WordPress

the missing manual®

The book that should have been in the box®

Matthew MacDonald

O'REILLY®

Beijing | Boston | Farnham | Sebastopol | Tokyo

WordPress: The Missing Manual

by Matthew MacDonald

Published by O'Reilly Media, Inc.,
1005 Gravenstein Highway North, Sebastopol, CA 95472.

O'Reilly books may be purchased for educational, business, or sales promotional use. Online editions are also available for most titles (*http://oreilly.com*). For more information, contact our corporate/institutional sales department: (800) 998-9938 or *corporate@oreilly.com*.

October 2012:	First Edition.
July 2014:	Second Edition.
October 2020:	Third Edition.

Revision History for the Third Edition:

    2020-09-02   First release

See *http://oreilly.com/catalog/errata.csp?isbn=9781492074168* for release details.

ISBN-13: 978-1-492-07416-8

[LSC]

Contents

The Missing Credits

ABOUT THE AUTHOR

 Matthew MacDonald is a science and technology writer with dozens of books to his name. Over the years, he's written about web design, programming, and nature's weirdest computing tool—the human brain. These days he's exploring quirky content for kids, including a free interactive guide for learning to write HTML and JavaScript. You can learn more about his new projects and his semi-regular publication, Young Coder, on his website, *http://prosetech.com*.

ABOUT THE CREATIVE TEAM

Amelia Blevins (editor) has been professionally editing for 10 years. When she's not editing, she's reading copiously and tending her garden in Philadelphia. She can be reached at *ablevins@oreilly.com*.

Katherine Tozer (production editor) lives in Boston, where she got her masters in publishing at Emerson and reads as much as she can get her hands on. Her email is *ktozer@oreilly.com*.

Joanne Sprott (indexer) has been a freelance book indexer, copyeditor, and proofreader since 1995. She currently enjoys working with her small partnership at Potomac Indexing LLC. Joanne lives and works in a lovely town in Oregon with outstanding views of the surrounding valley.

Sharon Wilkey (copyeditor) began her writing and editing career in daily newspaper journalism and has made her way back home to the tiny sandspit of Cape Cod, where her office is sometimes the beach. She can be reached at *sharonwilkey@comcast.net*.

James Fraleigh (proofreader) has operated a copyediting and proofreading practice since 2008. Contact him at *https://www.linkedin.com/in/jamesfraleigh*.

Jason Coleman (technical reviewer, third edition) has been pushing WordPress to its limits for years. Jason now leads development for Paid Memberships Pro, a membership-focused ecommerce plugin that powers many software as a service companies. Find Jason on Twitter: *@jason_coleman*.

Sallie Goetsch (technical reviewer, previous edition) (rhymes with "sketch") hand-coded her first website in HTML in 1995, but hasn't looked back since discovering WordPress in 2005. She works as an independent consultant and organizes the East Bay WordPress Meetup in Oakland, California. You can reach her at *www.wpfangirl.com*.

ACKNOWLEDGMENTS

No author could complete a book without a small army of helpful individuals. For this edition, I'm deeply indebted to Amelia Blevins, who kept everything relentlessly on track; Jennifer Pollock, who believed that writing one more edition was worth the effort; and Katherine Tozer, who managed the final transition to print. The content of this book was greatly improved by the feedback from tech reviewer Jason Coleman (in this edition) and Sallie Goetsch (in the previous one), as well as the sharp eyes and judicious touch of copyeditor Sharon Wilkey. Thanks also to the many folks who toiled behind the scenes, making this book better in ways I didn't even realize. I'm not sure who I can plausibly blame for the inevitable typos that remain, but it's definitely none of these people!

Finally, for the parts of my life that exist outside this book, I should thank my family members and credit their endless patience. They include my parents, Nora and Paul; my extended parents, Razia and Hamid; and—of course—my wife and daughters, Faria, Maya, Brenna, and Aisha. Thanks, everyone! I could have done it without you, but the results would've been ugly.

THE MISSING MANUAL SERIES

Missing Manuals are witty, superbly written guides to computer products that don't come with printed manuals (which is just about all of them). Each book features a handcrafted index.

For a full list of all Missing Manuals in print, go to *www.missingmanuals.com /library.html*.

Introduction

When you decide to create a website, there are several paths you can take. You could use a cookie-cutter website builder (like Wix, Weebly, or Squarespace). But if you do, you lose the ability to customize every detail to get what you *really* want. And because these companies host whatever you build, you also give up final control over your website.

You could hire an expert (or a team of experts) to build a website for you. Find the right person, and you'll get a great product. But how do you maintain it? How do you make changes, add content, and keep growing your site without asking your expert to make every change for you?

Or, you could choose a third path, and use the ridiculously popular publishing tool called *WordPress*. WordPress isn't perfect. But it does an amazing job of keeping things as simple or as complex as you want. You start out with free themes that have beautiful typography and layouts. You get a content publishing system that makes writing new pages as easy as typing in a word processor. You can snap in more advanced features when you need them, like social media integration, search optimization, and even ecommerce. And, best of all, you control everything.

At the time of this writing, WordPress powers roughly *one-third* of all the world's websites, according to the web statistics company W3Techs (see *http://tinyurl. com/3438rb6*). And while popularity can't tell you everything, WordPress's success has some great benefits:

- **Updates.** There's a small army of programmers at work on WordPress, guaranteeing it regular updates and new features. Living in the limelight also encourages people to find security vulnerabilities in the WordPress software, which the WordPress developers can fix with timely patches.

- **Support.** If you use WordPress, no one will ever say "that doesn't work here." These days, it's impossible to find a web host that doesn't support WordPress.

- **Community.** Millions of people are using WordPress, which means you can find plenty of friends who'll share tips, tricks, and plugins (the tiny bits of software that give WordPress more features).

Here's the bottom line: if you don't want to hire a black-belt web programmer (or become one yourself), WordPress is by far the easiest way to create a modern website. And because it's popular, you won't be left out on a limb with an abandoned product in a few years.

◾ About This Book

This book is for anyone who wants to learn how to make websites with WordPress. We assume you know your way around a computer (and the internet), but you haven't yet dipped your toes into the world of WordPress. You don't need to know HTML and CSS, although these web skills will come in useful if you get ambitious about theme customization (see Chapter 13).

In this book, you'll learn everything you need to create a new WordPress site, fill it with content, and make it look professional. Notice that we haven't yet used the word *blog*. Although WordPress is the world's premiere blogging tool, it's also a great way to create other types of websites, like sites that advertise businesses, report news, promote artists, and even sell products.

What You Need to Use This Book

Not much! The absolute bare minimum requirements are a proper computer (please don't try to build a WordPress site on your phone) and a web connection. It doesn't matter whether your computer runs Windows, macOS, or Linux, because you're going to do all your work online in your favorite browser.

One *recommended* ingredient is a web hosting account. (A *web host* is a company that rents out space on the web, so you can make your website available to the whole world.) This is likely to cost you a few dollars a month. One account is enough, because you can use a single account to create as many WordPress sites as you want. And it's easy, so you'll probably want to create a half dozen test sites to use while you practice the exercises in this book.

If you don't have a web host yet but would like to sign up, you'll get the lowdown on how that works in Chapter 2.

WordPress Without a Web Host

If you don't have a web host and don't want to sign up, never fear—there is another way! Chapter 3 explains how to use a fantastic WordPress testing tool called Local. With Local, you put a copy of WordPress on your computer and use that to create sites and make posts. Now, because Local is for your computer only, there's no way for anyone else to visit your site or see your work. But Local is a great way to practice your WordPress skills in private, before you're ready to sign up with a web host. Best of all, you don't need to pay anything to use it.

Not Recommended for WordPress.com

There's another option for folks who want to use WordPress but don't want to sign up with a web host. You can use the free WordPress.com hosting service to build a live WordPress website for free. Unfortunately, WordPress.com is not the blazing deal it seems. Many important features need paid upgrades, and there are fewer opportunities to customize your site with plugins and themes. And although WordPress.com works more or less the same as the install-it-yourself version of WordPress (called *self-hosted* WordPress), the management pages are a bit different.

This book focuses on self-hosted WordPress—the real, no-holds-barred, grown-up version, in all its glory. That said, once you become a master of classic WordPress, you'll have no problem stepping over to WordPress.com. But if you want to start out learning and practicing with WordPress.com, you may be better off with a different book. That's because the administrative interface—the pages you use to manage your website—is slightly different with WordPress.com.

◼ About the Outline

This book is divided into 14 chapters.

In the first few chapters, you'll start planning your path to WordPress web domination. First, you'll take a high-level look at the types of sites you can make with WordPress (Chapter 1). Then you'll learn how to get WordPress running on your web host (Chapter 2) or home computer (Chapter 3).

Next, you'll start writing content for WordPress, known as *posts* (Chapter 4). You'll pick a stylish theme (Chapter 5), and then you'll break out the formatting magic to make fancy posts (Chapter 6) and add pictures and videos (Chapter 7). Finally, you'll learn to create pages and menus (Chapter 8). At this point, you'll have all the skills you need to round out a basic, attractive WordPress website.

Then it's time to stretch your wings with plugins (Chapter 9), the extensions that allow you to add thousands of free new features from the WordPress community. Throughout the rest of this book, when there's a problem to solve, you'll often find a solution from a clever plugin.

As you continue, you'll learn how you can interact with other people on your site. First, you'll run discussions (and fight spam) through comments (Chapter 10). Next, you learn how to collaborate with a whole group of authors (Chapter 11), and how to attract boatloads of visitors (Chapter 12).

Finally, you'll take a look at some expert WordPress work. First, you'll crack open a WordPress theme and learn to change the way your site works by adding, inserting, or modifying the styles and code embedded inside the theme (Chapter 13). Then you'll go pro with a solid backup strategy, caching to boost performance, and a basic ecommerce shopping cart (Chapter 14).

■ About the Online Resources

As with all Missing Manuals, your learning doesn't end in these pages. You can visit the companion content for this book at *http://prosetech.com/wordpress*. There you'll find the following:

- **Sample sites.** This book is packed full of examples. Although you won't be able to take charge of the example site (modify it, manage comments, or do any other sort of administrative task), you can take a peek and see what it looks like. This is a handy way to witness features that are hard to experience in print—say, playing an embedded video or reviewing pictures in an image gallery.

- **Links.** Often this book will point you to a place on the web. It might be to learn more about a specialized WordPress feature, or to get background information on another topic, or to download a super-cool plugin. To save your fingers from the wear and tear of typing in all these web addresses, you can use this clickable list of links.

- **News and further reading.** Wondering where to go next once you've finished this book? We'll point you to some useful sites and resources where you can get more information, rub shoulders with other WordPressers, and follow the news.

Errata

To keep this book as up-to-date and accurate as possible, each time we print more copies, we'll make any confirmed corrections you suggest. We also note such changes on the book's website, so you can mark important corrections in your own copy of the book, if you like. Go to *http://oreilly.com/catalog/errata.csp?isbn=9781492074168* to report an error and view existing corrections.

The WordPress Landscape

Since you picked up this book, it's likely that you already know at least a bit about WordPress. You probably realize that it's a great tool for creating a huge variety of websites, from gossipy blogs to serious business sites. However, you might be a bit fuzzy on the rest of the equation—in other words, how WordPress actually works its magic.

This short chapter will introduce you to the WordPress way of life. First, you'll take a peek at the inner machinery that makes WordPress tick. If you're not already clear on why WordPress is so wonderful—and how it's going to save you days of work, years of programming experience, and a headful of gray hairs—this discussion will fill you in. After that, you'll take a quick look at the types of sites you can build with WordPress. As you'll see, WordPress began life as a blogging system but has since mutated into a flexible, easy-to-use tool for creating virtually any sort of site.

■ How WordPress Works

You probably already realize that WordPress isn't just a tool to build web pages. After all, anybody can create a web page—you just need to know a bit about HTML (the language that web pages are written in) and a bit about CSS (the language that formats web pages so they look beautiful). It also helps to have a good web page editor at your fingertips. Meet these requirements, and you'll be able to build a *static* website—one that looks nice enough, but doesn't actually *do* anything (*Figure 1-1*).

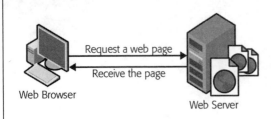

FIGURE 1-1

In an old-fashioned website, a web designer creates a bunch of HTML files and drops them into a folder on a web server. When you visit one of those pages, the web server sends the HTML to your browser. WordPress works a little differently—it builds its pages on the fly, as you'll see next.

With WordPress, you strike up a different sort of partnership. Instead of creating a web page, you give WordPress your raw content—that's the text and pictures you want published as an article, a product listing, a blog post, or something else. Then, when someone visits your site, WordPress assembles that content into a perfectly tailored page.

This system lets WordPress provide some useful features. For example, when visitors arrive at a WordPress blog, they can browse through the content in different ways. They can see the posts from a certain month, or on a certain topic, or tagged with a certain keyword. Although this seems simple enough, it requires a program that runs on a web server and puts the content together. If, for instance, a visitor searches a blog for the words "chocolate cake," WordPress needs to find all the appropriate posts, stitch them together into a web page, and then send the result back to your visitor's browser.

> **NOTE** Just in case your webmaster skills are a bit rusty, remember that a *web server* is the high-powered computer that runs your website (and, usually, hundreds of other people's websites too).

WordPress Behind the Scenes

In a very real sense, WordPress is the brain behind your website. When someone visits a WordPress-powered site, the WordPress software gets busy and—in the blink of an eye—delivers a hot-off-the-server, fresh new page to your visitor.

Two crucial ingredients allow WordPress to work the way it does:

- **A database.** This industrial-strength storage system sits on a web server. Think of it as a giant electronic filing cabinet where you can search and retrieve bits of content. In a WordPress website, the database stores all the content for its pages, along with extra information about the pages (like their categories and search tags), and all the comments that people have left. WordPress uses the MySQL database engine, because it's a high-quality, free, open source product, much like WordPress itself.

- **Programming code.** When someone requests a page on a WordPress site, the web server loads up a template and runs some code. It's the code that does all the real work—fetching information from different parts of the database, assembling it into a cohesive page, and so on. The code WordPress uses is written in a language called PHP.

Figure 1-2 shows how these two pieces come together.

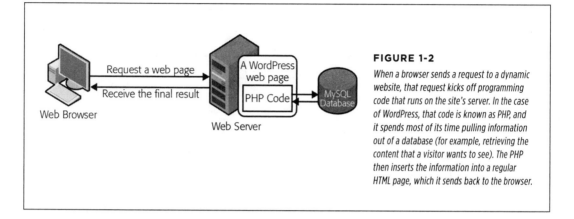

FIGURE 1-2

When a browser sends a request to a dynamic website, that request kicks off programming code that runs on the site's server. In the case of WordPress, that code is known as PHP, and it spends most of its time pulling information out of a database (for example, retrieving the content that a visitor wants to see). The PHP then inserts the information into a regular HTML page, which it sends back to the browser.

UP TO SPEED

The Evolution of Dynamic Sites

Dynamic websites are nothing new. They existed long before WordPress hit the scene. In fact, modern, successful websites are almost always dynamic, and almost all of them use databases and programming code. The difference is who's in charge. If you don't use WordPress (or another tool like it), it's up to you to write the code that powers your site. But if you use WordPress to build your site, you don't need to touch a line of code or worry about defining a single database table. Instead, you supply the content, and WordPress takes care of everything from storing it in a database to inserting it in a web page when it's needed.

Even if you *do* have mad coding skills, WordPress remains a great choice for site development. That's because using WordPress is a lot easier than writing your own software. It's also a lot more reliable and a lot safer, because every line of logic has been tested by a legion of genius-level computer nerds—and it's been firing away for years on millions of WordPress sites.

In short, the revolutionary part of WordPress isn't that it lets you build dynamic websites. It's that WordPress pairs its smarts with site-creation and site-maintenance tools that ordinary people can use.

WordPress Themes

One more guiding principle shapes WordPress—its built-in *flexibility*. WordPress wants to adapt itself to whatever design you have in mind, and it achieves that through a feature called *themes*.

Basically, themes let WordPress separate your content (which it stores in a database) from the layout and formatting details of your site (which it stores in a theme). Thanks to this system, you can tweak the theme's settings—or even swap in a whole new theme—without disturbing any of your content. *Figure 1-3* shows how this works.

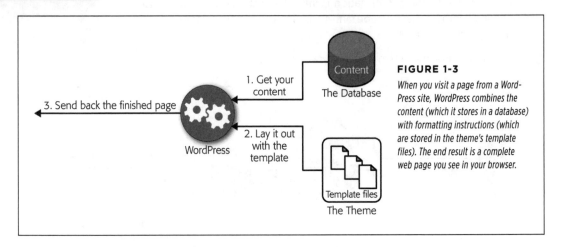

FIGURE 1-3

When you visit a page from a WordPress site, WordPress combines the content (which it stores in a database) with formatting instructions (which are stored in the theme's template files). The end result is a complete web page you see in your browser.

If you're still not quite sure how WordPress helps you with themes, consider an example. Imagine Jan decides to create a website so he can show off his custom cake designs. He decides to do the work himself, so he not only has to supply the content (the pictures and descriptions of his cakes), but also has to format each page the same way. Each page has two parts—a description of the cake and a picture of it—and he wants his pages to be consistent. But, as so often happens, a week after he releases his site, Jan realizes it could be better. He decides to revamp his web pages with a fresh, new color scheme and add a calorie-counting calculator in the sidebar.

Applying these changes to a non-WordPress website is no small amount of work. It involves changing the website's stylesheet (which is relatively easy) and modifying every single cake page, being careful to make *exactly the same change* on each (which is much more tedious). Even if Jan has a good HTML editing program, he'll still need to rebuild his entire website and upload all the new web pages.

With WordPress, these problems disappear. To get new formatting, you tweak your theme's style settings. To add the calorie counter, for example, you simply drop it into your theme's layout (and, yes, WordPress *does* have a calorie-counting plugin). And that's it. You don't need to rebuild or regenerate anything, go through dozens of pages by hand, or check each page to try to figure out which detail you missed.

■ What You Can Build with WordPress

There are many flavors of website, and many ways to create them. But if you want something reasonably sophisticated and don't have a crack team of web programmers to make that happen, WordPress is almost always a great choice.

That said, some types of WordPress websites require more work than others. For example, if you want to create an ecommerce site complete with a shopping cart and checkout process, you need to rely heavily on someone else's WordPress plugins. That doesn't necessarily make WordPress a poor choice for ecommerce sites, but it does present an extra challenge. (Chapter 14 has more information about ecommerce on WordPress.)

In the following sections, you'll see some examples of WordPress in action.

Blogs

As you probably know, a *blog* is a type of website that consists of separate, dated entries called *posts* (see *Figure 1-4*). Good blogs reflect the author's personality and are often informal.

When you write a blog, you invite readers to see the world from your viewpoint, whether the subject is work, art, politics, technology, or your personal experience. Blogs are sometimes described as online journals, but most blogs are closer to old-school newspaper editorials or magazine commentary. That's because a journal writer is usually talking to himself, while a half-decent blogger unabashedly addresses the reader.

Here are some common characteristics of blogs:

- **A personal, conversational tone.** Usually, you write blogs in the first person ("I bought an Hermès Birkin bag today" or "Readers emailed me to point out an error in yesterday's post"). Even if you blog on a serious topic—you might be a high-powered executive promoting your company, for example—the style remains informal. This gives blogs an immediacy that readers love.

- **Dated entries.** Usually, blog posts appear in reverse-chronological order, so the most recent post takes center stage. Often readers can browse archives of old posts by day, month, or year. This emphasis on dates keeps blogs looking current and relevant, assuming you post regularly. But miss a few months, and your neglected blog will seem old, stale, and seriously out of touch—and even faithful readers will drift away.

- **Interaction through comments.** Blogs aren't just written in a conversational way; they also "feel" like a conversation. Loyal readers add their feedback to your thoughts, usually in the form of comments appended to the end of your post (but sometimes through a ratings system or an online poll). Think of it this way: your post gets people interested, but their comments keep them invested, which makes them much more likely to come back and check out new posts.

The date the post was made

The post title. Click here to see the whole post and its comments.

FIGURE 1-4

Mr. Money Moustache writes a traditional (and wildly popular) blog about financial freedom, early retirement, and killing the habit of excessive consumerism. Scroll down the page (at http://www.mrmoneymustache.com) and you'll see a list of his most recent posts. And like all the site examples shown in this chapter, Mr. Money Moustache uses WordPress.

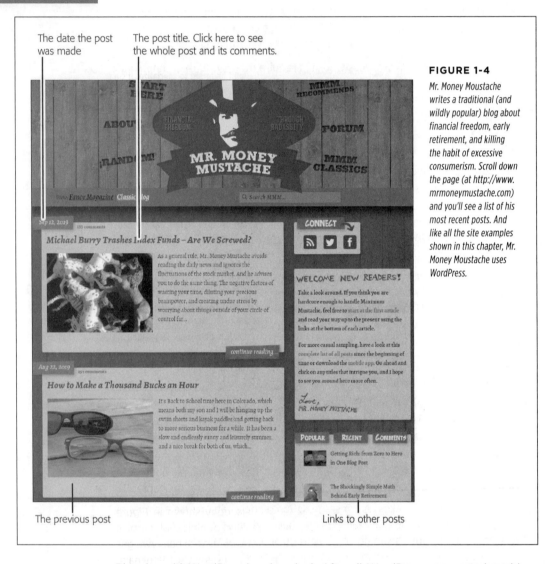

The previous post

Links to other posts

Blogging with WordPress is a slam dunk. After all, WordPress was created as a blogging tool (in 2003) and has since exploded into the most popular blogging software on the planet. If you plan to create a blog, there's really no good reason *not* to use WordPress. Although several other blogging platforms are out there, and they all work reasonably well, none of them has the near-fanatical WordPress community behind it, which is responsible for thousands of themes and plugins, and might even help you solve hosting and configuration problems (just ask your questions in the forums at *http://wordpress.org/support*).

Creating a Modern Blog

Perhaps the idea of writing a blog seems a bit boring to you. If so, you're probably locked into an old-fashioned idea about what a blog *is*.

Today's blogs aren't glorified online diaries. In fact, the best way to create an *un*successful blog is to stuff it full of meandering, unfiltered thoughts. Even your friends won't want to sift through that. Instead, follow these tips to get your blog on track:

- **Pick a topic and focus relentlessly.** People will seek out your blog if it's based on a shared interest or experience. For example, create a blog about your dining experiences around town, and foodies will flock to your pages. Talk up the challenges of taking care of a baby, and other new parents will come by and commiserate.

- **Add a clever title.** Once you choose your topic, give your blog a name that reinforces it, which will also help you stay on topic. The title "Pete Samson's Thoughts" is easy to ignore. "Mr. Money Moustache" definitely provokes a second look.

- **Don't be afraid to specialize.** It's a rule of the web that everything has been blogged before, so find a unique angle from which to attack your topic. You won't pique anyone's interest with yet another movie review site called My Favorite Movies. But throw a different spin on the subject with a blog that finds film flaws (In Search of Movie Mistakes) or combines your experience from your day job as a high-school science teacher (The Physics of Vampire Movies), and you just might attract a crowd.

- **Don't forget pictures and video.** Bloggers shouldn't restrict themselves to text. At a bare minimum, blogs need pictures, diagrams, comics, or some other visual element to capture the reader's eye. Even better, you can weave in audio or video clips of performances, interviews, tutorials, or related material. They don't even need to be your own work—for example, if you're discussing the avant-garde classical composer György Ligeti, it's worth the extra five minutes to dig up a performance on YouTube and embed that into your post.

Sites with Stories and Articles

WordPress makes a great home for personal, blog-style writing, but it's an equally good way to showcase the more polished writing of a news site, web magazine, short-story collection, scholarly textbook, and so on. WordPress also allows multiple authors to work together, each adding content and managing the site (as you'll see in Chapter 11).

Consider, for example, the Internet Encyclopedia of Philosophy shown in *Figure 1-5* (and located at *www.iep.utm.edu*). It's a sprawling catalog of philosophy topics amassed from about 300 authors and maintained by 25 editors, all with heavyweight academic credentials. Created in 1995, the site moved to WordPress in 2009 to make everyone's life a whole lot easier.

The Internet Encyclopedia of Philosophy is an interesting example for the sheer number and size of the articles it hosts. However, you'll also find WordPress at work in massive news sites, including TechCrunch, The New Yorker, TMZ, Salon, Boing Boing, and ThinkProgress. Even if you're the only one writing, it's easy to create a modern magazine-style layout, with grids of articles and pictures, polished sidebars, and pull quotes.

The current page (the
equivalent of a blog post)

Links to other pages,
organized by category

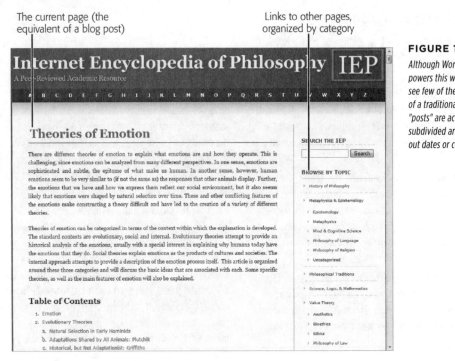

FIGURE 1-5

*Although WordPress
powers this website, you'll
see few of the hallmarks
of a traditional blog. The
"posts" are actually long,
subdivided articles, without dates or comments.*

How to Find Out If a Website Uses WordPress

Plenty of websites are built with WordPress, even if it's not always apparent. So what can you do if you simply *must* know whether your favorite site is one of them?

You could ask the website administrator, but if you're in a hurry, there are two easier ways. The first is the quick-and-dirty approach: right-click the page in your browser, choose View Source to bring up the page's raw HTML, and then hit Ctrl-F (Command-F on a Mac) to launch your browser's search feature. Hunt for text starting with "wp-". If you find *wp-content* or

wp-includes somewhere in the mass of markup, you're almost certainly looking at a WordPress site.

Another approach is to use a browser app, called a *sniffer*, that analyzes the markup. The advantage of this approach is that most sniffers detect other types of web-creation tools and programming platforms, so if the site isn't based on WordPress, you might still find out a bit more about how it works. One of the most popular sniffers is Wappalyzer (*http:// wappalyzer.com*), which works with the Mozilla Firefox and Google Chrome browsers.

Catalogs

WordPress is particularly well suited to websites stuffed full of organized content. For example, think of a website that has a huge archive of ready-to-make recipes (*Figure 1-6*). Or consider a site that collects classified ads, movie critiques, restaurant reviews, or custom products.

Because WordPress relies on a database, it's a wizard at organizing massive amounts of content. In a properly designed catalog site, people can find a review, recipe, product, or whatever else they want in multiple ways, such as searching by keyword or browsing by category.

FIGURE 1-6

Simply Recipes is a WordPress-powered site that features a giant catalog of recipes. What makes the site distinctly different from a blog is that it doesn't organize recipes by date, displayed one by one in reverse-chronological order. Instead, it orders them in common-sense categories, like Meat, Seafood, and Mushrooms. One thing that does make it like an ordinary blog is its support for comments at the end of each recipe.

It might occur to you that a catalog of products is nice, but a catalog of products that you can *sell* is even nicer. And, with a little work, you can use a WordPress site to not just showcase items, but let people add them to a virtual shopping cart as well.

What Makes a Catalog Site

Catalog sites are also known by many other names. Some people describe them as *content-based sites*; others call them *CMS sites* (for "content management system," because they manage reams of information). No matter what you call them, the sites share a few key characteristics:

- **They include lots of content.** If you want to create a recipe site with just four recipes, it probably wouldn't be worth the WordPress treatment.

- **The content is divided into distinct pages.** With a blog, the "pages" are actually blog posts. In a recipe site, each page is a recipe. And in the Internet Encyclopedia of Philosophy shown previously in *Figure 1-5*, each page is a lengthy scholarly article.

- **You browse the content by category.** In a blog, you often look at posts by date, and focus on the most recent ones. But in a catalog, everything is organized into categories, and you usually find what you want by clicking through a menu of choices. For example, you might choose the Dessert, Cake category to see what you can make for a birthday.

These criteria encompass a surprisingly huge range of modern-day websites. Examples include event listings for festivals, a portfolio of your work, and a list of products you sell. When you find a mass of text or pictures that needs to be categorized and presented to the world, you'll often find that WordPress is there, making itself useful.

Business Sites

WordPress isn't just a great tool for self-expression; it's also an excellent way to do business. The only challenge is deciding exactly *how* you want to use WordPress to help you out.

The first, and simplest, option is to take your existing business website and augment it with WordPress. For example, consider the Sony company. It has a custom-made site about its popular PlayStation at *http://playstation.com* and a WordPress site with news and trending stories at *http://blog.playstation.com*. Microsoft pulls off a similar trick with a standard site at *http://microsoft.com* and a WordPress-powered news hub at *http://news.microsoft.com*.

Other companies use WordPress to take charge of their entire websites. Usually, they're smaller sites, and often the goal is simply to promote a business and share its latest news. For example, you could use WordPress to advertise the key details about your new restaurant, including its location, menu, and recent reviews. Or imagine you need more detailed information for a tourist attraction, like the detailed website for the Guggenheim Museum (*Figure 1-7*).

FIGURE 1-7

*The Guggenheim website
has plenty of information
about the famous museum
and its collection. It looks
like the usual slick business site (and not at all
like a blog), but it's built
with WordPress.*

A greater challenge arises when a business doesn't want to just advertise or inform with its website, but also to *do business* over the web. For example, imagine you create a site for your family-run furniture store, like the one shown in *Figure 1-8*. You don't want to just advertise the pieces you offer; you want to take orders for them too, complete with all the trappings typical of an ecommerce website (such as a shopping cart, a checkout page, email confirmation, and so on). In this situation, you need to go beyond WordPress's native features and add a plugin to handle the checkout process.

For some small businesses, an ecommerce plugin offers a practical solution. But for many others, this approach isn't flexible enough. Instead, most ecommerce sites need a custom-tailored transaction-processing system that integrates with other parts of their business (like their inventory records, reporting system, or customer database). This functionality is beyond the scope of WordPress and its plugins. In other words, you won't be using WordPress to build the next Walmart.com.

FIGURE 1-8

On this WordPress-built furniture website, you can view the chairs for sale, their prices, and their dimensions. All this is possible with WordPress's standard features and a heavily customized theme. But if you want to allow online ordering, you need to use a specialized plugin, as you'll see in Chapter 14.

> **TIP** To see more examples of what you can do with WordPress, including plenty of business sites, visit the WordPress showcase at *http://wordpress.org/showcase*.

■ The Last Word

You've now finished your 10,000-foot WordPress flyover. You took your first look at how WordPress works and saw examples of the many types of sites you can make. Now it's time to dig in.

Where you go next depends on the type of setup you're working with:

- If you're creating your own WordPress websites on an internet web host, Chapter 2 covers everything you need to get started.

- If you prefer to play with WordPress in private, on your own personal computer, Chapter 3 shows you how to install the handy Local software to do just that.

- If you already have a WordPress setup—say, you're using an existing WordPress site at work and just want to start posting new content—you can skip straight to Chapter 4.

Happy trails!

Installing WordPress on Your Web Host

Before you create a WordPress website, you need to tick off a few boxes. The first requirement is to set up an account with a web host. In this chapter, you'll learn what you need and how to find a suitable web host (if you don't have one already). Then you'll learn the easy part—how to quickly install a fresh copy of WordPress on your website (or two).

If you don't have a web host account and aren't planning to get one just yet, there is a solution. You can start by putting a copy of WordPress on your home computer. Jump ahead to Chapter 3 to learn how that works, and come back here when you're ready to create a live site on the internet.

■ Choosing a Web Host

Maybe you already have a web host. If so, congratulations! You can soar right over this section and land at "Preparing For WordPress" (page 17). But if you don't have a web host yet and want to build a WordPress website that other people can see, you have a decision to make: who can you trust to do the job?

Choosing a good web host seems more daunting than it is. Technically, your host needs to meet just two requirements:

- It needs to be able to run PHP (ideally, version 7.3 or greater), the programming language that powers WordPress.

- It needs to recognize MySQL (version 5.6 or greater), which is the database that stores WordPress content.

These days, every web host meets these requirements. In fact, choosing a WordPress-friendly host is hard because so many hosts offer essentially the same thing. Most of the other selling points that web hosts advertise—the amount of disk space or bandwidth you get, for example—are less important. Even popular WordPress sites aren't likely to hit the web space and bandwidth limits of the average web host, unless you plan to host huge video files (and even then, you'll probably find it far easier to use a video-hosting service like Vimeo or YouTube).

Evaluating a Web Host

The best way to take a first look at a prospective web host is to visit its website. You'll usually see a page that divides various plans into a few basic categories, like the one in *Figure 2-1*.

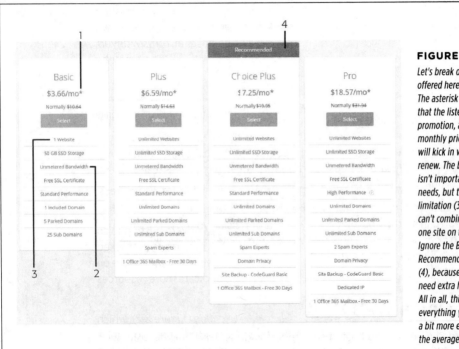

FIGURE 2-1

Let's break down the plans offered here on Bluehost. The asterisk (1) tells you that the listed price is a promotion, and the regular monthly price underneath will kick in when you renew. The bandwidth (2) isn't important for your needs, but the website limitation (3) means you can't combine more than one site on this account. Ignore the Best Value or Recommended highlights (4), because you don't need extra hand-holding. All in all, this host has everything you need but is a bit more expensive than the average on renewal.

Here are the technical details. If you're planning to make a starter site for yourself or a small business, you're looking for a *shared* hosting plan. You don't get the whole web server computer to yourself—other websites are on there too. This is the norm for small- to medium-sized sites, unless you want to pay a bundle of money.

Usually, you're looking at the cheapest plan that's available. But there's a catch. Some hosts limit their free plans to just a single website or domain. If you want to use one web host account to try out a bunch of websites (and why not, it's fun!) you might prefer a plan that lets you have several sites under one umbrella.

How much does it cost? To budget for WordPress, assume you'll pay $5 to $10 a month for web hosting. Then add the cost of a custom domain name (that's the web address that leads to your site), which you can typically get for an extra $10 to $35 per year.

The most important considerations in choosing a host aren't the amount of web space or bandwidth you get. Instead, they are reliability, security, and support—in other words, how often your website will be down because of technical troubles, how quickly you can get a real person to answer your questions, and whether your host will be in business several years into the future. These attributes are more difficult to assess, but before you sign up with a host, you should try contacting its support office (both by email and phone).

Don't trust website reviews, which are usually paid for. *Do* look up what other people say about the hosts you're considering on the popular forum Web Hosting Talk (*https://tinyurl.com/y5dw53ca*). And if you aren't comfortable doing your own research, you can consult a lineup of common WordPress hosts at *https://tinyurl. com/wphosts*. Just be warned that these hosts pay to be featured. They may be safe, but they aren't necessarily the cheapest choices.

> **NOTE** Here's the bottom line: WordPress has become so super-popular that virtually all web hosts embrace it, even in their most affordable web hosting plans.

Premium Performance and Managed Hosting

Although plenty of decent, cheap hosting options are available for WordPress, premium hosting plans cost several times more.

If you have a very popular website, you might choose *virtual private* hosting (also known as *semi-dedicated* hosting). This moves your site off the heavily trafficked web servers that host hundreds or thousands of other people's sites, and puts it on a computer that hosts fewer sites. That way, the server can dedicate more resources to handling *your* site and serving *your* visitors.

Another premium option for WordPress sites is *managed hosting* (*Figure 2-2*), which is like hotel concierge service for your website. With managed hosting, you get a domain name and web space (as usual), along with a bundle of extra WordPress services. These services might include the following:

- Daily backups, so you don't need to worry about doing it yourself

- Caching, which helps guarantee good performance

- Extra statistics, so you can analyze your web traffic

- Staging, so you can test changes to your site without making them live for the world to see

- WordPress freebies, like premium themes and ecommerce plugins

- Tools to promote your site
- A techy support person to install your plugins for you

You can accomplish all of these things on your own (and you'll learn how to in this book). But having a web host take care of the work is a nice convenience.

The drawback is cost. While basic WordPress hosting can be had for as little as $5 a month, managed hosting (and virtual private hosting) starts at around $30 per month. Heavily trafficked sites can pay hundreds of dollars a month.

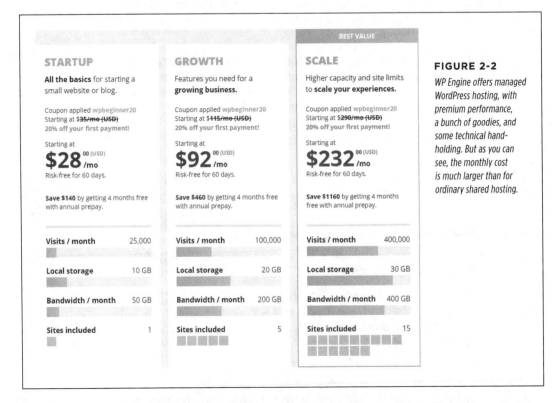

FIGURE 2-2

WP Engine offers managed WordPress hosting, with premium performance, a bunch of goodies, and some technical hand-holding. But as you can see, the monthly cost is much larger than for ordinary shared hosting.

You can learn more about managed hosting by checking out some of the web hosts that provide it. They're listed at *https://tinyurl.com/managedhosts.*

NOTE If you're just starting with WordPress, you won't yet know how well a particular web hosting plan will meet your needs. The best advice is to start with an affordable plan at a well-respected web host. You can always upsize or even transfer your site in the future. On the other hand, if you're setting up WordPress for an established business, managed hosting may be a worthwhile investment.

■ Preparing for WordPress

Once you have a web host, you're ready to set up a WordPress website. It's an easy task, but with a slight complication: different web hosts often have subtly different setup tools. The following sections will help you figure out where to start and will guide you through the setup process.

Using cPanel

Every web host has some sort of control panel where you can configure your website. Usually, these control panels are arranged in basically the same way. You log in through your web browser, using your super-secret web host username and password. Then you see a huge set of organized icons, each of which represents a feature or group of settings you can control, from email to security. Most of these features you'll never touch. But some of them you need in order to set up WordPress.

The most common control panel is called *cPanel*, and it's the one you'll see in this chapter (starting in *Figure 2-3*). Before you go ahead, make sure you can log in to your web host control panel. Often you'll use your domain name with *cpanel* tacked on the end (as in *www.HandMadePants.com/cpanel*). Your web host will have sent you the exact link in an email when you signed up.

FIGURE 2-3

The cPanel page is packed with website features. Each feature has an icon, and icons are grouped into boxes with titles like Files and Domains.

TIP If your web hosting provider isn't using cPanel, don't panic! You'll probably be able to find the same feature in your control panel. But if things look really different and weird, and you can't find the right icon to set up WordPress, you'll need to contact your web host's support people. Don't wait for an email response—try the phone or web-based chat.

Deciding Where to Put WordPress

Before you create your first WordPress site, you need to think a bit about how your web hosting account and your WordPress site will fit together.

When you sign up for a web hosting account, you typically get some space for your web pages and a domain name, which is the web address a visitor types into a browser to get to your site (like *www.HandMadePants.com*). If you've never had a website before, you probably assume that your domain name and your WordPress website address will be the same thing. But that's not necessarily true. In fact, you can choose to put WordPress in one of three places:

- **The root folder of your site.** This lets WordPress take over your entire site. For example, imagine you sign up for a site with the domain *www.Banana Republican.org* and then put WordPress in the root folder of that site. Now when visitors type that address into their browsers, they go straight to your WordPress home page.

- **A subfolder in your site.** This puts WordPress in a separate "branch" of your site. For example, if you bought the domain *www.BananaRepublican.org*, you might put WordPress in the subfolder *www.BananaRepublican.org/blog*. One reason you might use this arrangement is that you want your website to include a WordPress section (for example, a blog with news articles) and some traditional HTML web pages that you've already written. Another, less common reason is that you want to create more than one completely separate WordPress site on the same domain.

- **A subdomain in your site.** This is similar to the subfolder approach, but it uses a subdomain instead. To create a subdomain, you take your domain (say, *www. BananaRepublican.org*), remove the *www*, if it has it (now you have *Banana Republican.org*), and then put a different bit of text at the front, separated by a period (as in *social.BananaRepublican.org*). For example, you could have a traditional website at *www.BananaRepublican.org* and a news-style WordPress site with user feedback at *social.BananaRepublican.org*.

NOTE To see an example of the way companies separate sites, check out car-maker Toyota. It has a non-WordPress site that's all about selling cars at *http://toyota.com*, and a WordPress-powered news site at *http://pressroom.toyota.com*.

If you want to use either of the first two approaches, you don't need to do anything extra before you start installing WordPress. The WordPress autoinstaller will take care of everything.

If you plan to take the third approach and install WordPress on a subdomain, you need to create the subdomain before you go any further. The following section explains how.

Creating a Subdomain (If You Need One)

If you're planning to put WordPress in the root folder or in a subfolder of your website, skip this section—it doesn't apply. But if you're planning to host WordPress on a separate subdomain, you need to lay a bit of groundwork, so keep reading.

Creating a subdomain is a task that's quick and relatively straightforward—once you know how to do it. Unfortunately, the process isn't the same on all web hosts, so you may need to contact your host's support department to get the specifics. If your host uses the popular cPanel administrative interface, the process goes like this:

1. **In your browser, log in to the control panel for your web host.**

   Look for the Subdomains icon (usually, you'll find it in a box named Domains).

   You'll notice that cPanel piles dozens of icons into the same spot. If you can't find the Subdomains icon, you can use the search box that appears at the top of the cPanel page (*Figure 2-4*).

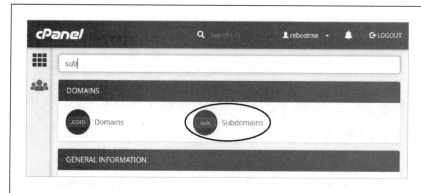

FIGURE 2-4

To jump right to the sub-domain feature you need, start typing the first few characters (that's "sub"). It will appear at the top of the page.

2. **When you find the Subdomains icon, click it.**

   This loads the Subdomains page (*Figure 2-5*).

3. **Choose the domain you want from a list of all the web addresses you own.**

   Some people have a web hosting account with just one domain, but others own dozens.

You create a subdomain here

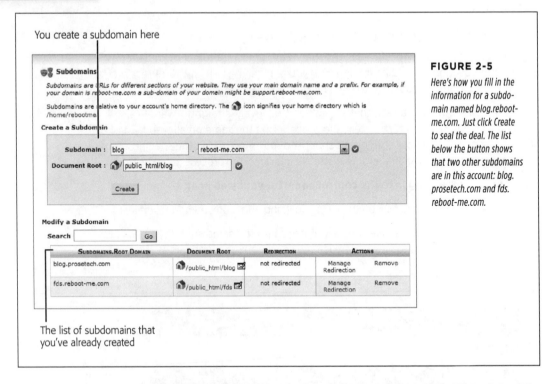

FIGURE 2-5

Here's how you fill in the information for a subdomain named blog.reboot-me.com. Just click Create to seal the deal. The list below the button shows that two other subdomains are in this account: blog.prosetech.com and fds.reboot-me.com.

The list of subdomains that you've already created

4. **In the Subdomain box, type in the prefix you want to use for the subdomain.**

   For example, if you want to create the subdomain *blog.reboot-me.com* on the domain *reboot-me.com* (*http://reboot-me.com*), you need to type "blog" in the Subdomain box.

5. **In the document Root box, pick the folder where you want to store the files for this domain.**

   Your web host will suggest something based on your subdomain. For example, it might be *public_html/blog* if you named the subdomain *blog*. You can use that if you're not sure what you want, or you can edit it to something you like better.

6. **Click the Create button to create your subdomain.**

   After a brief pause, you'll be directed to a new page that tells you your subdomain has been created. Click Go Back to return to the Subdomains page.

   You'll see your new subdomain in the list on the Subdomains page. Right now, it has no web files, so there's no point in typing the address into a browser. However, when you install WordPress, you'll put its files in that subdomain.

> **TIP** If you need to delete a subdomain, find it in the list and then click the Remove link. Technically, the subdomain folder still exists (the one you picked in step 5), but there's no way for someone to get there in a browser. If you want, you can delete that folder with cPanel's file management features or an FTP program.

Understanding the Administrator Account

Before you install WordPress, you need to decide what username and password you'll use for the all-powerful *administrator account*. This is the account you use to manage your site, and it gives you great powers (including the ability to nuke everything at a moment's notice).

Hackers, spammers, and other shady characters are very interested in your WordPress administrator account. If they get hold of it, they're likely to sully it with lurid ads (see *Figure 2-6*), phony software offers, or spyware.

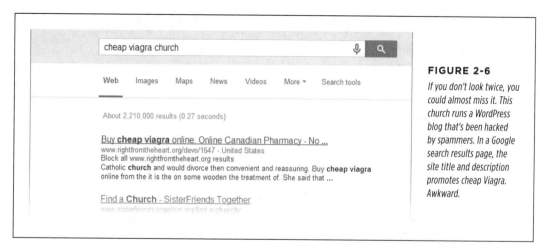

FIGURE 2-6

If you don't look twice, you could almost miss it. This church runs a WordPress blog that's been hacked by spammers. In a Google search results page, the site title and description promotes cheap Viagra. Awkward.

Your best protection against these attacks is to follow these rules when you create your administrator account:

- Make your username non-obvious (that means you should prefer AngryUnicorn to admin, user, or wordpress).

- Choose a strong, non-obvious password that includes a combination of letters and numbers (like bg8212beauty rather than bigbeauty). For guidelines on creating a secure password, see "A WordPress Password Is More Than a Formality" on page 27.

- Don't use the same password for your WordPress administrator account that you use to log in to your web host account.

Once you decide where you want to install WordPress and you pick a good username and password for your administrator account, you're ready to press on.

■ Installing WordPress

The easiest way to install WordPress is to use an *autoinstaller*, a special tool that installs programs on your site. Most web hosts offer an autoinstaller as part of their services.

There are several autoinstallers in the world. Two of the most popular are Softaculous and Fantastico, both of which you'll learn to use in this section. Other autoinstallers you might come across include Installatron and SimpleScripts.

> **NOTE** In an effort to please everyone, some web hosts support more than one autoinstaller. If that's the case for you, you can use either one. However, we prefer Softaculous, because it offers handy backup features that Fantastico doesn't. Page 32 has the scoop on those.

All autoinstallers work in more or less the same way: you sign in to your web hosting account and click the autoinstaller icon to see a catalog of the add-on software your host offers; look for WordPress and then start the installation. You need to supply the same basic pieces of information during the installation—most significantly, the website folder where you want to install WordPress, and the username and password you want to use for the WordPress administrator account (which your autoinstaller will create).

The following sections explain how to use Softaculous (first) and Fantastico (second). If your web host uses another autoinstaller, the steps are similar and you can follow along with a few adjustments.

Installing WordPress with Softaculous

How do you know if your host offers Softaculous? You could ask, but it's probably quicker to look for yourself:

1. **Log in to the control panel for your web host.**

   If you're already logged in to cPanel, make sure you're at the main page. You can go back to the main page by clicking the Home button in the top-left corner. (It's just under the "cPanel" heading, and it looks like a grid of squares.)

2. **Look for a Softaculous icon.**

   If your web host uses cPanel, you can use the search feature to hunt for Softaculous. *Figure 2-7* shows a successful search.

   If you can't find a Softaculous icon, you might luck out with one of the autoinstallers listed previously. Try searching for a Fantastico, Installatron, or SimpleScripts icon. If you find Fantastico, you can use the steps in the next section. If you find another autoinstaller, try following the steps listed here—just mentally replace "Softaculous" with the name of your autoinstaller.

FIGURE 2-7

Here, searching for Softaculous has brought up two useful icons. First is the Softaculous icon that launches the autoinstaller. Below that is a WordPress icon, which does almost the same thing—it launches Softaculous and preselects WordPress.

3. **Click the Softaculous icon.**

   Softaculous shows a large, colorful tab for each program it can install (*Figure 2-8*).

4. **Hover over the WordPress box and click the Install button.**

   This takes you to an all-in-one installation page that collects all the information WordPress needs. Most of these settings don't need to be changed. For example, Softaculous automatically selects the latest version of WordPress in the "Choose the version you want to install" box, which is the only safe choice. However, you need to review a few other settings. You'll take a look at them in the following steps.

5. **Pick a domain name and a directory.**

   This is the location where you want to put your WordPress website (*Figure 2-9*).

   The domain is the first part of the website address (*prosetech.com,* in this example). If you click the Domain box, you'll see all the domains that you own. You'll also see any subdomains you've created (like *news.prosetech.com*).

FIGURE 2-8

Along the left, Softaculous lists all the types of installation scripts it supports. But you won't need to hunt for the script that installs WordPress: it usually appears in the top position on the Softaculous page, because of its popularity.

You also need to decide whether to put your WordPress site in a subfolder. If you do, you need to fill the name in the Directory box. This example uses a folder named *teahouse*, which means the WordPress site will be created at *http://prosetech.com/teahouse*. It doesn't matter that the folder doesn't exist yet, because Softaculous will create it. If you leave the Directory box blank, WordPress uses the whole domain, and visitors can type an address like *http://prosetech.com* to go to your WordPress site.

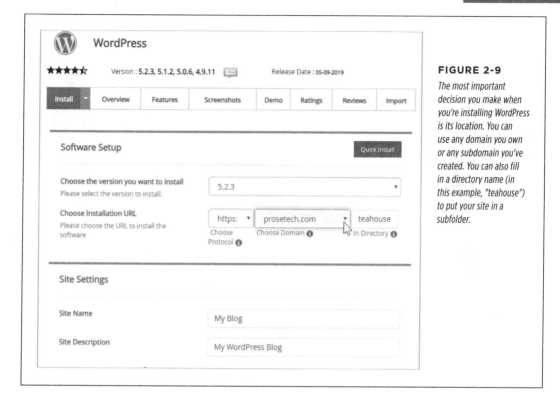

FIGURE 2-9

The most important decision you make when you're installing WordPress is its location. You can use any domain you own or any subdomain you've created. You can also fill in a directory name (in this example, "teahouse") to put your site in a subfolder.

6. **Choose a site name and description (*Figure 2-10*).**

   The site name is the title you want to give your WordPress site (like "Magic Tea Emporium"). It shows prominently on every page of your site.

   The description should be a short, one-sentence profile of your site. It appears in smaller text, just underneath the title on every page of your site.

   Don't worry about the Multisite feature just yet—you'll consider that in Chapter 11.

7. **Optionally, change the username, password, and email address for your administrator account.**

   Softaculous suggests *admin* for the username. You can do better. Pick something less obvious, which will be better hidden from attack.

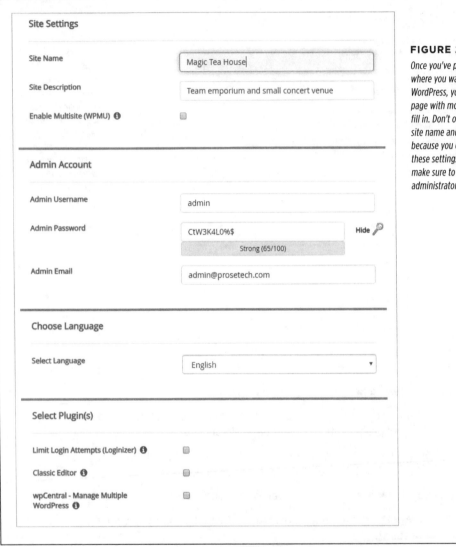

Site Settings

Site Name	Magic Tea House
Site Description	Team emporium and small concert venue
Enable Multisite (WPMU) ⓘ	☐

Admin Account

Admin Username	admin
Admin Password	CtW3K4L0%$ Hide 🔑
	Strong (65/100)
Admin Email	admin@prosetech.com

Choose Language

Select Language	English ▾

Select Plugin(s)

Limit Login Attempts (Loginizer) ⓘ	☐
Classic Editor ⓘ	☐
wpCentral - Manage Multiple WordPress ⓘ	☐

FIGURE 2-10

Once you've picked where you want to install WordPress, you have a page with more settings to fill in. Don't overthink the site name and description, because you can change these settings later. Do make sure to use a solid administrator password.

Remember, a good password is all that stands between you and a compromised WordPress site that's showing banner ads for timeshares. Do yourself a favor and follow the rules set out in the box on page 27 to defend your site properly.

The administrator email address is *your* email address. You'll get notifications from Softaculous at this email, and you'll use it if you need to reset your password.

A WordPress Password Is More Than a Formality

WordPress websites are commonly attacked by hackers looking to steal traffic or to stuff in some highly objectionable ads. The best way to avoid this danger is with a strong password.

With enough tries, web evildoers can guess any password by using an automated program. But most human WordPress hackers look for common words and patterns. If you use your first name (ashley), a string of close-together letters on the keyboard (qwerty, qazwsx), or a single word with a few number-fied or symbol-fied characters (like passw0rd and pa$$word), be afraid. These passwords aren't just a little bit insecure; they regularly make the list of the world's 25 most stolen passwords. (For the complete list of bad passwords, check out *https://tinyurl.com/badpassw*.)

That doesn't mean you need a string of complete gibberish to protect your site. Instead, you can deter casual hackers (who are responsible for almost all WordPress attacks) by taking a reasonably unique piece of information and scrambling it lightly. For example, you can use a favorite musician (HERBee-HANcock88), a movie title (dr.strangel*ve), or a short sentence with some vowels missing (IThinkThrforIM).

It's acceptable to write your password on paper and tuck it in a desk drawer—after all, you're not worried about family members or office colleagues; you're concerned with international spammers, who certainly won't walk into your office and rifle through your belongings. (However, it's still a bad idea to put your password in an email or text message.)

8. **Optionally, choose to install the Limit Login Attempts plugin.**

    Limit Login Attempts is a security-conscious plugin that temporarily closes down the administrative section of your site if it detects a potential intruder attempting to guess your username and password. This plugin is a good safeguard, but it's not immediately necessary for a new site. You'll learn more about the Limit Login Attempts plugin in Chapter 14.

    Don't worry about any other plugins—you don't need them. If you're curious, Classic Editor turns on the old-style WordPress post editor, and wpCentral is a tool for seeing multiple WordPress sites in one administrative interface.

9. **If you want to turn on Softaculous backups, click the Advanced Options heading. Otherwise, skip ahead to step 11.**

    The Advanced section has a few more Softaculous settings (*Figure 2-11*). The most useful settings are the ones for creating backups. Switch these on, and Softaculous automatically backs up your WordPress site, without requiring any work from you. (Technically, Softaculous works its magic by using *cron*, a scheduling tool that most web hosts support.)

10. **Fill in the backup settings: Backup Location, Automated Backup, and Backup Rotation.**

    Backup Location tells Softaculous where to store its backups. Usually, Local Folder is your only option, which means Softaculous stores its backups on your web server. (You can download them later.)

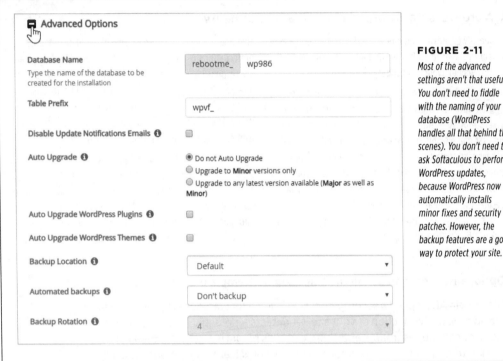

FIGURE 2-11

Most of the advanced settings aren't that useful. You don't need to fiddle with the naming of your database (WordPress handles all that behind the scenes). You don't need to ask Softaculous to perform WordPress updates, because WordPress now automatically installs minor fixes and security patches. However, the backup features are a good way to protect your site.

Automated Backup tells Softaculous how often it should make a backup (daily, monthly, or weekly).

Backup Rotation tells Softaculous how many old backed-up versions of your site to keep. For example, if you choose to keep four backups and use a weekly backup schedule, then on the fifth week Softaculous will discard the oldest backup to make way for the next one. Choose Unlimited to keep every backup you make, but be careful—daily backups of a large site can eventually chew up all your web hosting space. To avoid this problem, review your backups every once in a while and delete the old ones.

NOTE A regular backup schedule is a must for any WordPress site. However, you don't need to use Softaculous. Your web host may provide its own backup service, and plenty of WordPress plugins can perform regular backups, as you'll see in Chapter 14. If you need more time to think about your backup strategy, skip the Softaculous settings for now. You can always edit your Softaculous settings and switch on automatic backups afterward, as explained in the next section.

11. **Click Install to finish the job.**

Softaculous creates the folder you picked, copies the WordPress files there, and creates the MySQL database. After a few seconds, its work is done and you'll see a confirmation message (*Figure 2-12*).

Congratulations, the software was installed successfully

WordPress has been successfully installed at :
http://prosetech.com/magicteahouse
Administrative URL : http://prosetech.com/magicteahouse/wp-admin/

We hope the installation process was easy.

NOTE: Softaculous is just an automatic software installer and does not provide any support for the individual software packages. Please visit the software vendor's web site for support!

Regards,
Softaculous Auto Installer

Return to Overview

FIGURE 2-12

When Softaculous finishes creating your WordPress site, it gives you the address of your site and the address you use to get into the administration area (which you'll explore in Chapter 4).

Managing a Softaculous-Installed Site

Softaculous keeps track of the WordPress sites it installs. You can return to Softaculous to review this information and perform basic management tasks on your sites. Here's how:

1. **Log in to the control panel for your web host.**

2. **Find the Softaculous icon and click it.**

   This loads the familiar Softaculous page This time, turn your attention to the menu that runs down the left side of the page.

3. **Find the big WordPress icon. Click it, taking care not to click the Install or Demo buttons.**

   This brings you to the overview page with a description of WordPress ripped straight from the marketing copy. Scroll down and you'll find something much more useful—a list of all the places you've installed WordPress (*Figure 2-13*).

   If you accidentally click the Install button inside the Softaculous icon, the program assumes you want to install *another* WordPress site, and it opens the installation page. You can still get to the overview shown in *Figure 2-13*, but now you need to click the Overview button in the horizontal strip of buttons that appears just above the installation information.

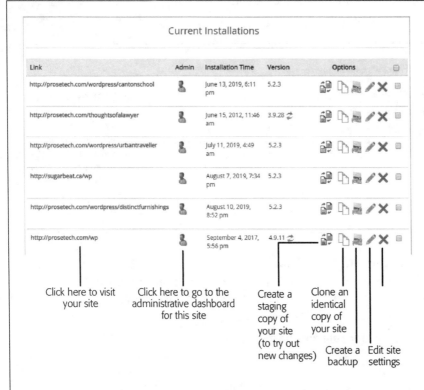

FIGURE 2-13

On this web hosting account, Softaculous has helped install WordPress in six places, and some of these installations are seriously in need of an update. Next to each site is information about the WordPress version and the icons that let you perform common tasks (such as updating, deleting, or backing up your site).

4. **Next to your site, click one of the icons in the Options column to perform a management task:**

   Clone creates an exact copy of your site, but in another folder. You could use this if you want to try out some extensive modifications before you make them a permanent part of your site.

   Backup lets you perform an immediate backup (rather than the more common scheduled backups, which Softaculous carries out automatically at a set time). When you click Backup, Softaculous asks you what you want to back up (*Figure 2-14*).

   Edit Details lets you change several of the settings you picked when you created the site. Don't change anything unless you know exactly what you're doing—changing the database name or WordPress folder at this point can confuse WordPress and break your installation. However, the Edit Details page is useful if you want to tweak the automatic backup settings.

Remove deletes your site. This removes all the WordPress files, the subfolder (if you installed WordPress in a subfolder), and the WordPress database. Once you take this step and confirm your choice, there's no going back.

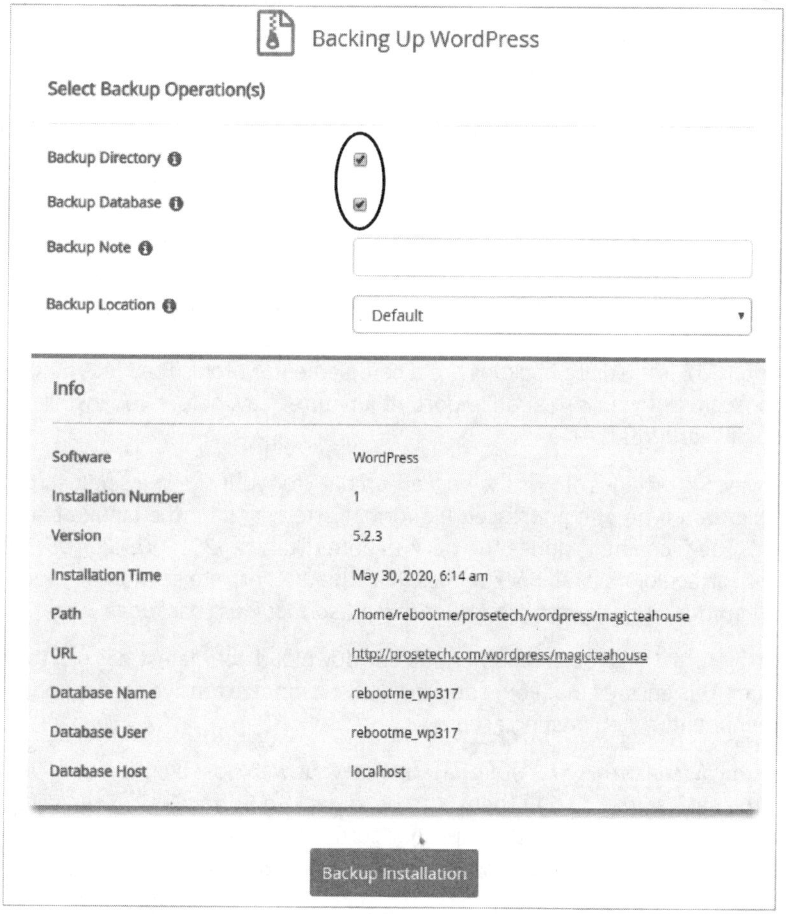

Backing Up WordPress

Select Backup Operation(s)

Backup Directory ℹ️   ☑

Backup Database ℹ️   ☑

Backup Note ℹ️

Backup Location ℹ️   Default   ▼

Info

Software	WordPress
Installation Number	1
Version	5.2.3
Installation Time	May 30, 2020, 6:14 am
Path	/home/rebootme/prosetech/wordpress/magicteahouse
URL	http://prosetech.com/wordpress/magicteahouse
Database Name	rebootme_wp317
Database User	rebootme_wp317
Database Host	localhost

Backup Installation

FIGURE 2-14

A WordPress site stores its text content in a database and stashes other supporting resources (like picture files) in the website directory. So a proper full backup includes a copy of both your database and your website directory. Make sure to select both checkboxes before you click Backup Installation.

NOTE Backups take place on your web host's web server, in the background. That means you can leave Softaculous and close your browser, and the program still makes scheduled backups. Softaculous will send you an email when it finishes a backup (using the administrative email address you supplied when you first installed the WordPress site).

5. **Optionally, you can browse directly to your site by clicking its URL.**

   Or click the head-and-torso Admin icon to visit the WordPress backend that controls your site. This interface (which is part of WordPress, not Softaculous) is where you manage your content, style your site, and take care of many more fine-grained configuration tasks. As with all WordPress sites, the administration page is your WordPress site's address with */wp_admin* tacked onto the end (for example, *www.prosetech.com/blog/wp_admin*).

Once your site is established, you probably won't visit Softaculous very often. But if you want to practice installing WordPress, fiddle with different installation choices, or just make a quick backup, it's a handy place to be.

Managing Softaculous Backups

Like every good web administrator, you need to regularly back up your site so you can recover from unexpected catastrophes (like the sudden bankruptcy of your web hosting company, or a spammer who defaces your site).

As you've already learned, Softaculous offers two backup options. You can set up scheduled backups, which take a snapshot of your site every day, week, or month. It's no exaggeration to say that every site should have a scheduled backup plan in place (if you don't want to use Softaculous, consider one of the plugins described in Chapter 14). Immediate backups are a complementary tool. They let you grab a quick snapshot of your site at an important juncture—say, before you install a new version of WordPress.

Either way, Softaculous stores the backed-up data for your site in a single large *.gz* file (which is a type of compressed file format often used on the Linux operating system). The filename includes the backup date (like *wp.1.2020-08-20_05-30-16.tarr.gz*). Softaculous stores backup files in a separate, private section of your web hosting account. Usually, it's in a folder named *softaculous_backups*.

For extra protection, you should periodically download your latest backup to your computer. This ensures that your site can survive a more extensive catastrophe that claims your entire web hosting account.

Although you can browse for your backups on your web host by using an FTP program, the easiest way to find them is to head back to Softaculous. Then click the Backups and Restore icon in the top-right corner of the page (*Figure 2-15*), which takes you to the Softaculous backup page (*Figure 2-16*).

> **NOTE**　An *FTP program* is a tool that can talk to a computer and exchange files with it over the internet. Using an FTP program, you can browse the files on your website and download them to your computer.

FIGURE 2-15

Click here to see all the backups Softaculous has made at your behest.

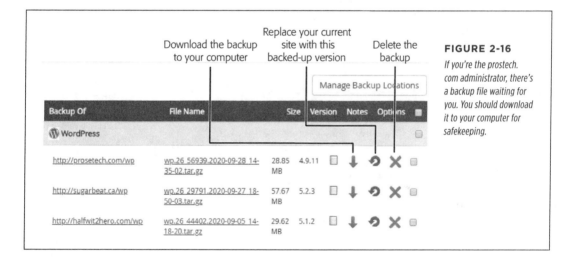

FIGURE 2-16

If you're the prostech. com administrator, there's a backup file waiting for you. You should download it to your computer for safekeeping.

TIP If you've set Softaculous to keep an unlimited number of backups, you'll eventually need to delete some of your oldest backups to free up more space. Otherwise, there's no reason to worry about the modest amount of space that a few WordPress backups will occupy.

If disaster strikes, you can restore your site by using the backup. From the Softaculous backups section, find the most recent backup, and then click the restore icon that appears next to it, which looks like an arrow that curves into a circle (see *Figure 2-16*).

You can also restore your site on a new web server—one that has Softaculous but doesn't have your backup file. First, upload your backup file to the *softaculous_backups* folder by using an FTP program. (Ask your web hosting company if you have trouble finding that folder.) Then, when you launch Softaculous and go to the backups section, you'll see your backup file waiting there, ready to be restored.

Installing WordPress with Fantastico

Fantastico is another popular autoinstaller. Like Softaculous, it replaces the aggravating manual installation process WordPress users once had to endure (in the brutish dark ages of a few years back) with a painless click-click-done setup wizard. Here's how to use it:

1. **Log in to the control panel for your web host.**

2. **Look for a Fantastico icon.**

   Remember, many control panels have a search feature that lets you type in the name of the program you want, rather than forcing you to hunt through dozens of icons (as shown in *Figure 2-4*).

3. **Click the Fantastico icon.**

   Fantastico's menu page appears, with a list of all the software it can install. Usually, you'll find WordPress near the top of the list, along with other site-building tools (*Figure 2-17*).

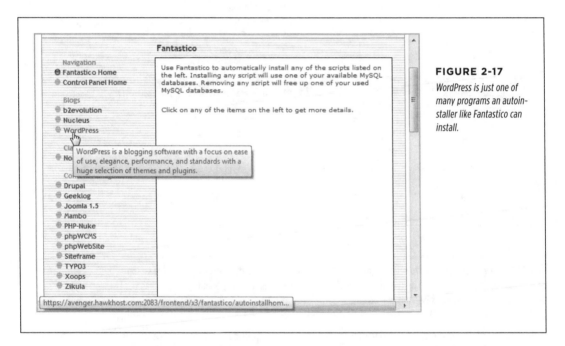

FIGURE 2-17

WordPress is just one of many programs an autoinstaller like Fantastico can install.

4. **Click WordPress.**

   Fantastico displays basic information about WordPress, including the version you're about to install and the space it will take up. Autoinstallers always use the latest stable version of WordPress, so you don't need to worry about these details.

5. **Click the New Installation link.**

Now Fantastico starts a three-step installation process.

6. **Pick a domain name and a directory (*Figure 2-18*).**

This is where you decide where to put WordPress and all its files. As you learned earlier (page 18), you have three basic options:

The root folder. This way, WordPress will run your entire site. To set this up, choose the domain name you registered for your website (in the first box) and leave the directory box blank.

A subfolder. This gives WordPress control over a section of your website. Choose your domain name in the first box and then fill in the name of the subfolder. The example in *Figure 2-18* uses the domain *reboot-me.com* (*http://reboot-me .com*) and a folder named *blog*. The autoinstaller will automatically create the folder you specify.

A subdomain. This is another way to give WordPress control over part of your website. To use this approach, you must have already created the subdomain (by following the steps on page 19). If you have, choose the subdomain name from the first box and leave the directory box blank.

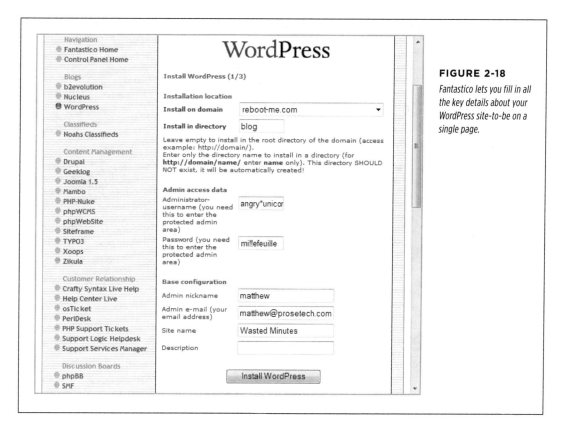

FIGURE 2-18

Fantastico lets you fill in all the key details about your WordPress site-to-be on a single page.

7. **Choose a username and password for your administrator account.**

Pick a name that's not obvious and a password that's difficult to crack (page 27). Doing otherwise invites spammers to hijack your blog.

8. **Fill in the remaining details in the "Base configuration" section.**

The administrator nickname is the name that WordPress displays at the end of all the posts and comments you write. You can change it later if you like.

The administrator email address is *your* email address, which becomes part of your WordPress user profile. It's also the email address you'll use for administration—for example, if you forget your administrator password and need WordPress to email you a password reset link.

The site name is the title you want to give to your WordPress site ("Wasted Minutes" per *Figure 2-18*). It shows prominently on every page of your site.

The description is a short, one-sentence summary of your site. WordPress displays it in smaller text just underneath the title on every page of your site.

9. **Click the Install WordPress button.**

The next screen summarizes the information you just typed in (*Figure 2-19*). For example, it displays the exact location of your new site and the name of the MySQL database that will hold all its content. You might want to double-check this info for accuracy, and then write down the details for safekeeping.

FIGURE 2-19

Here, Fantastico tells you what it's about to do. To hold all the data for this WordPress site, Fantastico will create a MySQL database named rebootme_wrdp1 (the name is based on the domain name www. reboot-me.com), and it will create the site at www.reboot-me.com/blog.

10. **Click "Finish installation" to move to the final step.**

Now Fantastico does its job—creating the folder you picked (in this case, *blog*), copying the WordPress files to it, and creating the MySQL database. When it finishes, you'll see a confirmation message. It reminds you of the administrator username and password you supplied, and lists the administration URL—the address you type into your browser to get to the admin area that controls your site. As with all WordPress sites, the administration page is your WordPress site's address with */wp_admin* tacked onto the end (for example, *www.reboot-me. com/blog/wp_admin*).

You can return to Fantastico to manage your WordPress installations anytime. However, Fantastico doesn't have the management features of Softaculous. Fantastico offers no way to modify, clone, or back up an existing WordPress site. Instead, it provides two links for each WordPress site you install: "Visit site" (which takes you there) and Remove (which deletes the site permanently).

Multiply the Fun with Multiple WordPress Sites

Most of the time, you'll install WordPress once. But you don't need to stop there. You can create multiple WordPress websites that live side by side, sharing your web hosting account.

The most logical way to do this is to buy additional web domains. For example, when you first sign up with a web hosting company, you might buy the domain *www.patricks-tattoos. com* to advertise your tattoo parlor. You would then install WordPress in the root folder on that domain. Sometime later, you might buy a second domain, *www.patrickmahoney.me*, through the same web hosting account. Now you can install WordPress for that domain too. (It's easy—as you'll see when you install WordPress, it asks you what domain you want to use.) By the end of this process, you'll have two distinct WordPress websites, two yearly domain name charges, but only one monthly web hosting fee.

Interestingly, you don't actually need to have two domains to have two WordPress sites. You could install separate WordPress sites in separate folders on the same domain. For example, you could have a WordPress site at *www.patrickmahoney. me/blog* and another at *www.patrickmahoney.me/tattoos*. This is a relatively uncommon setup (unless you're creating a bunch of WordPress test sites, as we do for this book at *http:// prosetech.com/wordpress*). However, it is possible, and there's no limit. That means no one is stopping you if you decide to create several dozen WordPress websites, all on the same domain. But if that's what you want, you should consider the WordPress multisite feature, which lets you set up a network of WordPress sites that share a common home but have separate settings (and can even be run by different people). Chapter 11 explains how that feature works.

■ Keeping WordPress Up-to-Date

No WordPress website should be left unprotected. If your site doesn't have the latest WordPress updates, it can become a target for hackers and spammers looking to show their ads or otherwise tamper with your site.

Fortunately, WordPress's creators are aware of the threat that outdated software can pose, and they designed the program for quick and painless upgrades. WordPress installs minor updates automatically, and it's quick to notify you about major updates so you can install them yourself. The following sections explain how these two updating mechanisms work.

Minor Updates

Since version 3.7, WordPress has included an autoupdate feature that downloads and installs new security patches as soon as they become available. So if you install WordPress 6.0 and the folks at WordPress.org release version 6.0.1, your site will grab the new fix and update itself automatically.

The autoupdate feature is a fantastic safety net for every WordPress site. However, it has an intentional limitation. It performs only *minor updates*, which are usually security enhancements or bug fixes. It doesn't attempt to install major releases—you need to do that yourself.

NOTE To spot the difference between a minor update and a major one, you need to look at the WordPress version number. Major releases change one of the first two digits in the version number (for example, 6.0.8 to 6.1.0 is a major update). Minor releases change the minor version number, which is the digit after the second decimal point (for example, 6.0.8 to 6.0.9).

Major Updates

A *major update* is a WordPress release that adds new features. Typically, WordPress puts out a major release every four months. You can find a list of recent and upcoming major releases at *http://wordpress.org/news/category/releases*.

You don't need to go out of your way to keep track of WordPress releases. Whenever you log in to the administration area (the control center you'll explore in Chapter 4), WordPress lets you know if there's a new version available.

To get to the admin area, take your WordPress site address (like *http://prosetech. com/magicteahouse* and add */wp_admin)* to the end (as in *http://prosetech.com /magicteahouse/wp_admin*). Initially, you start at the dashboard page. If WordPress detects that a newer version is available, it tries to grab your attention by adding a notification box to the top of this page (*Figure 2-20*).

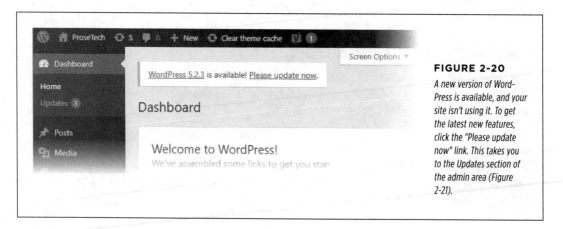

FIGURE 2-20

A new version of Word-Press is available, and your site isn't using it. To get the latest new features, click the "Please update now" link. This takes you to the Updates section of the admin area (Figure 2-21).

The Updates page is an all-in-one glance at everything that's potentially old and out-of-date on your site, including two types of WordPress extensions that you'll learn about later in this book: themes and plugins. Usually, the Updates page simply tells you that all is well. But when updates are available, you'll see something else. First, WordPress adds a black number-in-a-circle icon to the Updates command in WordPress's admin menu. The actual number reflects the number of website components that need updating. In *Figure 2-21*, that number is 3 because you need to update WordPress, a theme, and a plugin.

NOTE Themes and plugins are two ways you can enhance and extend your site. But if they contain flaws, hackers can use those flaws to attack your site. You'll learn more about themes in Chapter 5 and plugins in Chapter 9.

FIGURE 2-21

The Updates page explains that three components need updating: the WordPress software, the Akismet plugin, and the Mesmerize theme.

To install an update, use the buttons on the Updates page. If there's a new Word-Press update, click the Update WordPress button. If there's a newer theme or plugin, turn on the checkbox next to that theme or plugin, and then click Update Themes or Update Plugins.

> **TIP** Just to be safe, the recommended order is to update themes and plugins first, before you update the entire WordPress software. This lowers the chance of a conflict between a fancy new version of WordPress and an older component.

WordPress updates are impressively easy. There's no need to enter more information or suffer through a long wait. Instead, you'll see a quick welcome page that announces any new features or improvements. Your site will carry on functioning exactly as it did before.

Despite the rapid pace of new releases, WordPress's essential details rarely change. New versions may add new frills and change WordPress's administrative tools, but they don't alter the fundamental way that WordPress works.

■ The Last Word

This chapter introduced the essentials of web hosting with WordPress. First, you learned how to find a suitable web host to hold your site (and you saw how much it would cost). After that, you learned to install the WordPress software and create a new, blank site.

Installing WordPress is a relatively easy job. But if you've never done it before, it can seem intimidating at first. If you're just getting started, don't hesitate to create and delete a test site. Once you've done it a few times, you'll feel confident.

If you're ready to start using your site—filling it with content and making it look nice—skip to Chapter 4 next. But if you're curious about the idea of running a tiny, private WordPress server on your personal computer, continue on to Chapter 3 to learn about the handy tool that can make it happen.

NOTE Local Pro works only if you're hosting your WordPress site with a supported web host—at the time of this writing, that's Flywheel or WP Engine. The cost for Local Pro and Flywheel hosting with a single site hovers near $50 per month, so it isn't cheap. But Flywheel has pledged to keep Local Lightning free forever.

Installing Local

Now that you've learned what Local is, you're ready to try it for yourself. The first step is to download and install the Local program. There's a version for almost any operating system—Windows, Mac, and Linux. Here's how to get started:

1. **Fire up a browser and visit *https://localwp.com.***

   This is the dedicated page for the Local software. If you want to take a closer look at Flywheel, the web host that maintains Local, check out *https://getflywheel.com.*

2. **Click the Download button.**

   A window pops up that asks for some information (*Figure 3-1*).

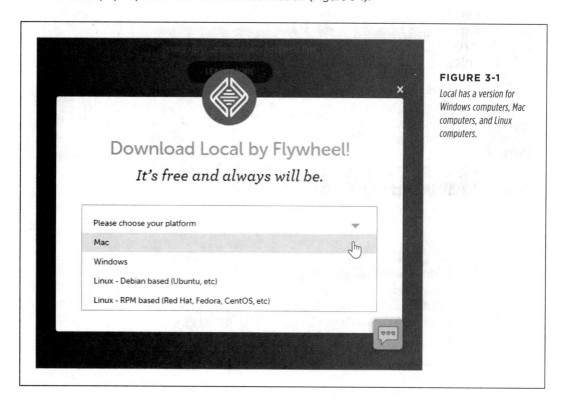

FIGURE 3-1

Local has a version for Windows computers, Mac computers, and Linux computers.

3. **Pick your operating system from the list and fill in your email address.**

   Once you pick your operating system, a few more text boxes pop into view, asking for your name (optional), phone number (optional), and email address (required).

4. **Click the Get It Now button.**

What happens next depends on your browser.

Some browsers, like Microsoft Edge, ask you what you want to do with the Local setup file before you start downloading it. (In this case, choose Open.)

Other browsers, like Google Chrome and Safari, start downloading the Local setup file immediately. In Chrome, you'll see it appear in a bar at the bottom of the window (*Figure 3-2*).

> **NOTE** The Local setup file is relatively big (between 200 and 500 MB, depending on your operating system). It has a name that includes the operating system and the Local version number, like *local-5-5-5-windows.exe* or *local-5-5-5-mac.dmg*.

FIGURE 3-2

Once you've down-loaded Local, a quick click launches the setup program.

Top: In Chrome, you'll find the setup file in the download bar.

Bottom: In Safari, you need to click the download arrow to see your down-loads, and then click the setup file.

5. **When the Local setup file is downloaded, open it.**

Your browser may do this automatically, or you might need to find the file and click it yourself (*Figure 3-2*). Either way, you'll find yourself looking at a fairly ordinary setup program.

6. **Use the setup program to install Local.**

On a Windows computer, the first question you get is about sharing (see *Figure 3-3*, top). If more than one person uses your computer, you can choose "Anyone who uses this computer" to let everyone use the program. Otherwise, choose "Only me" to keep it to yourself. Either way, click Next to continue and then click Install.

On a Mac computer, a window pops open that tells you how you should install Local (*Figure 3-3*, bottom). The best strategy is to click the Local icon in this window and then drag it over to the Applications icon. This pops open your Applications folder, where you can drop Local so it's always there when you need it.

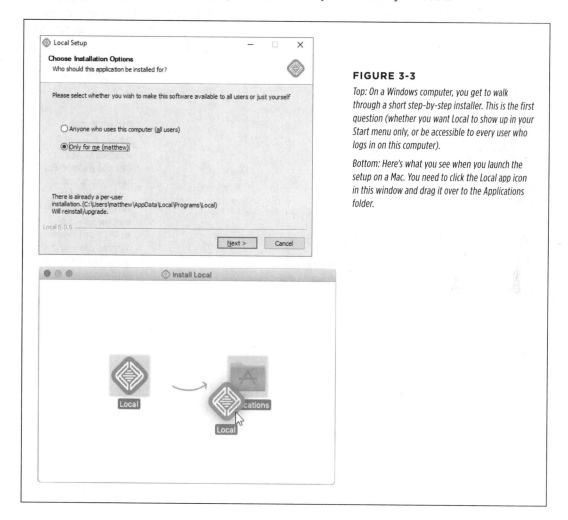

FIGURE 3-3

Top: On a Windows computer, you get to walk through a short step-by-step installer. This is the first question (whether you want Local to show up in your Start menu only, or be accessible to every user who logs in on this computer).

Bottom: Here's what you see when you launch the setup on a Mac. You need to click the Local app icon in this window and drag it over to the Applications folder.

Finally, you need to wait a short time while Local unpacks its files, moves in, and makes itself comfortable on your computer. You'll probably be asked to "grant permissions" to Local to make changes to your computer.

Creating a WordPress Site

Once you've installed Local, the next step is to create a WordPress site.

To get started, launch the Local program. You'll find Local in the usual place—the Start menu on a Windows computer or the Applications folder on a Mac. (Windows users also get a desktop shortcut.)

> **NOTE** When you run Local, you may see a warning message reminding you that you downloaded the product from the Web. Don't fear—you can safely carry on and allow Local to run.

The first time you start Local, you'll probably see a welcome message that promotes their Local Pro product. Click the X to close this message and to carry on to the main window. This is the heart of Local—the place where you see all the sites you've created (*Figure 3-4*).

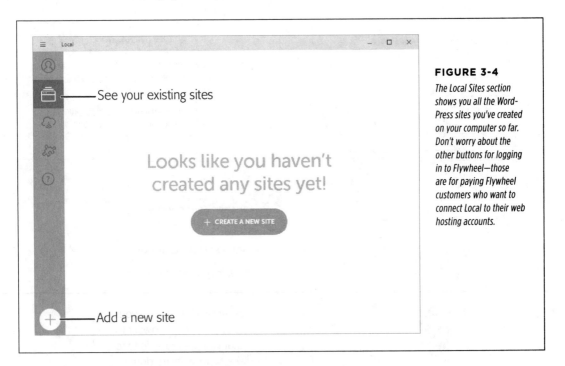

FIGURE 3-4

The Local Sites section shows you all the WordPress sites you've created on your computer so far. Don't worry about the other buttons for logging in to Flywheel—those are for paying Flywheel customers who want to connect Local to their web hosting accounts.

Here's how to make your first WordPress site with Local:

1. **Click Add Local Site.**

   The Add Local Site button is the plus-in-a-circle icon in the bottom-left corner of the Local window. Once you click it, Local asks for a bit more information (*Figure 3-5*).

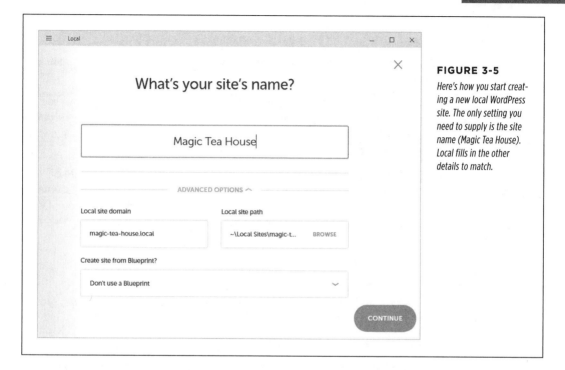

FIGURE 3-5

Here's how you start creating a new local WordPress site. The only setting you need to supply is the site name (Magic Tea House). Local fills in the other details to match.

2. **Choose a name for your site.**

This becomes the title at the top of every page of your site (although you can change it later).

3. **Optionally, you can tweak the settings in the Advanced Options section.**

None of them is particularly important right now, when you're just getting started with your first Local site. The settings include the following:

Local site domain is the web address you'll use to connect to the site on your computer. In the example in *Figure 3-5*, that means you'll point your browser to *https://magic-tea-house.local*. If you're creating a site that you plan to upload to a web host, you should change the local domain to match your website domain (like *www. magicteahouse.com*). That way, all your links will keep working when you transfer your site, with no extra fiddling required.

Local site path tells you where Local is going to put your website files on your computer. There's no need to change this setting.

Create site from Blueprint makes sense only if you've created a blueprint (which you haven't). Blueprints are a Local feature that lets you take a site you've already created and then create another, matching copy. You'll learn about that in the next section.

4. **Click Continue.**

Now Local asks for you to create an administrator account.

5. **Choose a username, password, and email address for your administrator account.**

The administrator account is what you use to log in to the administrative section of WordPress, change settings, and write posts. WordPress uses your email address to notify you about updates. You'll also need a working email to reset your password.

Because you're not creating a real WordPress website (just a test site on your own computer), you might think that it's OK to use the standard username (*admin*) and a flimsy password. And you're probably right! But if you plan to show others your site with the live link feature (page 54), or you just want to cultivate good habits, pick a reasonably complex password.

6. **Click Add Site.**

Now Local gets to work installing WordPress. It needs to transfer files, configure settings, and create the essential WordPress database. Along the way, you're likely to see several security warnings.

7. **Accept the security requests** (*Figure 3-6*).

Local needs to get through the tight security rules on modern computers that restrict what programs are allowed to communicate on the internet. If you're running on a Windows computer, you'll need to allow three programs through the firewall: Local, *nginx.exe* (the web server that Local uses), and *mailhog.exe* (the emailing tool that Local uses, so your site can send you email messages). On a Mac computer, you face the same requirements but get fewer details.

When you've clicked through all the security settings and Local has finished its work, you'll return to the Local Sites section in Local. But now you'll see the full details about your brand-new site (*Figure 3-7*).

8. **Click the View Site button to take a look at the site you created.**

To actually see your WordPress site, you need to open a web browser and type in the web address that you set in step 2. Local gives you a handy View Site button that does it for you. Click View Site to launch your browser and go right to your website's home page (*Figure 3-8*). Click the Admin button to open a browser window and go to the admin area of your site, which you'll learn to use in Chapter 4.

FIGURE 3-6

On a Windows computer, the Windows firewall asks you for permission several times to let Local's different apps communicate (top). On a Mac computer, you're likely to get a more generic warning (bottom). Either way, you should grant Local the permissions it needs.

FIGURE 3-7

The Local window is a sort of control panel where you can see all your sites and their details. Here, Local is running one website, called Magic Tea House. Use the View Site button to take a peek, or the Admin button to manage it.

FIGURE 3-8

Here's a blank and boring new WordPress website. It looks exactly the same as a new WordPress website created on a web host. If you look at the browser's address bar, you'll notice that this site uses the web domain that Local picked (in this case, magic-tea-house.local). This address works on your computer only. Behind the scenes, Local watches this domain for requests and handles them with its built-in web server.

Now that you know how to create a basic WordPress site with Local, you know enough to carry on to the next chapter and start exploring WordPress. But if you have a moment to linger, the next section introduces a few interesting Local features worth checking out.

◼ Managing Your Local Sites

If you're going to work with Local for longer than a few minutes, it's worth getting a little more comfortable with its environment. First, you should take a second look at the Local Sites page.

The Local Sites page gives you commands you can use to control your site—to shut it down, rename it, delete it, change the settings you picked when you created it, and so on. To perform any of these tasks, right-click the site name in the Local Sites list (*Figure 3-9*). A list pops up with handy commands.

The most important commands are self-explanatory. For example, you use Stop to temporarily pause your site—it's still there, but Local will ignore your browser if you try to visit it. The Start command resumes a stopped site, while Delete removes the site, its files, and all of its information from your computer. You better be sure you're really ready to move on before you click.

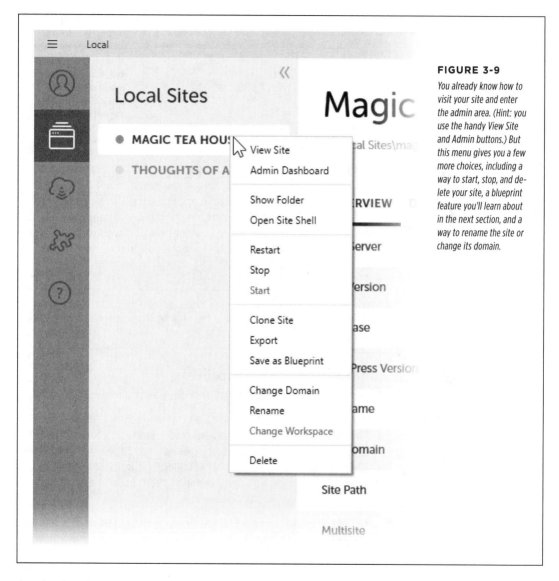

FIGURE 3-9

You already know how to visit your site and enter the admin area. (Hint: you use the handy View Site and Admin buttons.) But this menu gives you a few more choices, including a way to start, stop, and delete your site, a blueprint feature you'll learn about in the next section, and a way to rename the site or change its domain.

Another handy trick is viewing the *files* for your site. In a real, live WordPress site, you need to use an FTP program to connect to your web server and see its files. But WordPress sites in Local are easier to inspect, because they live on your computer.

To take a look, you can use the Show Folder command, or click the tiny arrow under your site title (*Figure 3-10*).

FIGURE 3-10

Local uses your personal documents folder to store all the sites you create. So, if you're logged in to your computer with the username maya, Local finds that folder, creates a Local Sites folder inside, and then creates a folder inside that for your website. In this example, the WordPress website folder is c:\Users\maya\Local Sites\magic-tea-house, and you can take a look inside by clicking the tiny arrow.

Right now, you won't be able to make much sense of what's in your WordPress website folder. But as you continue through this book and get more experience, this window onto WordPress will help you figure out how it all works. For example, in Chapter 13 you'll dig into the theme files in the website folder to change the way your site works.

NOTE WordPress uses your website folder to store your themes, your plugins, and the pictures and other files you upload—ingredients you'll learn about throughout this book. WordPress doesn't store the posts and pages you write, because that information goes into its database.

Making Multiple Sites and Blueprints

Local makes it easy to create as many WordPress sites as you want.

If you want a new site, just click the Add Local Site icon and follow the same instructions you did the first time around. All the sites you create show up in the Local Sites list. (If you look carefully at *Figure 3-9*, for example, you'll see two WordPress sites—one for the Magic Tea House, and another called Thoughts of a Lawyer.) Local hosts all of your sites at once. You can even load them up at the same time in separate browser tabs.

Sometimes you might want to make a change or customization to a site without replacing the current version. For example, maybe you have an idea that you want to try out, but you're not sure if you'll like the results. Or maybe you're a fancy web design company that wants to use a similar starter site for different people (and then customize it to suit each client).

Local lets you do this with a feature it calls *blueprints*. The idea is that you create a site, customize it however you want, and create a blueprint of your work. You can then use that blueprint to make perfectly cloned copies of the site. Best of all, the blueprint stores your settings, your theme, and all the data. It even keeps all the plugins that you've installed.

Here's how to make a blueprint:

1. **Right-click your site in the Local Sites list and choose Save as Blueprint.**

   Local asks for a bit more information (*Figure 3-11*).

2. **Give your blueprint a name.**

   Your name might be based on the site ("Magic Tea House") or on the type of site it represents (say, "Basic Blog Starter").

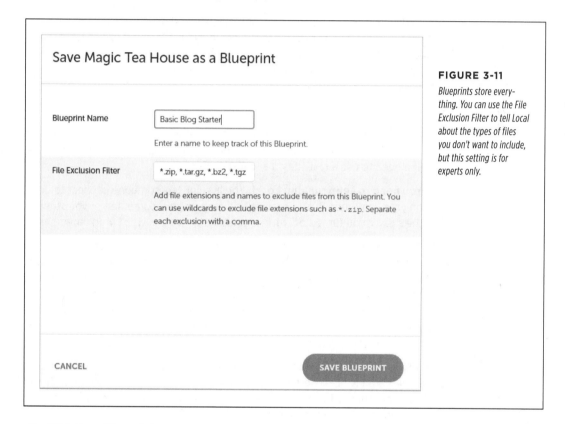

FIGURE 3-11

Blueprints store every-thing. You can use the File Exclusion Filter to tell Local about the types of files you don't want to include, but this setting is for experts only.

3. **Click Save Blueprint.**

   Wait a moment. Local needs to transfer your entire site into this blueprint, so there's a fair bit of copying to do. It compresses all the files to save space.

Now your blueprint is ready to go. The next time you create a new site (using the familiar Add Local Site button), you can choose your blueprint from the "Create Site from Blueprint" list (page 47). Right now, that won't be any different from creating an ordinary WordPress site. But once you start creating and customizing your sites, blueprints give you a way to reuse a significant chunk of work.

Using Live Links

One of Local's most innovative features (and its best kept secret) is *live links*. The idea is simple. Imagine you're a highly paid web developer polishing off a WordPress site for a new customer. You're using Local to do all your work on your personal computer. But once the design is perfect, you want to show your hard work to your client—and they want to see it before it goes live on their web servers.

In this situation, the usual course of action is to upload your hard work to a test website. Your test site will be live on the web, but it won't be advertised to anyone and it won't show up in a Google search. This approach works well—but what if you don't have your own web host?

The live link feature is a handy shortcut that lets you share your work without a test website. In fact, you don't even need a web hosting account. Instead, you open a temporary door to your Local-hosted website by using a secret link. Anyone who uses the link can go through the Flywheel web servers to find your computer and see the site you've shared.

Before you use live links, you need to understand a few limitations:

- **It's temporary.** Your link will expire after a short time, possibly just an hour, and certainly not more than a day. If you want to keep your site around longer, you need to use a test site on a real web host.

- **You need to keep running Local.** The live link lets someone connect to Local on your computer. But if Local isn't running on your computer, or you stopped the site, or you turned off your computer, the live link won't work anymore.

- **There are security risks.** You are allowing other people to connect to your computer. Yes, there is protection in place. Local acts like a traditional web server, which means it lets people access your WordPress site but not your personal files or computer hardware. However, there's always the potential for security flaws in the Local software that we don't know about. Fortunately, the short lifetime of each live link makes it very unlikely that a hacker will take advantage of it to attack your computer.

If you want to use the live link feature, it's easy. At the bottom of the Local Sites page, there's a Live Link setting with an Enable button (*Figure 3-12*). Click that to get your link. Then, when you're finished sharing your work and you want to close your site off to the public, click Disable to remove the link.

NOTE   Every time you enable live links, you get a new link. So if you disable your live link and then enable it again, you'll get a new link that you'll need to share with your peeps all over again.

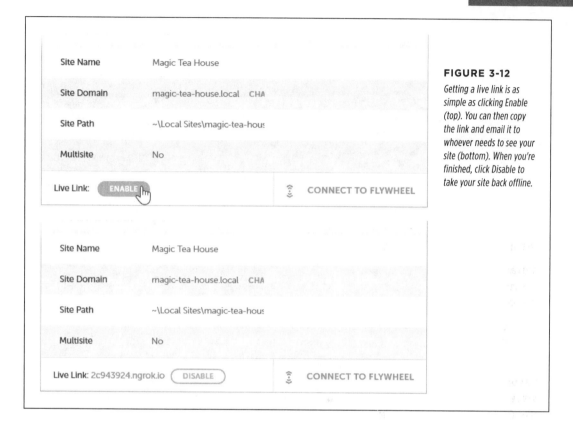

FIGURE 3-12

Getting a live link is as simple as clicking Enable (top). You can then copy the link and email it to whoever needs to see your site (bottom). When you're finished, click Disable to take your site back offline.

◼ The Last Word

WordPress is designed to run on a web server. But in this chapter, you learned about handy tools that let your computer *pretend* to be a web server, so you can create completely private websites without a web host.

Next, you saw how to install Local—the most straightforward choice—and make as many sites as you can fit on your computer. You even saw how to show your test sites on the web (temporarily) by using live links.

Because Local is running WordPress, you can use a Local-powered website to do pretty much everything you'll learn about in this book. To really get started, your next step is Chapter 4, where you'll learn to use the WordPress admin area and create your first post.

Creating Posts

N ow that you've installed the WordPress software on your web host (Chapter 2) or your personal computer (Chapter 3), you're ready to get started publishing on the web. In this chapter, you'll go to your fledgling WordPress site and start posting *content*, which can be anything from bracing political commentary to cheap celebrity gossip. Along the way, you'll learn several key WordPress concepts.

First, you'll get comfortable in WordPress's administration area—the central cockpit from which you pilot your site. Using the admin area, you'll create, edit, and delete the posts that appear on your site.

Next, you'll learn how to classify your posts by using categories and tags, so you can group them in meaningful ways. WordPress calls this art of organization *taxonomy*, and if you do it right, it gives your readers a painless way to find the content they want.

Finally, you'll take a hard look at the web address (URL) that WordPress generates for every new post. You'll learn how to take control of your URLs, making sure they're meaningful, memorable, and accessible to search engines.

■ Introducing the Admin Area

The administration area (*admin area* for short) is the nerve center of WordPress management. When you want to add a new post, tweak your site's theme, or review other people's comments, this is the place to go.

The easiest way to get to the admin area is to take your WordPress website address and add */wp-admin* to the end of it. For example, if you host your site here:

http://magicteahouse.net

you can reach the site's admin area here:

http://magicteahouse.net/**wp-admin**

This URL trick works exactly the same if you're running WordPress with a virtual web server on your own computer, using a tool like Local (from Chapter 3). Just take your local address and add */wp-admin* on the end.

When you go to the admin area, WordPress asks for your username and password. Once you log in, you'll see a page like the one in *Figure 4-1*.

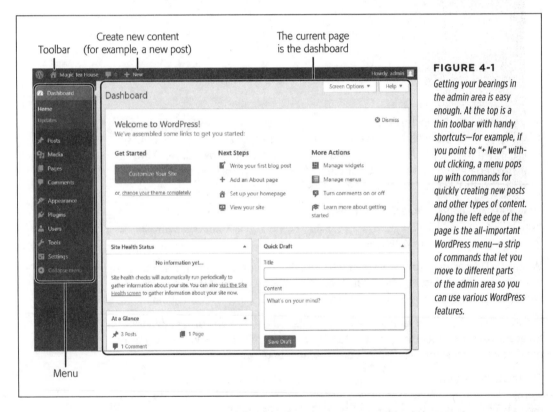

Toolbar   Create new content (for example, a new post)   The current page is the dashboard

Menu

FIGURE 4-1

Getting your bearings in the admin area is easy enough. At the top is a thin toolbar with handy shortcuts—for example, if you point to "+ New" without clicking, a menu pops up with commands for quickly creating new posts and other types of content. Along the left edge of the page is the all-important WordPress menu—a strip of commands that let you move to different parts of the admin area so you can use various WordPress features.

When you finish working in the admin area, it's a good idea to log out. That way, you don't need to worry about a smart-alecky friend hijacking your site and adding humiliating posts or pictures while you're away from your computer. To log out, click your username in the top-right corner of the toolbar and then click Log Out.

The Admin Menu

To browse around the admin area, you use the menu—the panel that runs down the left side of the page. It has a link to every administrative feature WordPress offers.

WordPress groups the menu commands into submenus. To see a submenu, hover over one of the menu headings (like Posts) and it will pop open (*Figure 4-2*).

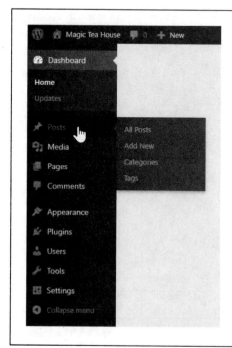

FIGURE 4-2

WordPress's menu packs a lot of features into a small strip of web page real estate. Initially, all you see are first-level menu headings. But point to one of the items without clicking, and WordPress opens a submenu. (The exception is the Comments heading, which doesn't have a submenu. You just click it to review comments.)

When you click a menu command, you go to that part of the admin area. For example, say you choose Posts→Add New. (In other words, you mouse over to the left-side menu, hover over Posts until its submenu appears, and then click the Add New item.) Now you'll see a slick editor where you can type your new post.

You can also click a menu heading directly (for example, Posts). If you do, you go to the first item in the corresponding submenu, as shown in *Figure 4-3*.

The menu has another trick—it can shrink itself down to save space. To see how this works, click the "Collapse menu" arrow at the very bottom of the menu. When the menu is compressed, it shows only icons without any text (*Figure 4-4*). Click the arrow button again to restore the menu to its full glory.

NOTE WordPress will also collapse the menu automatically if you resize your browser window to be very narrow. Size your browser back up, and the full menu reappears.

FIGURE 4-3

If you click the heading Posts, you actually go to the submenu item Posts→All Posts. And if you lose your bearings in the admin area, just look for the bold text in the menu to find out where you are. In this example, that's All Posts.

FIGURE 4-4

In its collapsed state, the menu removes the text of the menu headings, leaving just the tiny icons. Hover over one of these icons (like Posts, shown here), and the submenu appears, but with a helpful difference: now WordPress displays the menu name as a title at the top.

The Dashboard

Your starting place in the admin area is a densely packed home page called the *dashboard*. You can get back to this page at any time by clicking Dashboard in the menu.

The dashboard may seem like a slightly overwhelming starting point, because it's crowded with boxes. Each box handles a separate task, as shown in *Figure 4-5*.

TIP You don't *need* to use the dashboard—you can manage all the same tasks by using the menu and going to different parts of the admin area. But the dashboard collects some of the most important information and links in one spot so you can get a quick overview of your site.

The first time you visit the dashboard, you'll see a Welcome to WordPress box at the top of the page. Although the links in the welcome box work perfectly well, it's probably a better idea to get used to finding what you need using the menu. Before you continue, scroll down past the welcome box or click the Dismiss link to remove the box altogether, so you can see the rest of the dashboard page.

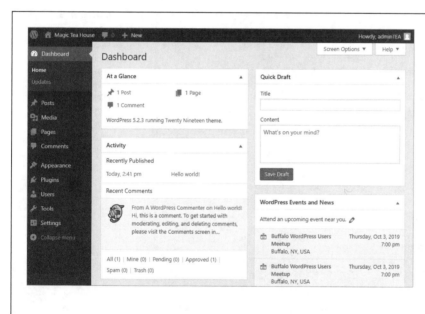

FIGURE 4-5

At the top of the dashboard, the At a Glance box displays your site's vital signs—including how many posts, pages, and comments it has. To the right is a Quick Draft box that lets you create a new post in a hurry. Below that, you'll find boxes with information about recent posts (articles you've written), recent comments (which other people have left in response to your posts), and links to WordPress news.

Don't be surprised to find that your brand-new WordPress site contains content. WordPress starts off every new site with one blog post, one page, and one comment, all of which are dutifully recorded in the At a Glance box. Once you learn to create your own posts, you'll see how to delete these initial examples.

NOTE WordPress continually evolves. When you use the latest and greatest version, you may find that minor details have changed, such as the exact wording of links or the placement of boxes. But don't let these details throw you, because the underlying WordPress concepts and procedures have been steady for years.

■ Administration Practice

Now that you understand how the admin area works, why not try performing a couple of basic administrative tasks? In the following sections, you'll review a few useful WordPress settings and then take a closer look at the theme you're using.

Changing Basic Settings

The following steps show you some settings that you might want to change before you start posting. Even if you don't want to change anything yet, this task will help you get used to clicking your way through the admin menu to find what you need:

1. **In the menu, choose Settings.**

   You'll see a page of tweakable settings (*Figure 4-6*).

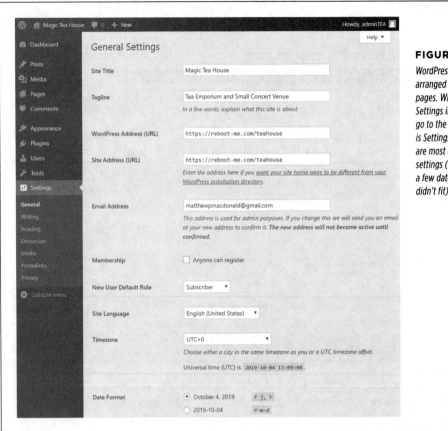

FIGURE 4-6

WordPress settings are arranged into several pages. When you click Settings in the menu, you go to the first page, which is Settings→General. Here are most of the general settings (not including a few date options that didn't fit).

2. **If you'd like, you can change some settings.**

   Here are some suggestions:

Site Title and Tagline. In a basic WordPress site, every page has a header section at the top. WordPress puts the site title and tagline there (*Figure 4-7*). The site title also shows up at the top of the browser window, and, if a visitor decides to bookmark your site, the browser uses the site title as the bookmark text. You shouldn't change these details often, so it's worthwhile to double-check that you have a clear title and catchy tagline right now.

Timezone. This tells WordPress where you are, globally speaking. (For example, UTC-4 is the time zone for New York.) If WordPress doesn't have the right time zone, it will give posts and comments the wrong timestamp. For example, it might tell the world that a comment you left at 8:49 p.m. was actually recorded at 3:49 a.m. If you're not sure what your time zone offset is, don't worry, because WordPress provides a list of cities. Pick the city you live in, or another city in the same time zone, and WordPress sets the offset to match.

Date Format and **Time Format.** Ordinarily, WordPress displays the date for every post you add and the time for every comment made. These settings control how WordPress displays the date and time. For example, if you want dates to be short—like "2020/12/18" rather than "December 18, 2020"—the Date Format setting is the one to tweak.

Week Starts On. This tells WordPress what day it should consider the first day of the week in your country (typically, that's Saturday, Sunday, or Monday). This setting changes the way WordPress groups posts into weeks and the way it displays events in calendars.

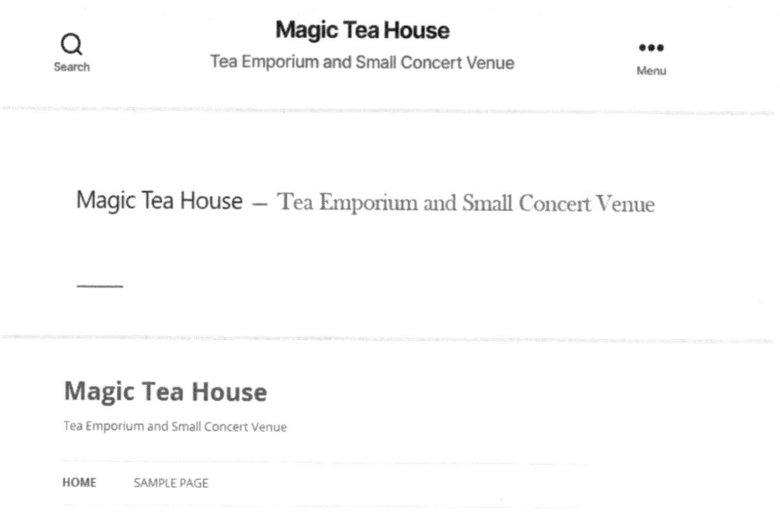

FIGURE 4-7

A WordPress header includes the site's title and tagline. The exact way this information is presented depends on the theme. Here's the same header in the themes Twenty Twenty (top), Twenty Nineteen (middle), and Twenty Sixteen (bottom).

3. **Click Save Changes to make your changes official.**

   WordPress takes a fraction of a second to save your changes and then shows a "Settings saved" message at the top of the page. You can now move to a different part of the admin area.

NOTE There are plenty more WordPress settings to play around with in the Settings page. As you explore various WordPress features in this book, you'll return to these settings to customize them.

Choosing a Starter Theme

Every WordPress site has a theme that sets its layout and visual style. As you begin to refine your site, you'll take a closer look at themes, starting in Chapter 5. But before you even get to that point, you need to pick a good starter theme for your site.

WordPress includes a few themes with every new installation. Somewhat awkwardly, it names each theme after the year it released the theme. At the time of this writing, a WordPress installation includes the Twenty Twenty, Twenty Nineteen, and Twenty Seventeen themes. (The folks at WordPress didn't get around to creating a theme in 2018.)

It's natural to assume that the best theme is the latest one, that the older year themes are obsolete, and that later themes should replace earlier ones in new sites. However, the creators of WordPress take a different approach. They want each new theme to showcase a current, popular style. The Twenty Seventeen theme, for example, suits a site with large, attractive photos (think restaurant or gallery). That's because it features edge-to-edge images and a scrolling trick that makes the content float past the pictures in the background. By comparison, Twenty Twenty is a streamlined theme that has modern typography, assertive headings, and a clean, simple layout.

NOTE There's no shame in using an older year theme, like Twenty Seventeen, if it suits your site. The WordPress designers want their themes to stay useful, even years later.

You can use any theme to learn the fundamentals of WordPress. However, you'll have an easier time using a standard year theme, rather than a theme developed by another company that adds its own features and design rules. The examples in this chapter use Twenty Twenty.

To start yourself out right, here's how to change your site's theme to Twenty Twenty:

1. **In the menu, choose Appearance→Themes.**

   You'll see a small gallery of preinstalled themes. Usually, there won't be many themes here yet, but some WordPress installers bundle in a few extra themes to start you out. The number of themes depends on your web host, but it doesn't matter too much, because you'll eventually want to hunt for your own perfect theme.

2. **Find the box for the Twenty Twenty theme and hover over it (*Figure 4-8*).**

   If you see a Customize button, your site is already using this theme, and there's nothing more to do. If you see an Activate button, you're currently using a different theme.

3. **Click Activate.**

   WordPress immediately applies this theme to your site. Click your site name (near the top-left corner of the page) to view your site and see the results.

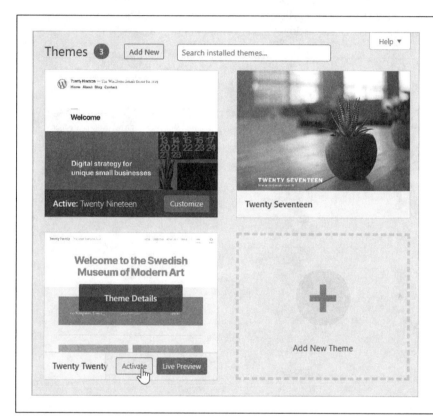

FIGURE 4-8

Here the Twenty Nineteen theme is currently active. You know that because it's the first theme in the list, and the one with the Customize button underneath it. Hover over another theme (like Twenty Twenty, bottom-left) and you'll see the Activate and Live Preview buttons appear.

◼ Adding Your First Post

Comfortable yet? As you've seen, the WordPress admin area gives you a set of relatively simple tools to manage your website. In fact, a good part of the reason WordPress is so popular is that it's so easy to take care of. (And as any pet owner knows, the most exotic animal in the world isn't worth owning if it won't stop peeing on the floor.)

But to really get going with your website, you need to put some content on it. So it makes perfect sense that one of the first tasks that every WordPress administrator learns is *posting*.

Creating a New Post

To create a new post, follow these steps:

1. **In the admin menu, choose Posts→Add New.**

   The Add New Post page appears (*Figure 4-9*). The first time you visit this page, you may get a pop-up message welcoming you to the "wonderful world of blocks." If so, click the tiny X icon to close the pop-up.

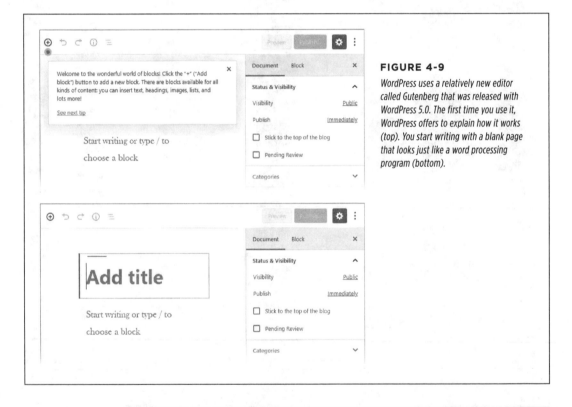

FIGURE 4-9

WordPress uses a relatively new editor called Gutenberg that was released with WordPress 5.0. The first time you use it, WordPress offers to explain how it works (top). You start writing with a blank page that looks just like a word processing program (bottom).

> **TIP** Want to start a new post in a hurry? Once you're logged in to your site, you'll see a thin black toolbar that stays at the top of the page, no matter where you go on your WordPress site. Click New→Post in the toolbar to go straight to the Add New Post page.

2. **Start by typing a post title into the "Add title" box at the top of the page.**

   A good post title clearly announces what you're going to discuss. Often, visitors will come across your post title before getting to your post text. For example, they might see the title in a list of posts or on a search engine results page. A good title communicates your subject and entices the reader to continue on to the post. A lousy title might be cute, clever, or funny, but fails to reflect what the post is about.

Here are some good post titles: "Angela Merkel Struggles in Recent Poll," "Game of Thrones Is Officially Off the Rails," and "My Attempt to Make a Chocolate-Bacon Soufflé." And here are weaker titles for exactly the same content: "Polls, Polls, and More Polls," "GOT Recap," and "My Latest Kitchen Experiment."

3. **Press Enter when you've finished your title to move down to the content area (*Figure 4-10*).**

A basic blog post consists of one or more paragraphs. After each paragraph, press the Enter key (once) to start the next paragraph. WordPress automatically adds a bit of whitespace between paragraphs, so they don't feel too crowded. Resist the urge to sign your name at the end, because WordPress automatically adds this information to the post.

Technically, each paragraph you write is called a *block*. When you move the mouse over a block, a mini toolbar appears with more formatting choices. There's a lot you can do to stylize your blocks, and you'll learn all about that in Chapter 6.

NOTE Don't worry if you're not yet feeling inspired. It's exceptionally easy to delete blog posts, so you can add a simple post just for practice and then remove it later (see page 75).

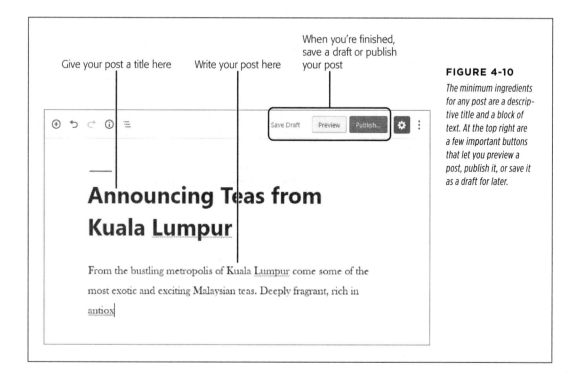

FIGURE 4-10

The minimum ingredients for any post are a descriptive title and a block of text. At the top right are a few important buttons that let you preview a post, publish it, or save it as a draft for later.

4. **Double-check your post.**

A post with typographic errors or clumsy spelling mistakes is as embarrassing as a pair of pants with a faulty elastic band. Before you inadvertently reveal yourself to the world, it's a good idea to double-check your writing.

Most browsers have built-in spellcheckers. You'll see red squiggly lines under your mistakes, and you can right-click misspelled words to choose the right spelling from a pop-up menu.

5. **When you finish writing and editing, click the Preview button (at the top right).**

Your post preview opens in a new browser tab or new browser window. It shows you a perfect rendition of what the post will look like on your site, with the current theme. However, your post isn't live just yet. When you've finished looking at the preview, close the tab to get back to the Add New Post page.

6. **If you like what you saw in the preview, click Publish.**

WordPress shows extra publishing options in the sidebar on the right (you'll get more information on those a bit later in this chapter) and asks "Are you ready to publish?"

7. **Click Publish again.**

A message box appears at the top of the page, confirming that your post has been published. Your post is now live on your site and visible to the world. Click the "View post" link to see the published post (*Figure 4-11*).

If you're not quite done but need to take a break, click Save Draft instead of Publish. WordPress holds onto your post so you can edit and publish it later. Returning to a draft is easy—head to the list of posts (Posts→All Posts), find your post in the list (it'll be near the top), hover over it, and click the Edit link.

If you've decided you don't like your post and want to toss it away, WordPress makes you work a little harder. As you work on a post, WordPress periodically saves it as a draft. So even if you click away to another part of the admin area (or close your browser), WordPress keeps a copy of your unpublished post. To delete it, you need to use the post list described on page 75.

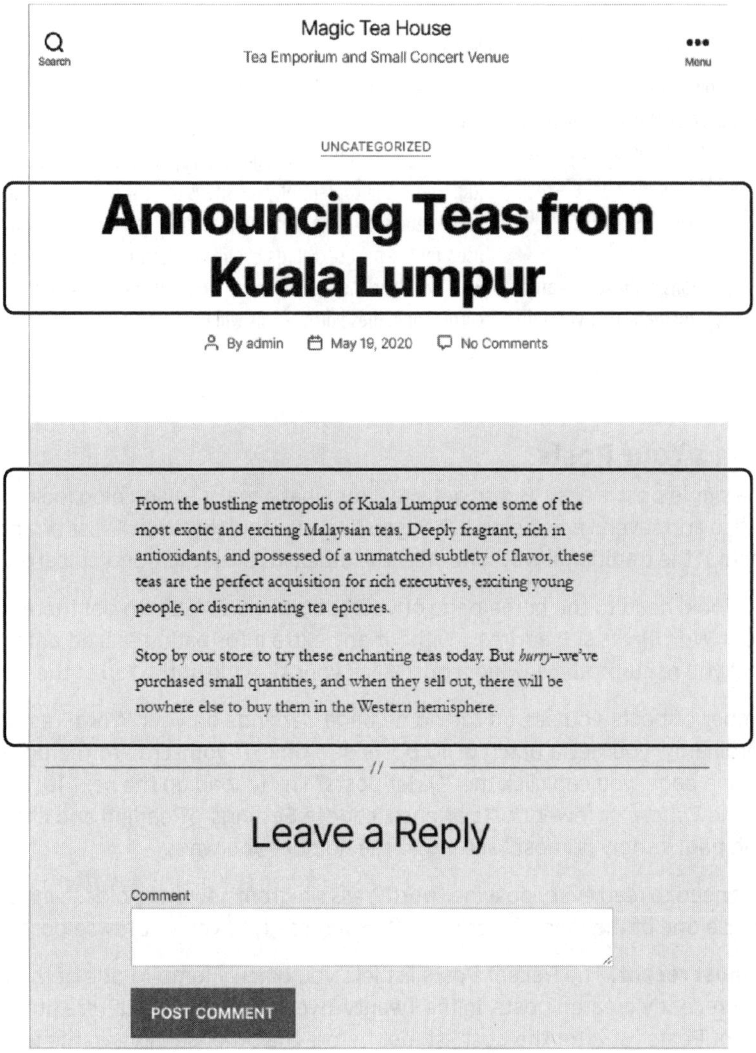

Magic Tea House

Tea Emporium and Small Concert Venue

Q
Search

•••
Menu

UNCATEGORIZED

Announcing Teas from Kuala Lumpur

By admin   May 19, 2020   No Comments

From the bustling metropolis of Kuala Lumpur come some of the most exotic and exciting Malaysian teas. Deeply fragrant, rich in antioxidants, and possessed of a unmatched subtlety of flavor, these teas are the perfect acquisition for rich executives, exciting young people, or discriminating tea epicures.

Stop by our store to try these enchanting teas today. But *hurry*—we've purchased small quantities, and when they sell out, there will be nowhere else to buy them in the Western hemisphere.

Leave a Reply

Comment

POST COMMENT

FIGURE 4-11

Here's the finished post, transplanted into the stock layout of your WordPress site. The two outlined sections represent the content you contributed. WordPress has added extra details, like the category and date information above your post. You'll learn to take charge of these details in this chapter and the next.

Why Your Post Might Look a Little Different

If you try out these steps on your own WordPress site (and you should), you might not get exactly the same page as shown in *Figure 4-11*. For example, the date information, the author byline, and the link that lets you jump to the previous post may be positioned in different spots or have slightly different wording.

You might assume that these alterations represent feature differences, but that isn't the case. Instead, this variabil-

ity is the result of different themes, plugins, and WordPress settings.

The best advice is this: don't get hung up on these differences. Right now, the content of your site is in your hands, but the other details (like the placement of the sidebar and the font used for the post text) are beyond your control. In Chapter 5, when you learn how to change to a new theme or customize a current one, these differences will begin to evaporate.

Browsing Your Posts

Adding a single post is easy. But to get a feel for what a real, thriving blog looks like, you need to add several new posts. When you do, you'll find that WordPress arranges your posts in the traditional way: one after the other, in reverse-chronological order.

To take a look, head to the home page of your blog (*Figure 4-12*). To get there, just enter your WordPress site address, without any extra information tacked onto the end. Or, if you're currently viewing a post, click your site title at the top of the page.

The number of posts you see on the home page depends on your WordPress settings. Ordinarily, you get a batch of 10 posts at a time. If you scroll to the bottom of the home page, you can click the "Older posts" link to load up the next 10. If you want to show more or fewer posts at once, choose Settings→Reading and change the "Blog pages show at most" setting to the number you want.

You don't need to read every post in a WordPress site from newest to oldest. Instead, you can use one of the many other ways WordPress gives you to browse posts:

- **By most recent.** The Recent Posts list lets you quickly jump to one of the five most recently created posts. In the Twenty Twenty theme, WordPress puts the Recent Posts list after the post list, but before the other sections in the footer.

- **By month.** Using the Archives list (which you'll also find in the Twenty Twenty footer), you can see a month's worth of posts. For example, click "June 2020" to see all the posts published that month, in reverse-chronological order. Some WordPress blogs also include a calendar for post browsing, but if you want that, you'll need to add it yourself (see page 132 to learn how).

- **By category or tag.** Later in this chapter, you'll learn how to place your posts in categories and add descriptive tags. Once you take these steps, you'll have another way to hunt through your content, using either the Categories list in

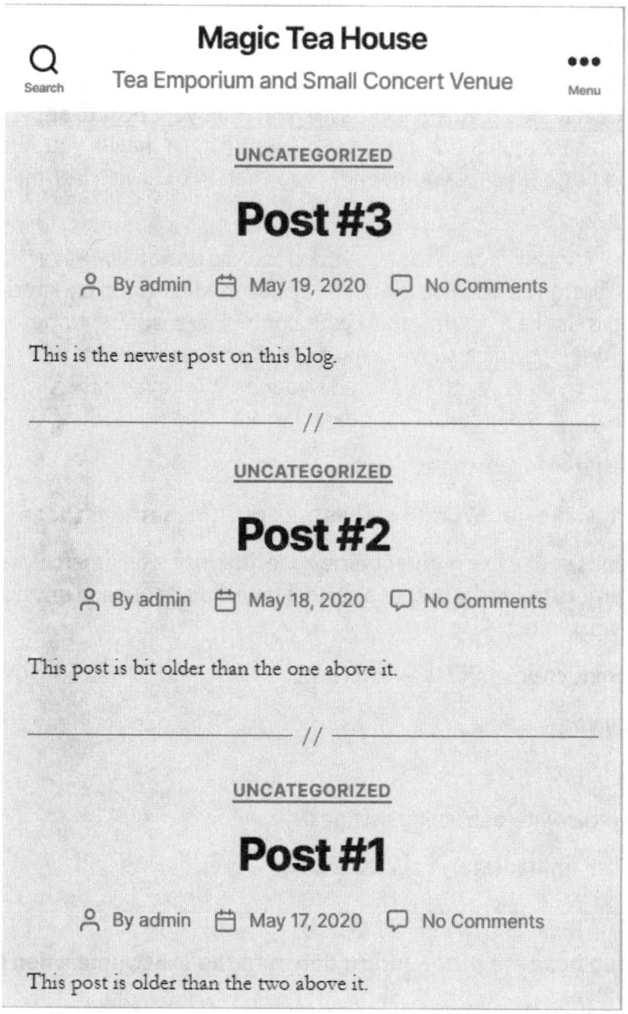

Magic Tea House

Tea Emporium and Small Concert Venue

Q Search

••• Menu

UNCATEGORIZED

Post #3

By admin   May 19, 2020   No Comments

This is the newest post on this blog.

———————— // ————————

UNCATEGORIZED

Post #2

By admin   May 18, 2020   No Comments

This post is bit older than the one above it.

———————— // ————————

UNCATEGORIZED

Post #1

By admin   May 17, 2020   No Comments

This post is older than the two above it.

FIGURE 4-12

When you visit the home page of a blog, you start out with a reverse-chronological view that puts the most recent post first. In this example, the most recent post is named Post #3. The oldest visible post is Post #1.

the footer or the category and tag links that WordPress adds to the end of every post.

- **By author.** If your site has posts written by more than one person, WordPress automatically adds a link with the author name at the top or bottom of every post. Click that link and you'll see all the posts that person has created for this site, in reverse-chronological order (as always). You can't use this feature just yet, but it comes in handy when you create a blog that has multiple authors, as you'll learn to do in Chapter 11.

- **Using a search.** To search a blog, type a keyword or two into the search box, which appears at the top of your footer, and then press Enter. WordPress searches the title and body of each post and shows you a list of matching posts.

Delaying Publication

Sometimes, you might decide your post is ready to go, but you want to wait a little before putting it on the web. For example, you might want your post to coincide with an event or product announcement. Or maybe you want your post to appear at a certain time of day, rather than the 2 a.m. time you wrote it. Or maybe you simply want to add a bit of a buffer in case you get new information or have a last-minute change of heart.

In all these situations, you can choose to save your post as a draft (click Save Draft) and publish it later. That gives you complete control over when the post appears, but it also forces you to make a return trip to your computer. A different approach is to use *delayed publishing*, which allows you to specify a future publication time. Before that time arrives, you may return and edit your post (or even cancel it). But if you do nothing, the post will magically appear, at exactly the time you specified.

To use delayed publishing, follow these steps:

1. **Before you start, make sure WordPress has the right time settings (page 63).**

   If WordPress thinks you're in a different time zone, its clock won't match yours, and when you tell it to publish a post at a certain time, it will appear a few hours before or after you expect.

2. **To write your post, choose Posts→Add New.**

   Write your post in the usual way.

3. **Click the Publish button.**

   This shows a sidebar with publishing settings.

4. **Click the "Publish immediately" section.**

   A calendar pops into view (*Figure 4-13*).

5. **Use the provided boxes to pick a future date and the exact time when the post should go live.**

   At this point, the Publish button turns into a Schedule button.

6. **Click the Schedule button to commit to publishing the post.**

   WordPress will wait until the time you specify and then publish your work.

   If you decide you don't actually want to publish the post at the time you set, you can edit the post (as described in the next section) and put the scheduled time to a very distant future date.

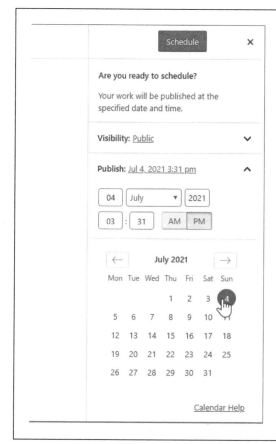

FIGURE 4-13

WordPress lets you schedule content for future publication down to the minute.

TIP You can use the same technique to create a post with an *older* publication date. To do that, just type in a date and time that falls in the past. Although no one wants to make their content look old, this trick is occasionally useful. For example, it's handy if you publish several posts at the same time and want to change their order or spread them out. It also makes sense if you're republishing content that appeared on another site and want to use the true, first-publication date.

Editing a Post

Many people assume that posting on a blog is like sending an email message: you compose your thoughts, write your content as best you can, and then send it out to meet the world. But the truth is that you can tinker with your posts long after you publish them.

WordPress gives you two easy ways to do it.

If you're logged in and viewing a post, an Edit Post link appears in the tiny black toolbar at the top of the page. Click that link, and WordPress takes you to the Edit Post page, which looks almost identical to the Add New Post page. In fact, the only difference is that the Publish button has been renamed Update. Using the Edit Post page, you can change any detail you want, from correcting a single typo to replacing the entire post. When you finish making changes, click the Update button to commit your edit.

Another way to pick a post for editing is to use the admin area. First, choose Posts→All Posts, which shows you a list of all the posts you've published (*Figure 4-14*). Find the post you want to edit, hover over it, and then click the Edit link to get to the Edit Post page, where you can make your changes.

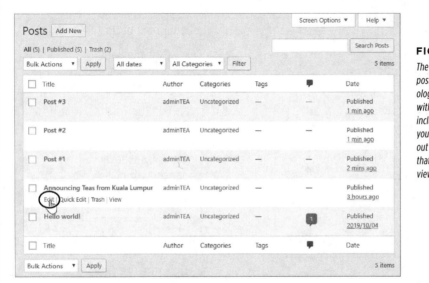

FIGURE 4-14

The Posts page lists your posts in reverse-chronological order, starting with the most recent, and including any drafts. When you point to a post (without clicking), you see links that let you edit, delete, or view the post.

Along with the Edit link, the Posts page includes a Quick Edit link. Unlike Edit, Quick Edit keeps you on the Posts page but pops open a panel that lets you edit some of the post details. For example, you can use Quick Edit to change a post's title, but you can't use it to change the content of the post.

Being able to edit in WordPress is a nearly essential feature. Eventually, even the best site will get something wrong. There's no shame in opening up an old post to correct an error, clean up a typing mistake, or even scrub out a bad joke.

NOTE Unlike some blogging and content management systems, WordPress doesn't display any sort of timestamp or message about when you last edited a post. If you want that, you'll need to add it as part of your edit. For example, you might tack an italicized paragraph onto the bottom of a post that says, "This post edited to include the full list of names" or "Updated on January 25th with the latest survey numbers."

Deleting a Post

As you've just seen, you can edit anything you've ever written on your WordPress website, at any time, without leaving any obvious fingerprints. You can even remove posts altogether.

The trick to deleting posts is to use the Posts page (*Figure 4-14*). Point to the post you want to vaporize and then click Trash. Or, on the Edit Post page, click the Move to Trash link that appears in the Publish box. You can delete posts you've published and drafts (incomplete posts) that never saw the light of day.

TIP Now that you know how to remove a post, try out your new skill with the "Hello world!" example post that WordPress adds to every new blog. There's really no reason to keep it.

Trashed posts aren't completely gone. If you discover you removed a post that you actually want, don't panic. WordPress gives you two ways to get your post back.

If you realize your mistake immediately after you trash the post, look for the message "Item moved to Trash. Undo." It appears in a box at the top of the Posts page. Click the Undo link, and your post returns immediately to both the Posts list and your site.

If you want to restore a slightly older trashed post, you need to dive into the Trash. Fortunately, it's easy (and not at all messy). First, click Posts→All Posts to get to the Posts page. Then click the Trash link that appears just above the list of posts (*Figure 4-15*). You'll see every post that's currently in the trash. Find the one you want, hover over it, and then click Restore to resurrect it (or click Delete Permanently to make sure no one will find it again, ever).

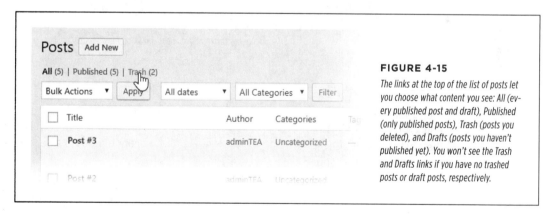

FIGURE 4-15

The links at the top of the list of posts let you choose what content you see: All (every published post and draft), Published (only published posts), Trash (posts you deleted), and Drafts (posts you haven't published yet). You won't see the Trash and Drafts links if you have no trashed posts or draft posts, respectively.

NOTE Of course, removing posts from your blog and scrubbing content from the web are two vastly different things. For example, if you post something impolite about your boss and remove it a month later, the content can live on in the cache that search engines keep and in internet archival sites like the Wayback Machine (*http://web.archive.org*). So always think before you post, because WordPress doesn't include tools to reclaim your job or repair your online reputation.

Creating a Sticky Post

As you know, WordPress orders posts by date on the home page, with the most recent post occupying the top spot. But you might create an important post that you want to feature at the top of the list, regardless of its date. For example, you might write up a bulletin that announces that your business is temporarily closing for renovations, or answers frequently asked questions ("No, there are no more seatings available for this Sunday's Lobster Fest"). To keep your post at the top of the list so it can catch your readers' eyes, you need to turn it into a *sticky post*.

NOTE WordPress displays all your sticky posts before all your normal posts in the post list. If you have more than one sticky post, it lists the most recent one first.

You can designate a post as sticky when you first write it (on the Add New Post page) or when you edit it later (on the Edit Post page). Either way, you use the settings in the sidebar on the right. Here's what to do:

1. **If no sidebar is showing on the right, click the gear icon in the top-right corner to show it.**

2. **At the top of the sidebar, click Document (*Figure 4-16*).**

3. **Put a checkmark in the "Stick to the top of the blog" setting.**

4. **Click Publish or Update to confirm your changes.**

FIGURE 4-16

The sidebar has two tabs: Document and Block. The Document tab has the sticky setting.

The only caveat with sticky posts is that they stay sticky forever—or until you "unstick" them. The quickest way to do that is to choose Posts→All Posts, find the sticky post in the list, and then click the Quick Edit link underneath it. Turn off the "Make this post sticky" checkbox and then click Update.

UP TO SPEED

The Path to Blogging Success

There's no secret trick to building a successful blog. Whether you're recording your thoughts or promoting a business, you should follow a few basic guidelines:

- **Make sure your content is worth reading.** As the oft-reported slogan states, *content is king*. The best way to attract new readers, lure them in for repeat visits, and inspire them to tell their friends about you, is to write something worth reading. If you're creating a topical blog (say, putting your thoughts down about politics, literature, or gourmet marshmallows), your content needs to be genuinely *interesting*. If you're creating a business blog (for example, promoting your indie record store or selling your real estate services), it helps to have content that's truly *useful* (say, "How to Clean Old Records" or "The Best Chicago Neighborhoods to Buy In").

- **Add new content regularly.** Nothing kills a site like stale content. Blogs are particularly susceptible to this problem because posts are listed in chronological order, and each post prominently displays the date you wrote it (unless you remove the dates by editing your theme files; see Chapter 13).

- **Keep your content organized.** Even the best content can get buried in the dozens (or hundreds) of posts you'll write. Readers can browse through your monthly archives or search for keywords in a post, but neither approach is convenient. Instead, a good blog is ruthlessly arranged using *categories* and *tags* for the posts (see the next section).

■ Organizing Your Posts

WordPress gives you two complementary tools for organizing your posts: categories and tags. Both work by grouping related posts together. In the following sections, you'll learn how to use them effectively.

Understanding Categories

A *category* is a short text description that describes the topic of a group of posts. For example, the Magic Tea House uses categories like *Tea* (posts about teas for sale), *Events* (posts about concert events at the tea house), and *News* (posts about other developments, like renovations or updated business hours).

Categories are really just text labels, and you can pick any category names you want. For example, the categories Tea, Events, and News could just as easily have been named Teas for Sale, Concerts, and Miscellaneous, without changing the way the categories work.

In a respectable WordPress site, every post has a category. (If you don't assign a category, WordPress automatically puts your post in a category named Uncategorized, which presents a bit of a logical paradox.) Most of the time, posts should have just one category. Putting a post in more than one category is a quick way to clutter up the structure of your site, and confuse anyone who's browsing your posts one category at a time.

TIP A good rule of thumb is to give every post exactly one category. If you want to add more information to make it easier for people to find posts that are related to each other, add tags, which you'll learn about next.

You don't need to create all your categories at once. Instead, you can add them as needed (for example, when you create a new post that needs a new category). Of course, you'll have an easier time organizing your site if you identify your main categories early on.

It's up to you to decide how to categorize posts and how many categories you want. For example, the Magic Tea House site could just as easily have divided the same posts into more categories, or into different criteria, as shown in *Figure 4-17*.

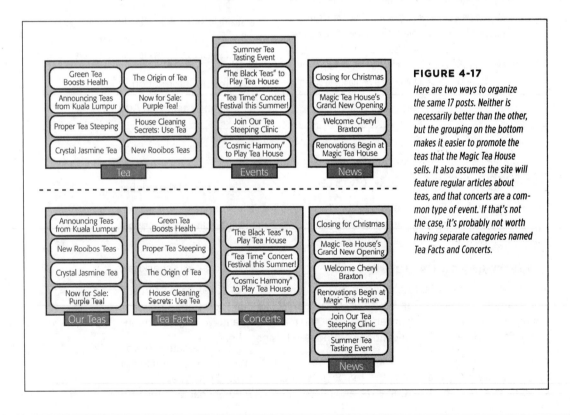

FIGURE 4-17

Here are two ways to organize the same 17 posts. Neither is necessarily better than the other, but the grouping on the bottom makes it easier to promote the teas that the Magic Tea House sells. It also assumes the site will feature regular articles about teas, and that concerts are a common type of event. If that's not the case, it's probably not worth having separate categories named Tea Facts and Concerts.

Categorizing Posts

You can easily assign a category to a post when you first add the post. Here's how:

1. **Choose Posts→Add New to start a new post.**

   Or you can start editing an existing post and then change its category. The Add New Post and Edit Post pages work the same way, so it doesn't make a difference.

2. **Look to the sidebar on the right. At the top of it, click Document.**

   If you've accidentally closed the sidebar, click the gear icon in the top-right corner to show it.

   The Document tab has the options for adding extra information to your post. The Block tab has options for formatting individual parts of your post. (You'll take a closer look at that in Chapter 6.)

3. **Click the Categories heading in the sidebar.**

   You'll see a list with all the categories you've created (*Figure 4-18*).

   If the category you want exists, skip to step 6.

   If your post needs a new category, one that you haven't created yet, continue on to step 4.

4. **At the bottom of the Categories section, click Add New Category.**

   This expands the Categories section so you can enter category information.

5. **Enter the category name in the New Category Name box and then click Add New Category.**

   Don't worry about the Parent Category box—you'll learn to use that on page 82, when you create subcategories.

   Once you add your category, it appears in the Categories section of the sidebar so you can use it.

6. **Find the category you want to use in the list and then turn on the checkbox next to it.**

   When you add a new category, WordPress automatically turns on its checkbox, because it assumes this is the category you want to assign to your post. If it isn't, simply turn off the checkbox and pick something else.

7. **Carry on editing your post.**

   That's it. When you publish your post, WordPress assigns it the category you chose (*Figure 4-19*). If you didn't choose a category, WordPress automatically puts it in a category named (paradoxically) Uncategorized.

FIGURE 4-18

Here there are three custom categories to choose from (Events, News, and Tea), along with the fallback category (Uncategorized).

TIP Ordinarily, the category named Uncategorized is WordPress's *default category*—that means WordPress uses it for new posts unless you specify otherwise. However, you can tell WordPress to use a different default category. Simply choose Settings→Writing and pick one of your categories in the Default Post Category list.

TEA

Announcing Teas from Kuala Lumpur

👤 By admin 📅 May 20, 2020 💬 No Comments

From the bustling metropolis of Kuala Lumpur come some of the most exotic and exciting Malaysian teas. Deeply fragrant, rich in antioxidants, and possessed of a

FIGURE 4-19

This post is in the Tea category. Click the link (the word "Tea" above the post), and you'll see all the posts in that category. If your post is in more than one category, you'll see a separate link for each one.

UP TO SPEED

How to Choose Good Categories

To choose the right categories, you need to imagine your site, up and running, several months down the road. What posts does it have? How do people find the content they want? If you can answer these questions, you're well on the way to choosing the best categories.

First, you need to choose categories that distribute your posts *well*. If a single category has 90% of your posts, you probably need new—or different—categories. Similarly, if a category accounts for less than 2% of your posts, you may have too many categories. (Although there are exceptions—perhaps you plan to write more on that topic later, or you want to separate a very small section of special-interest posts from the rest of your content.)

You may also want to factor in the sheer number of posts you plan to write. If your site is big and you post often, you may want to consider more categories. For example, assuming the Magic Tea House has two dozen posts, a category split like this works fine: Tea (70%), Concerts (20%), News (10%). But if you have hundreds of posts, you'll probably want to subdivide the big Tea group into smaller groups.

It also makes sense to create categories that highlight the content you want to promote. For example, if you're creating a site for a furniture store, you'll probably create categories based on your products (Couches, Sofas, Dining Room Tables, and so on). Similarly, the Magic Tea House can split its Tea category into Our Teas and Tea Facts to better highlight the teas it sells (as shown previously in *Figure 4-17*).

Finally, it's important to consider how your readers will want to browse your information. If you're a lifestyle coach writing articles about personal health, you might decide to add categories like Good Diet, Strength Training, and Weight Loss, because you assume that your readers will zero in on one of these subjects and eagerly devour all the content there. Be careful that you don't split post categories too small, however, because readers could miss content they might otherwise enjoy. For example, if you have both a Good Diet Tips and Superfoods category, a reader might explore one category without noticing the similar content in the other. This is a good place to apply the size rule again—if you can't stuff both categories full of good content, consider collapsing them into one group or using subcategories.

Using Subcategories

If you have a huge site with plenty of posts and no shortage of categories, you may find that you can organize your content better with *subcategories*. The idea behind subcategories is to take a large category and split it into two or more smaller groups.

However, rather than make these new categories completely separate, WordPress keeps them as subcategories of the original category, which it calls the *parent category*. For example, the Magic Tea House site could make Tea a parent category and create subcategories named Black Tea, Green Tea, Rooibos, and Herbal Tea.

Done right, subcategories have two potential benefits:

- **Visitors can browse posts by subcategory or parent category.** People using the Magic Tea House site can see all the tea posts at once (by browsing the Tea category) or can drill down to the subcategory of tea that interests them the most.

- **You can show a category tree.** A category tree shows the hierarchy of your categories. In a complex site with lots of categories, most readers find that this makes it easier to browse the categories and understand how the topics you cover are related. You'll learn how to show a category tree in Chapter 5.

You can create subcategories by using the Categories box—in fact, it's just as easy as creating ordinary categories (*Figure 4-20*). The only requirement is that you create the parent category first. Then enter the subcategory name, pick the parent category in the Parent Category list, and click Add New Category.

Yes, you can also create subcategories inside subcategories. But try to avoid complicating your life in this way unless you really need to (for example, your site has dozens of categories that you want to group together).

WordPress displays categories hierarchically in the Categories box, so you'll see your subcategories (like Green Tea) displayed underneath the parent category (Tea). However, there's an exception—when you first add a new subcategory, WordPress puts it at the top of the list, and it stays there until you refresh the page or add a new category. Don't let this quirk worry you; your new category is still properly attached to its parent.

> **NOTE** When you assign a post to a subcategory, make sure you pick the subcategory only, *not* the parent category. For example, if you want to add a post about green tea, you should turn on the checkbox next to the Green Tea box, but *not* the Tea box. Because Tea is the parent category, people who browse the Tea category will automatically see your Green Tea posts.

When you start adding subcategories to your site, you might be disappointed by the way they appear in the Categories list. The standard list of categories is a flat, one-dimensional list in alphabetical order (*Figure 4-21*). You can't see the relationships between parent categories and subcategories. But you can change the way this list works with the Categories widget, as you'll see in Chapter 5.

Categories
Events
News
Tea
Rooibos
Black Tea
Herbal Tea
Uncategorized
Add New Category
New Category Name
Green Tea
Parent Category
Tea
Add New Category

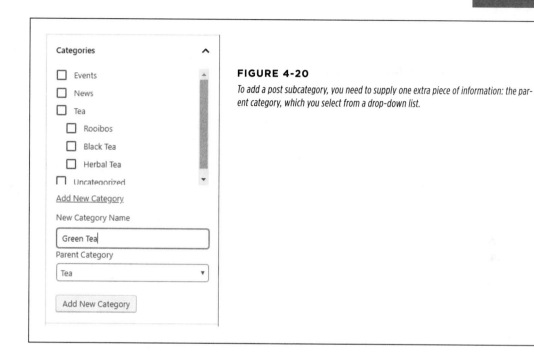

FIGURE 4-20

To add a post subcategory, you need to supply one extra piece of information: the parent category, which you select from a drop-down list.

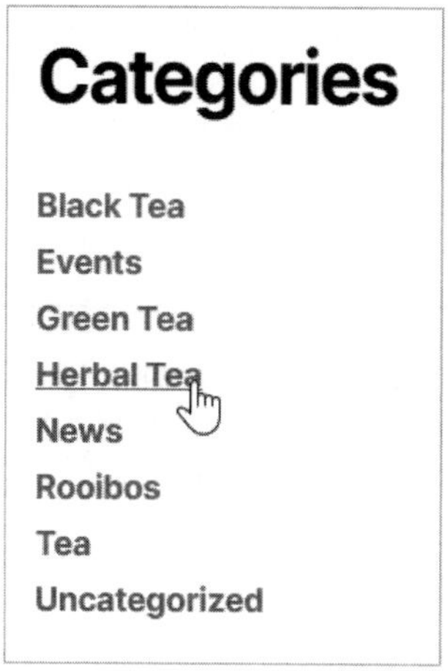

Categories
Black Tea
Events
Green Tea
Herbal Tea
News
Rooibos
Tea
Uncategorized

FIGURE 4-21

Ordinarily, the Categories list shows all the categories that you use. If a category doesn't have any posts, it won't appear in the Categories list.

Managing Categories

As you've seen, you can create a category whenever you need one, right from the Add New Post or Edit Post page. However, the WordPress admin area also includes a page specifically for managing categories. To get there, choose Posts→Categories, and you'll see a split page that lets you add to or edit your categories (*Figure 4-22*).

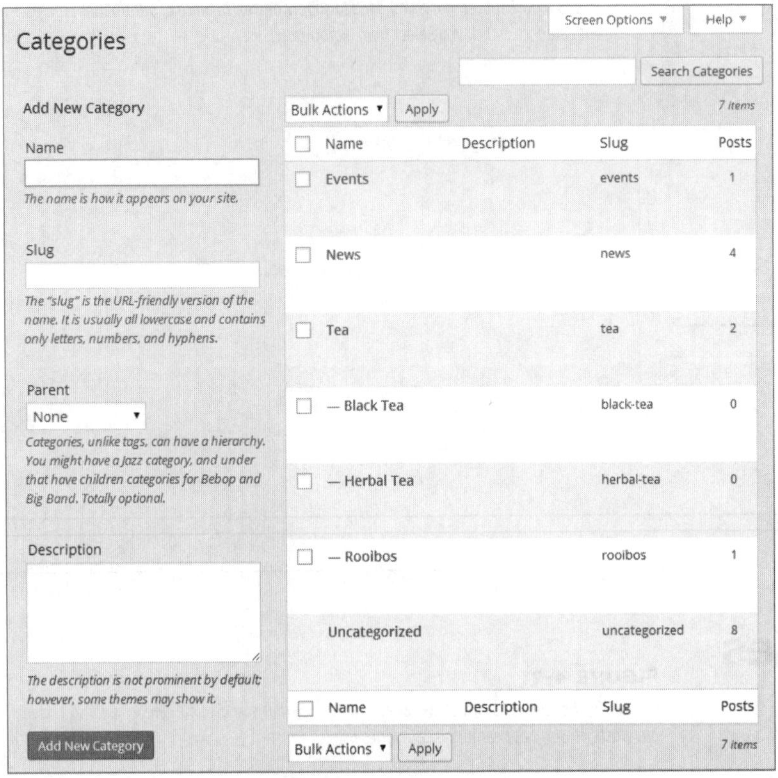

FIGURE 4-22

The Categories page includes a section on the left for adding new categories and a detailed list of all your categories on the right. The categories list works in much the same way as the list of posts on the Posts page. Point to a category without clicking, and you get the chance to edit or delete it.

The Categories page lets you perform a few tasks that aren't possible from the lowly Categories box:

- **Delete categories you don't use.** When you take this step, WordPress reassigns any posts in the category to the default category, which is Uncategorized (unless you changed the default in the Settings→Writing page).

- **Edit a category.** For example, you might want to take an existing category and rename it, or make it a subcategory by giving it a parent.

- **Enter extra category information.** You already know that every category has a name and, optionally, a parent. In addition, categories have room for two pieces of information that you haven't used yet: a slug and a description. The *slug* is a simplified version of the category name that appears in the web address when you're browsing the posts in a specific category (as explained on page 96). The *description* explains what the category is all about. Some themes display category descriptions in their category-browsing pages, but otherwise it isn't important.

Understanding Tags

Like categories, *tags* are text labels that add bits of information to a post. But unlike categories, a post can (and should) have multiple tags. For that reason, the process of applying tags is less strict than the process of putting your post in the right category.

Tags are often more specific than categories. For example, if you write a review of a movie, you might use Movie Reviews as your category and the movie and director names as tags.

Follow these guidelines when you use tags:

- **Don't over-tag.** Instead, choose the best 5 to 10 tags for your content.

- **Keep your tags short and precise.** Pick "Grateful Dead" over "Grateful Dead Concerts."

- **Reuse your tags on different posts.** Once you pick a good tag, put it to work wherever it applies. After all, tags are designed to help people find related posts. And never create a similarly named tag for the same topic. For example, if you decide to add the tag "New York Condos," and then you use the tags "NY Condos" and "Condo Market," you've created three completely separate tags that won't share the same posts.

- **Consider using popular tags.** If you're trying to attract search engine traffic, you might consider using hot search keywords for your tags.

- **Don't duplicate your category with a tag of the same name.** WordPress treats categories and tags in a similar way, as bits of information that describe a post. Duplicating a category with a tag is just a waste of a tag.

> **TIP** Here's some advice to help you get straight about categories and tags. Think of the category as the *fundamental grouping* that tells WordPress how a post fits into the structure of your site. Think of a tag as a *searching convenience* that helps readers hunt for content or find a related post.

Tagging Posts

Adding a tag to a post is even easier than assigning it to a category. When you create a post (in the Add New Post page) or edit a post (in the Edit Post page), look for the Tags section in the sidebar on the right. It appears just under the Categories section.

To add a tag, type the tag into the text box and press Enter or the comma (,) key. This adds the first tag, but you can keep typing to add as many tags as you want (*Figure 4-23*). Tags can have spaces in them (like "store event"), but that's less common than one-word tags (like "sale").

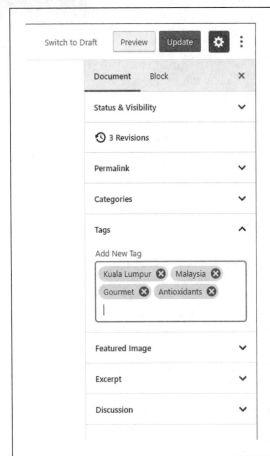

FIGURE 4-23

Right now, this post has four tags. Keep typing to add another. If you change your mind, you can remove a tag by clicking the X icon that appears next to it.

Different themes put category and tag information in different places. In Twenty Twenty, the category is always displayed above the post title. The tags appear at the end of the post (*Figure 4-24*). You can click a tag to see a list of similarly tagged posts on your site. Many blogs also use a *tag cloud*, a cluster of tag links, sized in proportion to how often you use them (in other words, in proportion to how many posts feature that tag). The default WordPress site layout doesn't use a tag cloud, but you can add one easily by using the Tag Cloud widget. You'll learn how in Chapter 5.

As with categories, tags have their own management page, which you can see by choosing Posts→Tags. There you can add tags, remove tags, and edit the tag slugs.

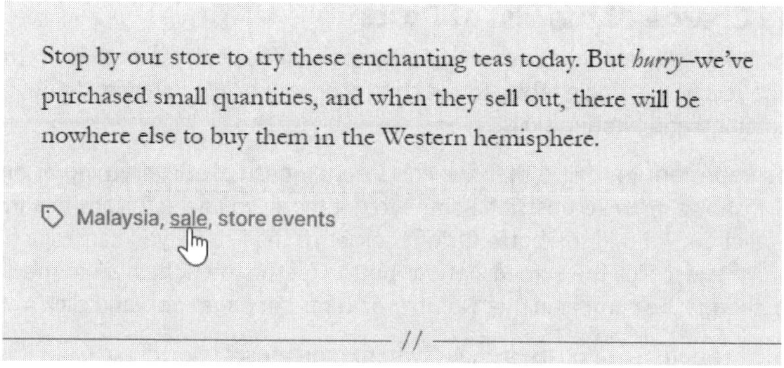

Stop by our store to try these enchanting teas today. But *hurry*—we've purchased small quantities, and when they sell out, there will be nowhere else to buy them in the Western hemisphere.

🏷 Malaysia, sale, store events

FIGURE 4-24

The exact look of a post byline depends on your theme, but most list the categories and tags you added to a post. In Twenty Twenty, the tags are at the end of a post.

POWER USERS' CLINIC

Being in Two Places at Once

One potential problem with the admin area is that it lets you view only one page at a time. This limitation can become awkward in some situations. For example, imagine you're in the middle of creating a post when you decide you want to review a setting somewhere else in the admin area. You *could* save the post as a draft, jump to the settings page, and then return to continue with your post. And, for many people, this approach works just fine. But if you're the sort of power user who's comfortable with browser shortcuts, another approach may appeal to you: opening more than one browser page at a time, with each positioned on a different part of the admin area.

It all works through the magic of the Ctrl-click—a nifty browser trick where you hold down the Ctrl key (Command on a Mac)

while clicking a link. This causes the target of the link to open in a new tab, which appears next to the page you're already looking at.

For example, imagine you're at the Add New Post page and want to review your post display settings. To open the settings page in a new tab, hover over Settings in the menu and Ctrl-click the Reading link. Keep in mind, however, that if you change something in one tab that affects another, you might not see the results of your change right away. For example, if you add a category on the Categories page while the Add New Post page is open, you won't see it in the Add New Post page unless you refresh the page.

■ Working with Several Posts at Once

So far, you've worked on one post at a time. If you plan to change a post's title or edit its text, this is the only way to go. But if you want to manipulate several posts in the same way—for example, change their category in one step—the Posts page lets you carry out your work in bulk.

You've already seen the Posts page, which lists the posts on your site. Now you'll learn how to use it to change or remove groups of posts. But first you need to learn how to search for the posts you want.

Taking Charge of the List of Posts

As your site grows larger, it becomes increasingly difficult to manage everything on one page and in a single table. To get control of your posts, you need to develop your searching and filtering skills.

First, it's important to realize that the Posts page doesn't list everything at once. Instead, it shows up to 20 posts at a time—to get more, you need to click the arrow buttons that show up in the bottom-right corner of the list. Or you can adjust the 20-post limit: just click the Screen Options button (at the top right, next to the Help button), change the number in the "Number of items per page" box, and click Apply.

Changing the number of posts is one way to fit more posts into your list, but it isn't much help if you want to home in on a specific batch of posts that might be scattered throughout your site. In this situation, WordPress has another set of tools to help you out: its filtering controls. Using the drop-down lists at the top of the table of posts (*Figure 4-25*), you can choose to show only posts that were made during a specific month (for example, "June 2020") or that belong to a specific category (say, "Green Tea").

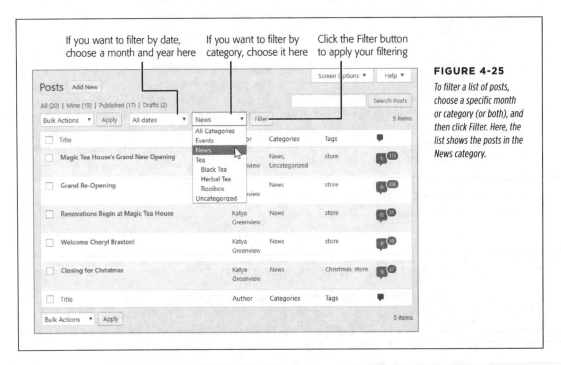

FIGURE 4-25

To filter a list of posts, choose a specific month or category (or both), and then click Filter. Here, the list shows the posts in the News category.

Ordinarily, WordPress displays posts in the same reverse-chronological order that they appear on your site. But you can change that by clicking one of the column headings. Click Title to sort alphabetically by headline, or Author to sort alphabetically by writer. If you click the column heading a second time, WordPress *reverses* the sort order. So if you sort a list of posts by Title, clicking Title a second time shows the posts in reverse-alphabetical order. And if you sort posts by Date, so that the newest posts appear first, clicking Date again displays the posts with the oldest ones on top.

The last trick that the Posts window offers is the search box in the top-right corner, above the posts list. You can search for all the posts that have specific keywords in their titles or text. For example, to show posts that talk about veal broth, you would type in *"veal broth"* (using quotation marks if you want to turn both words into a single search term), and then click Search Posts.

Deleting a Group of Posts

The easiest bulk action is deleting. And while it might seem unlikely that you'd want to wipe out a bunch of your work at once, bulk deletes are useful if you need to clean up a bunch of old drafts.

To send a batch of posts to the trash, follow these simple steps:

1. **Choose Posts→All Posts in the menu.**

   That takes you to the familiar Posts page.

2. **Turn on the checkbox next to each post you want to remove.**

   If you have a huge collection of posts, you can use searching or filtering to home in on just the posts you want.

3. **In the Bulk Actions list, choose Move to Trash.**

   The Bulk Actions list appears in two places: just above the list of posts and just underneath it. That way, it's easily accessible no matter where you are.

4. **Click Apply.**

   WordPress moves all the selected posts to the trash.

Editing a Group of Posts

You can also use a bulk action to make certain post changes. This can be a lifesaver if you need to make a big, sweeping reorganization. For example, you can add tags to a whack of posts in one go, change some posts from one author to another, reassign categories, or publish a group of drafts at the same time. If you need to perform an operation like this, bulk editing will save you plenty of mouse clicks.

To edit a batch of posts, follow these steps:

1. **Choose Posts→All Posts in the menu.**

2. **Turn on the checkbox next to each post you want to edit.**

   Use searching and filtering if you need it to find the right posts.

3. In the Bulk Actions list, choose Edit and then click Apply.

WordPress opens a panel at the top of the post list with editing options (*Figure 4-26*).

4. Manipulate the details you want to change.

Using the Bulk Edit panel, you can add tags (type them in) or apply a category. However, you can't remove tags or *change* the category. That means that if you apply a new category, your posts will actually have two categories, which probably isn't what you want. (Sadly, you have to remove the category you don't want individually, post by post.)

The Bulk Edit panel also lets you change the post's author (if your site has more than one), its status (for example, turning a draft into a published post), and a few other settings that you'll explore in the coming chapters.

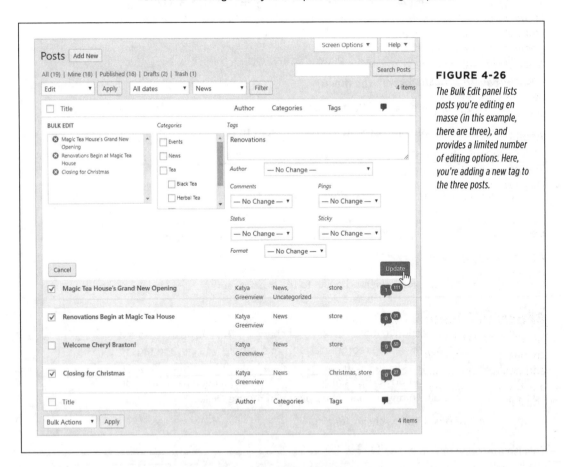

FIGURE 4-26

The Bulk Edit panel lists posts you're editing en masse (in this example, there are three), and provides a limited number of editing options. Here, you're adding a new tag to the three posts.

5. Click Update in the Bulk Edit panel.

Or you can always back out by clicking Cancel.

How to Get High-Quality Web Addresses

Every post you put on a WordPress site has its own unique web address, or URL. So far, you haven't really thought about what those URLs look like. After all, nobody *needs* to type in a web address to read a post. Instead, people can simply visit the front page of your site and click through to whatever content interests them.

However, seasoned web designers know that web addresses matter—not just for the front page of your site, but for each distinct bit of content. One reason is that search engines pay attention to the keywords in a URL, so they treat something like *http://wastedminutes.com/best_time_wasters* differently from *http://wastedminutes.com/post/viewer.php?postid=3980&cat=83*. All other factors being equal, if someone searches for "time wasters" in Google, the search engine is more likely to suggest the first page than the second.

Another important detail is the *lifetime* of your URLs. Ideally, a good web address never changes. Think of an address as a contract between you and your readers. The promise is that if they bookmark a post, the web address will still work when they return to read it, even months or years later (assuming your entire site hasn't gone belly-up in the interim). WordPress takes this principle to heart. In fact, it calls the unique web address that's assigned to every post a *permalink*, emphasizing its permanent nature.

As you'll soon see, WordPress will happily give your posts meaningful web addresses that last forever, but you might need to help it out a bit. Before you can do that, you need to understand a little more about how the WordPress permalink system works.

Choosing a Permalink Style

Here's the good news: WordPress lets you choose the permalink style. If you don't change this setting, WordPress opts into a permalink naming system called "Day and name." That means it adds date and name information into your post URLs, along with a simplified version of the post title. So if you publish a post on July 28, you'll get a URL like this:

```
http://magicteahouse.net/2020/07/28/announcing-teas-from-kuala-lumpur
```

This isn't terrible, but unless you're running a news-oriented site, it's probably not exactly what you want. For one thing, it makes URLs quite long. And it also emphasizes the date, which can make older content feel out-of-date. If someone sees a link in this style and it's a few years old, they might not bother to visit the page, even though the content may still be accurate and up-to-date.

Usually, you'll want to use the "Post name" style, which doesn't include any date information. This gives you cleaner, simpler URLs like this:

```
http://magicteahouse.net/announcing-teas-from-kuala-lumpur
```

The only quirk is that if you give more than one post the same name (please don't), WordPress will tack a number on the end of the permalink to make each post unique.

To choose a different permalink style, follow the next steps. However, be warned that changing the permalink style will change the links for any posts you've already published.

> **WARNING** If you want to tweak the way your WordPress site generates permalinks (and you almost certainly do), it's best to make that change as soon as possible. Otherwise, changing the permalink style can break the web addresses for old posts, frustrating readers who have bookmarked them. Think twice before tampering with the URL structure of an established site.

1. **In the admin menu, choose Settings→Permalinks.**

   The Permalinks Setting page appears.

2. **Choose a new permalink style:**

   Plain. This style uses brief but obscure permalinks that feature WordPress's internal post ID, like *http://magicteahouse.net/?p=13*. WordPress used to use this system for new sites, but changed its ways because almost no one likes these links. After all, can you tell the difference between *http://lazyfather .com/blog/?p=13* (a post about cute family pets) and *http://lazyfather.com /blog/?p=26* (a post about the coming Mayan apocalypse)?

   Day and name. This default style includes the year, month, and date, separated by slashes. At the end is the more descriptive post name, as in *http://magic teahouse.net/2020/07/28/announcing-teas-from-kuala-lumpur*.

   Month and name. This style is similar to "Day and name," except that it leaves out the date number, giving you a slightly more concise permalink, like *http:// magicteahouse.net/2020/07/announcing-teas-from-kuala-lumpur*.

   Numeric. Like the Plain style, this style uses the post ID. But instead of including the *?p=* part, it adds the text */archives*, as in *http://magicteahouse.net /archives/13*. This type of permalink is still as clear as mud.

   Post name. This style omits all the date information, using just the post name, as in *http://magicteahouse.net/announcing-teas-from-kuala-lumpur*. The advantages of this system are that the permalinks it creates are concise and easy to remember and understand. They don't emphasize the date the content was created, which is important if you have timeless content that you want to refer to months or years later.

Custom structure. This is an advanced option that lets you tell WordPress exactly how it should cook up permalinks. The most common reason to use a custom structure is that you want the post category to appear in your permalink (as explained in the next section).

TIP If you don't want to emphasize dates, the "Post name" style is a great choice. If you're concerned about clashing titles, "Month and name" is safer, and if you want to emphasize the exact date of your posts—for example, if you write time-sensitive or news-like content—"Day and name" is a good choice.

---

3. **Click Save Changes.**

WordPress applies the permalink change. As a side effect, the links of all your existing, published posts change to the new style, and your old links will no longer work. This could be a big problem if you have an established site and its content has been bookmarked by visitors and indexed by search engines.

Creating a Custom Permalink System

If you're ambitious, you can make deeper customizations to the way WordPress generates post permalinks. To do that, you need to choose the Custom Structure permalink type and then fill in your permalink "recipe" with the right codes. But how do you build the right recipe?

You might have already noticed that when you choose a permalink style other than Default, WordPress automatically inserts the matching codes in the Custom Structure box. For example, if you choose "Day and name," these codes appear:

%year%/%monthnum%/%day%/%postname%

Think of this as a recipe that tells WordPress how to build the permalink. Each code (that's the bit of text between percentage signs, like *%year%* and *%monthnum%*) corresponds to a piece of information that WordPress will stick into the web address.

In this example, four codes are separated by three slashes. When WordPress uses this format, it starts with the site address (as always) and then adds the requested pieces of information, one by one. First it replaces the *%year%* code with the four-digit year number:

http://magicteahouse.net/**2020**/%monthnum%/%day%/%postname%

Then it replaces the *%monthnum%* code with the two-digit month number:

http://magicteahouse.net/2020/**07**/%day%/%postname%

It carries on until the permalink is complete:

http://magicteahouse.net/2020/07/**28/the-origin-of-tea**

WordPress recognizes 10 codes. More than half are date-related: *%year%*, *%monthnum%*, *%day%*, *%hour%*, *%minute%*, and *%second%*. Additionally, there's a code for the category slug (*%category%*) and the author (*%author%*). Finally, every permalink must end with either the numeric post ID (represented by *%post_id%*) or post name (*%postname%*), because this is the unique detail that identifies the post.

Often, WordPress gurus use a custom permalink structure that adds category information. They do so because they feel that the permalinks are aesthetically nicer—in other words, clearer or more meaningful—or because they think that this increases the chance that search engines will match their post with a related search query.

Here's an example of a custom permalink structure that creates category-specific permalinks:

```
%category%/%postname%
```

Now WordPress creates permalinks that include the category name (in this case, Tea) and the post name (The Origin of Tea), like this:

```
http://magicteahouse.net/tea/the-origin-of-tea
```

This type of URL doesn't work well if you assigned some of your posts more than one category. In such a case, WordPress picks one of the categories to use in the web address, somewhat unpredictably. (Technically, WordPress uses the category that has the lower category ID, which is whichever one you created first.) But otherwise it's a neat way to put extra information in your URL.

Changing a Post's Permalink

Most WordPress fans prefer *pretty permalinks*—web addresses that include the post title. You get pretty permalinks in every permalink style except Default and Numeric.

However, pretty permalinks aren't always as pretty as they should be. The problem is that a post title doesn't necessarily fit well into a web address. Often, it's overly long or includes too many dashes. In this situation, you can help WordPress out by explicitly editing the *slug*—the version of the post name that WordPress uses in your permalinks.

You can change the slug when you add or edit a post. Here's how:

1. **Publish or preview your post.**

   If you've just written a brand-spanking-new post, WordPress hasn't actually picked a slug yet. It does that when you publish the post. (WordPress has a reason to wait until the last second. It knows you might change your post title, and it wants to choose the best possible permalink.)

   If you want to change your permalink *before* you publish a post, you have to trick WordPress into picking the slug early. To do that, you preview your post by clicking the Preview button. You don't need to actually take a look at the preview—just the act of forcing WordPress to create the preview also makes it choose a slug.

2. **Click the post title.**

A box appears around the title. The permalink appears at the top of this box.

WordPress creates the slug automatically, once you type in the post title and start entering the post content. After this point, the slug doesn't change, even if you edit the title, unless you edit it explicitly.

3. **Click the Edit button next to the permalink.**

WordPress converts the portion of the permalink that holds the slug into a text box (*Figure 4-27*). You can then edit to your heart's content, so long as you stay away from spaces and special characters, which aren't allowed in URLs. The best permalinks are short, specific, and unlikely to be duplicated by other posts. (Although WordPress is smart enough to add a number to your slug if you've already assigned it to another post.)

> **NOTE** If you see a Change Permalinks button next to the permalink where the Edit button should be, you're using the Plain permalink style. Follow the instructions in the previous section to change that.

4. **Click Save to make your change official.**

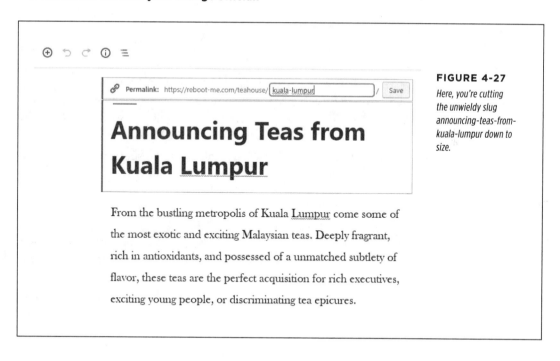

FIGURE 4-27

Here, you're cutting the unwieldy slug announcing-teas-from-kuala-lumpur down to size.

HOW TO
GET HIGH-
QUALITY WEB
ADDRESSES

How WordPress Handles Permalinks That Change

WordPress is very conscientious about dealing with old permalinks. Imagine you create a post with one slug (*tea-sale*) and publish it. Then, you change it to use a different slug (*spring-tea-sale*). If you type in the old permalink using the old slug (*http://magicteahouse.net/tea-sale*), WordPress will send you to the new address (*http://magicteahouse.net/spring-tea-sale*). Nice! But remember, this redirection is for changed slugs only. WordPress won't work its redirection magic if you change the permalink *style* of your entire site.

WordPress's automatic redirection makes sure old links don't break. But this convenience can cause problems. Consider the tea sale permalinks. What happens if, a month later, you decide you want to create a new post that uses the old, original slug (*tea-sale*)? This won't work, because WordPress recognizes that the old slug is in use—it's being redirected to another

address. Instead, when you try to use *tea-sale* in your new post, WordPress will change it to *tea-sale-2*.

There's no built-in way to tell WordPress you don't need it to redirect a post, and you want to get your favorite permalink back. To resolve this sort of problem, you could delete the original post. Another option is to use the Redirection plugin (*https://tinyurl.com/redirectio*) to create your own redirects that override WordPress. For example, using the Redirection plugin, you could create a redirect for */sale* that points to the post at */spring-sale-2021*. Later in the year, you could change your redirect so */sale* points to a different post at */end-of-year-sale-2021*. This way, you're in charge, and there are no changed slugs or automatic redirects. But before you try any of this sorcery, you need to learn about plugins in Chapter 9.

Browsing Categories and Tags Using a Web Address

Earlier, you saw how category links let you see a list of posts for any category. For example, click the Herbal Tea link above a post (or in the Categories widget in the footer) and—presto!—you see the posts about brewing your favorite dried leaves.

WordPress works this category-browsing magic by using a specific form of web address. If you understand it, you can use category web addresses yourself, wherever you need them. First, you start with the site address:

    http://magicteahouse.net

Then, you add */category/* to the end of the address, like this:

    http://magicteahouse.net**/category/**

Finally, you add the bit that identifies the category you want to use. This isn't the category name (after all, it might have spaces and other special characters that don't belong in a web address). Instead, it's the category *slug* you saw earlier, which looks like this:

    http://magicteahouse.net/category/**herbal-tea**

WordPress cooks up the slug based on your category name, using the same process it follows to pick the slug for a new post. First, WordPress replaces every uppercase letter with a lowercase one. Next, it replaces spaces with hyphens (-). Lastly, it strips out forbidden special characters, if you used them. As a result, the category Herbal Tea gets the slug *herbal-tea*.

Remember, you can modify the slug when you edit a category in the Posts→Categories page (shown previously in *Figure 4-22*). For example, you can shorten the preceding address by replacing *herbal-tea* with the simpler slug *herbal*.

Tags work the same way as categories, except the */category/* portion of the web address becomes */tag/*. So, to browse the posts that use a specific tag, you need an address like this:

```
http://magicteahouse.net/tag/kuala-lumpur
```

You can tweak tag slugs in the Tags page. However, it's far less common to tailor tag slugs than it is to edit post and category slugs.

GEM IN THE ROUGH

Posting with Your Phone or Tablet

The admin area is stocked full of power-user tools. But it's impossible to beat the convenience of posting far from your computer, wherever you are, using a few swipes and taps on your favorite mobile device. In the past, developers created plugins that made mobile posting possible. Today, WordPress itself has taken over that role, and it offers an impressive range of free mobile apps at *http://wordpress.org/mobile*. You'll find apps that work with iPads, iPhones, and Android devices. All the apps are polished, professional, and free.

◼ The Last Word

In this chapter, you made the jump from novice to true WordPress user. If you've never touched WordPress before (except to install it), this was a chapter of firsts—your first post, first look at the admin area, first time changing settings. And your first time learning about some of WordPress's most important concepts, including these:

- **Categories.** The essential tool for organizing WordPress posts, so it's a catalogued arrangement of articles instead of a mere pile of posts.

- **Tags.** A way to attach descriptive details to your post, so visitors can search for related content.

- **Permalinks.** The secret to giving every post a short, memorable address.

- **Themes.** The formatting blueprint that controls your site's layout, fonts, and formatting.

In the next chapter, you'll go deeper into themes. You'll learn how to find different themes, change their style settings, and add extra widgets into your layout.

Choosing and Polishing Your Theme

Using the skills you picked up in the previous chapters, you can create a WordPress site and stuff it full of posts. However, your site will still come up short in the looks department. That's because every new WordPress site starts out looking a little drab and pretty much the same as everyone else's freshly created WordPress site. If that sounds colossally boring to you, keep reading, because this chapter shows how to inject some style into your site.

The key to a good-looking WordPress site is the theme you use. Essentially, a *theme* is a set of files that control how WordPress arranges and styles your content, transforming it from text in a database into beautiful web pages. You can think of a theme as a *visual blueprint* for your content. Themes tell WordPress how to lay out the components of your site, what colors and fonts to use, and how to integrate pictures and other graphical details.

Every WordPress site starts out with a standard theme. Right now, yours probably uses the straightforward Twenty Twenty theme (assuming you followed the instructions on page 64). In this chapter, you'll learn how to enhance the Twenty Twenty theme or pick a stylish new one from the thousands that WordPress offers. You'll also see how to choose widgets and arrange them on your site. Master these skills, and you can start the transformation of your site from standard-issue plainness to eye-popping pizzazz.

■ How Themes Work

One of WordPress's most impressive tricks is the way it generates web pages *dynamically*, by pulling content out of a database and assembling it into just the right web page. Themes are the key to this process.

In an old-fashioned website, you format pages before you upload them to your web server. If you want your site to look different in any respect, you have to update your pages and re-upload the whole site. But in a WordPress site, your content and formatting information are separate, with your theme handling the formatting. As a result, you can change the way WordPress styles your pages by editing or changing your theme, without needing to touch the content. The next time someone requests a page, WordPress grabs the same content and quietly applies the latest formatting instructions.

So how does a theme work its content-formatting magic? Technically, a theme is a package of files. Most of them are *templates* that set out the structure of your pages. For example, the template file *header.php* determines how the header at the top of every page on your site looks (see *Figure 5-1*), and the template file *single.php* assembles the content for a single post.

Each template includes a mixture of HTML markup and PHP code. (If you're fuzzy on these web basics, know that HTML is the language in which all web pages are written, and PHP is one of many web programming languages you can use to create dynamic content.) You won't actually touch the template files in this chapter, but you'll learn how to edit them later, in Chapter 13.

Along with your template files, WordPress uses a stylesheet, named *style.css*, that supplies formatting information for virtually every heading, paragraph, and font on your site. This stylesheet uses the CSS (Cascading Style Sheets) standard, and it formats WordPress pages in the same way that a stylesheet formats almost every page you come across on the web today. There's no special WordPress magic here, but you can edit your theme's stylesheet to add special effects.

Understanding the WordPress Year Themes

Every new WordPress site starts out with a default theme. Odds are it's a year theme, one of a few standard themes officially sanctioned by the team of WordPress developers. We call them *year themes* because WordPress releases them annually (occasionally skipping a year) and names them after the year in which they appear. The first year theme was named Twenty Ten. It was followed by the Twenty Eleven theme, the Twenty Twelve theme, and so on. The fine folks at WordPress plan to continue this release pattern for the foreseeable future.

The year themes are also called the *default themes*, because most sites start out using one of them. WordPress comes with the most recent year themes preinstalled—currently, those are the ones described in Table 5-1.

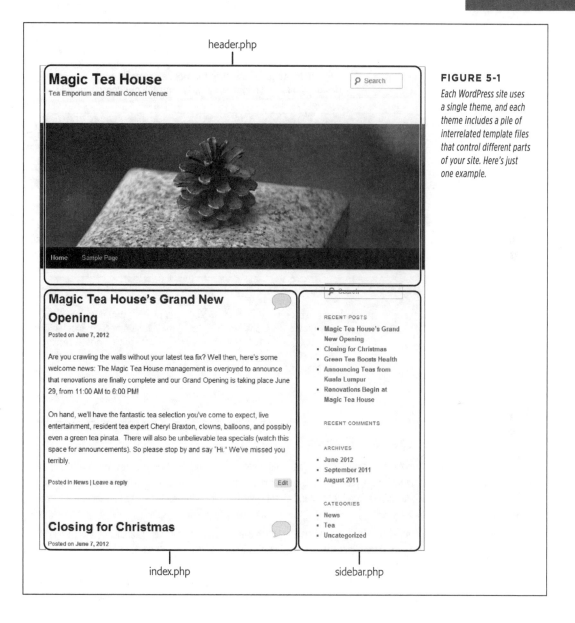

header.php

Magic Tea House

Tea Emporium and Small Concert Venue

Home   Sample Page

Magic Tea House's Grand New Opening

Posted on June 7, 2012

Are you crawling the walls without your latest tea fix? Well then, here's some welcome news: The Magic Tea House management is overjoyed to announce that renovations are finally complete and our Grand Opening is taking place June 29, from 11:00 AM to 6:00 PM!

On hand, we'll have the fantastic tea selection you've come to expect, live entertainment, resident tea expert Cheryl Braxton, clowns, balloons, and possibly even a green tea pinata. There will also be unbelievable tea specials (watch this space for announcements). So please stop by and say "Hi." We've missed you terribly.

Posted in News | Leave a reply   Edit

Closing for Christmas

Posted on June 7, 2012

RECENT POSTS

- Magic Tea House's Grand New Opening
- Closing for Christmas
- Green Tea Boosts Health
- Announcing Teas from Kuala Lumpur
- Renovations Begin at Magic Tea House

RECENT COMMENTS

ARCHIVES

- June 2012
- September 2011
- August 2011

CATEGORIES

- News
- Tea
- Uncategorized

index.php   sidebar.php

FIGURE 5-1

Each WordPress site uses a single theme, and each theme includes a pile of interrelated template files that control different parts of your site. Here's just one example.

TABLE 5-1 *The WordPress year themes*

THEME	WHAT IT LOOKS LIKE
Twenty Twenty	Twenty Twenty is a modern theme with a magazine-like layout. It pads pages with plenty of space on the sides, but lets quotes, pictures, and other elements expand into the margins. It uses high-contrast text (very heavy headlines with much lighter text), a tinted background, and a menu that adjusts itself for mobile devices. There's also one big design choice you might not like: Twenty Twenty doesn't include a sidebar.
Twenty Nineteen	Twenty Nineteen is a minimalist theme, with heavy black typography on bare white pages. Its clean, simple layout makes it suitable for a variety of types of websites, but its built-in customization options are somewhat limited. Like Twenty Twenty, Twenty Nineteen doesn't include a sidebar.
Twenty Seventeen	Twenty Seventeen is a slick theme that's good at showcasing big header pictures. It supports some interesting frills, like customized sections on the home page and video banners. You also get plenty of whitespace (no crowding here). Although Twenty Seventeen is suitable for anything, WordPress experts often recommend it for business sites.
Twenty Sixteen	Twenty Sixteen uses a streamlined classic layout (straightforward header, sidebar with neatly aligned sections, simple menu), but adds some extra touches. For example, it lets pictures and pull quotes expand into the blank whitespace on the left side of the page, which is a nice layout detail that fits newsy sites or serious blogs. It also has the option of a solid border around the page, which is its one slightly retro touch.

UP TO SPEED

What's in a Name: The Year Themes

The year themes often confuse WordPress newcomers. The naming system seems to imply that older year themes are out-of-date and that newer year themes are their natural successors. But that's not true.

First, it's important to understand that each year theme is a new creation. That means Twenty Nineteen didn't evolve into Twenty Twenty, and Twenty Twenty doesn't replace Twenty Nineteen.

Similarly, newer year themes don't replace older ones. WordPress clearly states that the goal of each year theme is to be different and to showcase new features. The goal is *not* to provide a single all-purpose theme that satisfies everyone. Not everyone will stick with Twenty Twenty, for example, because it lacks the ability to show a sidebar.

When WordPress releases its next year theme, it will offer yet another new style that doesn't replace the themes that have gone before. This change won't affect *you*, however, because you're about to learn how to make any theme work for you—and how to find even more specialized themes in WordPress's expansive theme catalog.

Making Your Theme Suit Your Site

Now that you have a basic idea of how themes work, you're ready to start improving the look of your site. There are several paths you can take to change the appearance of your theme, depending on how dramatic your alterations are. Your choices include these:

- **Changing to a different theme.** To give your site a dramatic face-lift, you can pick a completely different theme. With that single step, you get new fonts, colors, and graphics, a new layout, and—sometimes—new features. WordPress offers thousands of choices.

- **Tweaking your theme settings.** WordPress gives you a number of useful ways to personalize your theme. The options depend on the theme, but you can usually alter a theme's color scheme, change the header picture, and shift the layout.

- **Customizing your widgets.** Most WordPress themes include one or more sections you can customize, like a sidebar or footer you can stock with bits of content (like pictures, links, menus, and more). These bits of content are called *widgets*, and you can change them, rearrange them, and add new ones.

- **Editing your theme.** Advanced WordPress fans can crack open their theme files, work on the code with help from this book, and make more substantial changes. The simplest modification is to fine-tune the CSS styles that format your pages. More ambitious theme hackers can change virtually every detail of their sites.

In the rest of this chapter, you'll start to explore the first three tasks in this list. But you won't get to the last, most ambitious point (hardcore theme editing) until Chapter 13.

■ Choosing a Theme

While you're learning how to use WordPress, you might decide to stick with the year themes. That way, you won't get confused about which features are part of WordPress and which details are added by your theme. But by the end of this book, you'll probably decide to move on to a different theme.

Thousands of other themes work with WordPress, and they're called *third-party themes*. (The name reflects the fact that WordPress didn't make the theme, and you didn't make the theme, but a third person did.) Although WordPress doesn't own third-party themes, it helps distribute them through the theme gallery—as you'll see in a moment.

Choosing a theme is the Big Choice you make about your site's visual appearance. Themes determine important design details—for example, the way WordPress uses graphics, fonts, and color across your pages, and the overall layout of your pages. It also determines the way WordPress presents key ingredients, like the date of a post, the post's category and tag information, and the links that guests use to browse through your archives. An ambitious WordPress theme can rework almost every visual detail of your site (*Figure 5-2*).

FIGURE 5-2

The Greyzed theme creates a grungy effect that looks like lined paper on stone. Every design detail—from the look of the menu to the widget titles—differs from a plain-vanilla WordPress site.

WordPress tailors some themes for specific types of content. You can find custom themes for travel blogs, photo blogs, and magazine-style news blogs. There are even themes that lean heavily toward specific topics, including one for coordinating and celebrating a wedding (called Forever) and one that looks like the old-school Mac desktop (called Retro MacOS). *Figure 5-3* shows the latter.

Themes also influence the way your site works, in ways subtle and profound. For example, some themes tile your posts with pictures instead of putting them in a top-to-bottom list (which is great if you want your site to show a portfolio of work rather than a list of articles). Or your theme may include a fancy frill, like a slideshow of featured posts.

Even if you're happy with the standard WordPress theme, it's worth trying out a few others, just to open your mind to new possibilities. As you'll see, although changing a theme has a profound effect on the way your site looks, doing so is almost effortless. And most themes are free, so there's no harm in exploring.

TIP If you're a style-conscious site designer who wants a distinctive theme, and you aren't put off by a bit of hard work, you can customize any theme. Chapter 13 explains how.

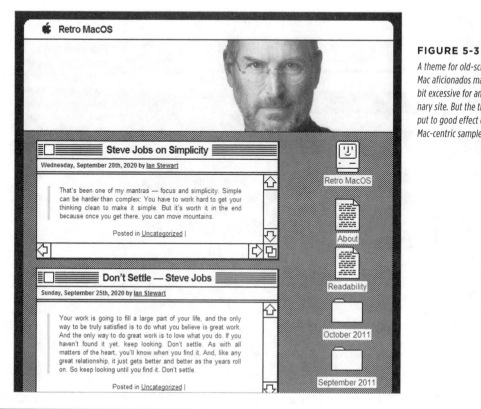

FIGURE 5-3

A theme for old-school Mac aficionados may be a bit excessive for an ordinary site. But the theme is put to good effect on this Mac-centric sample site.

One limitation with WordPress themes is that somewhere in cyberspace, there are sure to be plenty of other websites using the same theme as you. This isn't a huge problem, provided that you're willing to customize your site in little ways—for example, by changing colors, choosing a suitable header picture, and so on. Some themes are more flexible in this respect than others, and allow you to make your site's appearance truly unique.

Visiting the Theme Gallery

To change your theme, you need to visit the theme gallery (*Figure 5-4*). To get there, log in to the admin area and choose Appearance→Themes in the menu.

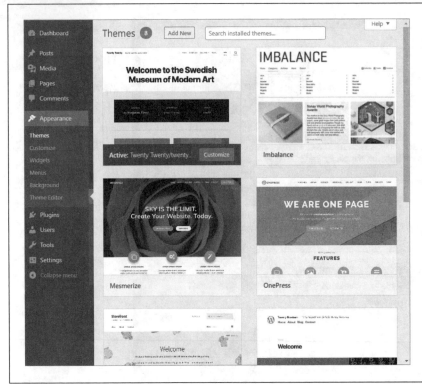

FIGURE 5-4

On the Themes page, WordPress displays all the themes currently installed on your site. In this WordPress site, eight themes are installed, as indicated by the number next to the Themes title at the top of the page. The first theme in the gallery (here, that's Twenty Twenty) is the one that your site currently uses.

At first, you'll find the theme gallery fairly sparse. A standard WordPress installation includes the latest year themes and nothing else. You need to add any other theme you want to use. Fortunately, there's no reason to fear this process—new themes take up very little space, installing them takes mere seconds, and you can do it all without leaving the admin area.

> **NOTE** Some web hosts preinstall extra WordPress themes. If you use one of these hosts, your theme gallery won't start out quite as empty as just described. However, you'll probably still need to install new themes to find the one you *really* want.

Installing a New Theme

Before you can switch to a new theme, you need to install the theme itself. To see the themes you can add, click the Add New button next to the Themes heading at the top of the Themes page. This takes you to the Add Themes page, where you can browse WordPress's theme repository to find exactly what you want (*Figure 5-5*).

Initially, the Add Themes page shows you a list of *featured* themes, which are new and particularly noteworthy designs that the folks at WordPress think might interest you. If they don't, you can browse the most popular themes (click Popular) or the most recent ones (click Latest).

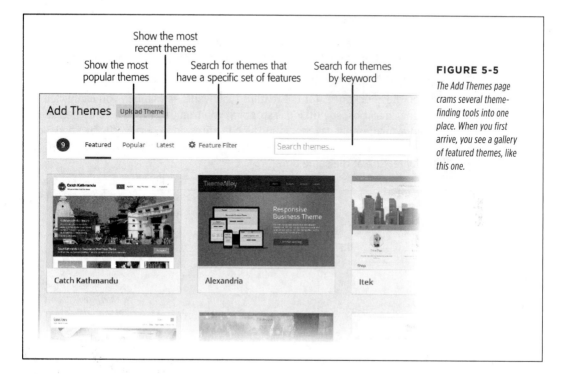

Show the most recent themes

Show the most popular themes

Search for themes that have a specific set of features

Search for themes by keyword

FIGURE 5-5

The Add Themes page crams several theme-finding tools into one place. When you first arrive, you see a gallery of featured themes, like this one.

Browsing is fine, but the most powerful way to hunt down new themes is with a search. The Add Themes page gives you two ways to do that: by keyword and by feature.

To search by keyword (for example, "magazine" or "industrial" or "professional"), type the word in the search box and then click Search. WordPress shows you themes that have that search term in their names or descriptions.

To search by feature, click Feature Filter. WordPress displays a long list of check-boxes, representing different features and theme characteristics (*Figure 5-6*). The checkboxes are organized into three sections:

- **Subject.** This category lets you hunt for themes that are designed for specific types of sites. For example, choose Photography to find the themes that are good for showcasing your pictures. (They probably have a graphical layout with some sort of gallery feature.) Or use Education to find a school theme, News to find a theme that suits text-heavy articles, Portfolio to find a theme that can showcase your work, or something else altogether.

- **Features.** This category lets you find themes that offer specific features. The list is short but includes useful details. For example, use Custom Colors if you want a theme that has a flexible color scheme, and Full Width Template for a theme that lets some pages widen to fill up more space. You won't know what most of the features mean, because we haven't yet explored things like featured images and image sliders.

- **Layout.** This category lets you choose themes that organize their pages the way you like. For example, you can pick a layout with a specific number of columns, or ask for a layout with a left or right sidebar.

Turn on the checkboxes for the features you want. Remember, you can pick more than one checkbox, so nothing is stopping you from launching a combined search that finds themes that have news and blog features (just check both News and Blog). Click Apply Filters to perform the search. However, you can't search by keyword and filter by feature at the same time.

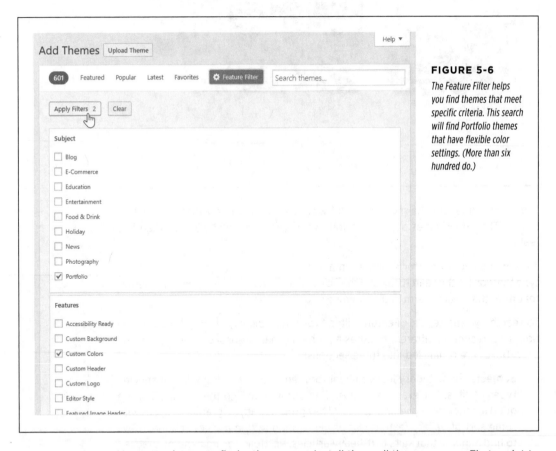

FIGURE 5-6

The Feature Filter helps you find themes that meet specific criteria. This search will find Portfolio themes that have flexible color settings. (More than six hundred do.)

No matter how you find a theme, you install them all the same way. First, point to your new theme without clicking the mouse. The Install and Preview buttons appear (*Figure 5-7*, top). Click Preview to see a sample site that uses the theme, and to read a brief description of the theme and the person or company that created it. Click Install to copy the theme to your website so it's ready to use.

TIP Don't be afraid to install a theme that you might not want to use—all the themes in the WordPress repository are guaranteed to be safe and spyware-free. And don't worry about downloading too many themes—not only are they tiny, but you can easily delete those you don't want.

Once WordPress installs a theme, it gives you three choices (*Figure 5-7*, bottom):

- **Live Preview.** This opens a window showing you what your site would look like if WordPress applied the chosen theme. Think of it as a test-drive for a prospective theme. You can read posts, search your site, and click your way around your content, secure in the knowledge that you haven't changed the real, live version of your site.

- **Activate.** Click this to start using the theme.

- **Return to Theme Installer.** This takes you back to the Add Themes page (*Figure 5-6*), where you can search for another theme.

FIGURE 5-7

Top: When you're ready to take a closer look at a theme, point to it and click Preview. Or click Install to copy it to your website so it's ready to use.

Bottom: WordPress has finished installing the latest version of the Alexandria theme on your site. Click Activate to start using it.

To see all the themes you've installed on your site so far, choose Appearance→Themes. This returns you to the theme gallery. To take a closer look at a theme, point to it. If you point to the theme you currently use, you'll see a Customize button, which launches the theme customizer that you'll learn about in the next section. If you point to one of the other themes, you'll see two familiar buttons: Live Preview, which lets you test-drive the theme; and Activate, which applies it to your site.

TIP Don't be shy! You can always switch back to your previous theme. Later, when you've performed extensive theme customizations, you'll need to be more cautious. Changing a theme can lose some theme-specific customizations, like the colors you've picked, the header graphics you're using, and your arrangement of widgets. But the content in your site (the posts, pages, categories, menu items, and so on) are always safe because they don't depend on the theme.

Once you activate a theme, the best way to get familiar with it is to poke around. Try adding a post, viewing it, and browsing the list of posts on the home page, just as you did in Chapter 4. Check out the way your theme formats the home page and presents individual posts.

Once you familiarize yourself with your new theme, you'll want to check out its options and consider tweaking it. That's the task you'll tackle next.

NOTE There's one other way to add a theme: by uploading it from your computer to your site, using the Upload Theme button on the Add Themes page. You'll use this option only if you find a theme on another site, buy a premium theme from a third-party company, or build the theme yourself.

UP TO SPEED

The Proper Care of Your Themes

When you install a theme, you get the current version of it. If the theme creator releases a new version later, your site sticks with the old version.

That makes some sense, because WordPress has no way to be sure that the new version of the theme won't make drastic changes that break your site. But this design also raises the risk that you might ignore an important security update for one of your themes, thereby exposing your site to an attack. (Remember, themes contain templates, and templates contain PHP code, so themes can create security vulnerabilities.)

To minimize your risk, check regularly for theme updates. Go to Dashboard→Updates and look at the Themes section. If a theme has been updated, WordPress lists it here, and compares the version number of your copy with the version number of the latest release. If your site has multiple themes installed, WordPress shows the updates for *all* of them—it makes no difference whether you currently use the theme or not. To protect yourself, download all the available updates. To do that, turn on the checkbox next to each update and then click Update Themes.

Finally, if you install a theme, try it out, and decide you aren't ever likely to use it, it's a good idea to delete it. Even if a theme isn't active, its template files are accessible on your site, allowing attackers to exploit any security flaws the templates may have. This isn't a common method of attack, but it has happened to unsuspecting site owners before.

◼ Tweaking Your Theme

The first step in getting the look you want is choosing the right theme, but your site-styling doesn't end there.

Every theme lets you customize it. In fact, you *need* to customize your theme to make sure it meshes perfectly with your site. Your page's header image is a perfect example—if your theme includes one, you'll almost certainly want to replace the stock image with a picture that better represents your site. Other basic theme-customization tasks include shuffling around your widgets and setting up your menus.

The best place to customize most themes is in the theme customizer. Here's where things go a little wonky. Every theme is slightly different. Most modern themes have lots of options available in the theme customizer, but some older themes have just a few. If you can't find what you want in the theme customizer, you'll need to hunt around for more options in the Appearance menu.

The Theme Customizer

The *theme customizer* is a relatively recent addition to the WordPress world. It provides a central dashboard where you can quickly change the settings in your theme. The theme customizer fits in all these settings by using a multitabbed page. But its most impressive feature is a live preview that shows what your changes will look like on your site, even before you've officially applied them.

To open the customizer, choose Appearance→Customize. Or, when viewing the theme gallery (Appearance→Theme), point to the current theme and click the Customize button. Either way, WordPress displays a preview of your site with a sidebar of theme-customization options. *Figure 5-8* shows the Twenty Twenty theme under the microscope in the theme customizer.

> **NOTE** The theme customizer doesn't always look as it does in *Figure 5-8*. Some themes offer more customizable sections than others. However, the basic concept remains the same: you configure common settings while watching a constantly updated preview of your work.

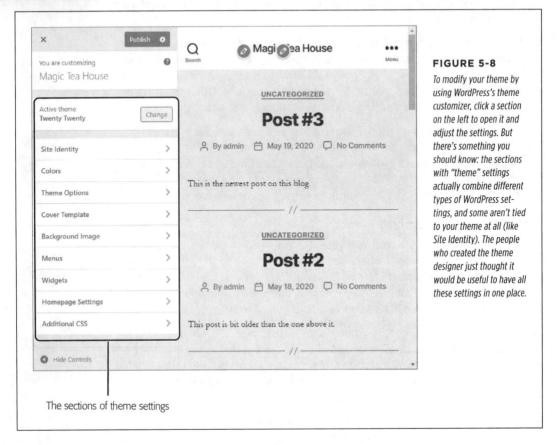

FIGURE 5-8

To modify your theme by using WordPress's theme customizer, click a section on the left to open it and adjust the settings. But there's something you should know: the sections with "theme" settings actually combine different types of WordPress settings, and some aren't tied to your theme at all (like Site Identity). The people who created the theme designer just thought it would be useful to have all these settings in one place.

The sections of theme settings

So what can you do with your site in the theme customizer? Here's a quick rundown:

- **Site Identity.** This section lets you change the title of your site and the subtitle (known as a *tagline*). These aren't actually theme-specific settings. (You set them when you installed WordPress, and you can also change them through the Settings→General section of the admin area.) Some themes also use this section for header pictures and logos.

- **Colors.** This section lets you tweak the colors in your theme. Themes are all over the map when it comes to color options—some give you fine-grained control over individual design elements (like buttons and headings), while others limit you to a couple of prebuilt color schemes (and some themes have no color options at all). Twenty Twenty falls somewhere in between—it lets you change the background and accent color, but doesn't let you colorize much else.

- **Theme Options.** This is a custom section that some themes have and others don't. Twenty Twenty uses it for a few miscellaneous settings. The most important is the ability to use excerpts, which you'll see on page 122.

- **Cover Template.** This is a special section that you'll see only if you're using Twenty Twenty. It controls how the Cover template works, which you can use to make a fancy, graphical home page. You'll try it out after you start building pages, on page 234.

- **Background Image.** This section lets you use a picture for your background. That picture then shows up on every page on your site, whether you're browsing through a list of posts or reading a single one.

- **Menus.** You use this section to assemble a menu. But before you can use menus, your site needs to have some pages. You'll cover pages and menus in Chapter 8.

- **Widgets.** This section lets you add prebuilt bits of content to various places in your site, such as its sidebar or footer. You'll learn to configure widgets later in this chapter.

- **Homepage Settings.** This section lets you use a custom page for your site's home page, provided you've already created that page. You'll learn to make a custom home page (and other types of pages) in Chapter 8.

- **Additional CSS.** This section lets you add your own CSS settings to override some of the visual details in your theme. This is a much more advanced theme customization technique, because you need to know not only the CSS standard, but also a bit about how your theme is designed. You'll explore this feature in Chapter 13.

To take a closer look at these settings, click the section title. You can then tweak whatever you say, safe in the knowledge that you aren't actually changing your site. If you like what you see in the preview, click the Publish button at the top of the panel to make your changes permanent. Or click the X icon in the top-left corner of the panel to exit the theme customizer and back out of any changes you haven't published.

Let's try it out.

Colors

It's up to your theme to decide what color settings it offers in the theme customizer. Twenty Twenty lets you set two types of background colors. It uses one background color for headers and footers (initially, it's white) and the other behind your post content (initially, that's beige). To change a color, click the Select Color box (*Figure 5-9*).

Twenty Twenty also lets you pick a primary color, which is better described as the *highlight* color. The primary color is used wherever the theme needs to accent something. To change the primary color, you use a more limited slider (at the bottom of the sidebar) that lets you pick a color but not its intensity. In Twenty Twenty, the primary color sets the background color of buttons and the text color of links.

In Twenty Nineteen, the primary color is used to tint featured images, provided you tick the "Apply a filter to featured images" box. (Featured images are described on page 188.)

NOTE The Twenty Nineteen filter feature looks neat in a monochrome kind of way, but it's unlikely that a real site would want every picture to be tinted exactly the same. Truthfully, this feature screams "for cool demo purposes only."

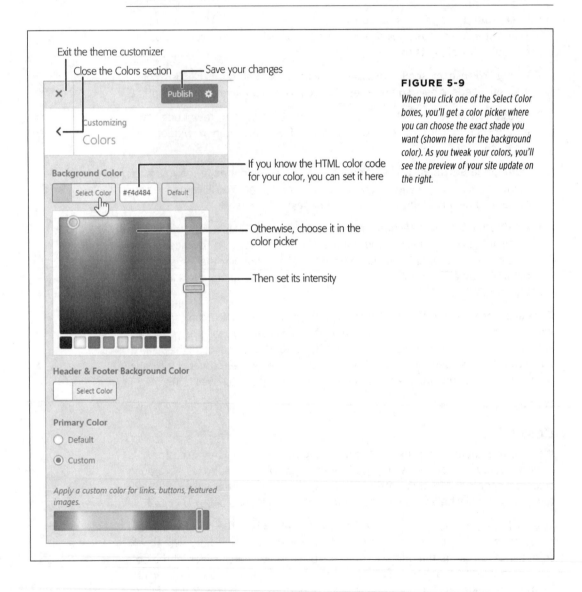

FIGURE 5-9

When you click one of the Select Color boxes, you'll get a color picker where you can choose the exact shade you want (shown here for the background color). As you tweak your colors, you'll see the preview of your site update on the right.

Some themes force you to choose between preset color schemes.

If you click the Colors section in the Twenty Seventeen theme, for example, you'll find three options (*Figure 5-10*): Light (the standard black text on a white background), Dark (for a dramatic white on black color scheme), and Custom (which lets you pick your preferred colors). However, the Custom setting in Twenty Seventeen doesn't behave like the Custom setting in Twenty Twenty. In Twenty Twenty, you change a highlight color that applies to buttons and a few other things; in Twenty Seventeen, you pick the color that's used to tint text. For example, pick bright red and you'll get red-tinged black text.

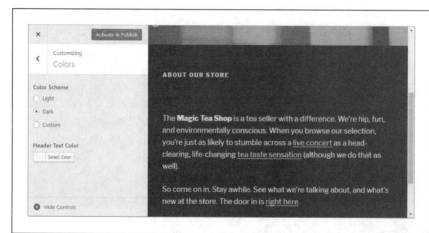

FIGURE 5-10

Different theme, different color choices. The Twenty Seventeen theme lets you choose from two prebuilt color schemes or set a custom text color. You can also choose a different color for the site title and tagline by clicking the Select Color button under Header Text Color.

If you change colors in the older Twenty Sixteen theme, you'll get yet another slightly different set of color choices. You'll find separate options for colorizing various design elements, including the text, the links, the page background, and the border around the page.

NOTE This is part of the wackiness of themes. Every theme has the ability to define its own settings. This design is great for flexibility, but it means there's often a learning curve when changing from one theme to another.

Header Pictures and Logos

Many themes allow you to replace the ordinary text title of your site with some sort of image. But every theme does it a bit differently.

Depending on the theme, the header text may be displayed above the image, beneath the image, or on top of it, or you may choose to hide the header text altogether (*Figure 5-11*).

Magic Tea House

Tea Emporium and Small Concert Venue

HOME   **POSTS**   TEA CONCERT SERIES

Magic Tea House's Grand New Opening   CATEGORIES

FIGURE 5-11

Top: The Magic Tea House is a bit plain with a text-only header.

Middle: Every theme makes its own decision about where to place the header image. In this theme, the header image appears under the menu.

Bottom: Hiding the header text results in a cleaner look.

Magic Tea House

Tea Emporium and Small Concert Venue

HOME   **POSTS**   TEA CONCERT SERIES

Magic Tea House's Grand New Opening   CATEGORIES

HOME   **POSTS**   TEA CONCERT SERIES

Magic Tea House's Grand New Opening   CATEGORIES

Leave a reply   Events

In recent years, WordPress has been less interested in header images. The Twenty Twenty and Twenty Nineteen themes don't support header images. Instead, they substitute a similar feature that lets you add a much smaller header logo at the top of your site.

If you're interested in adding a header picture, here are the steps that will do it on most themes:

1. **In the admin area, choose Appearance→Customize.**

   This loads the theme customizer, if you aren't already there.

2. **Choose the Header Image section.**

   Here's where things get tricky. In some themes, this section is named Header Media, or just Header. In Twenty Twenty, you use the Site Identity section to use the similar header logo feature. But if you can't find any sort of header-related section and Site Identity doesn't have any picture-adding options, your theme probably doesn't support headers.

3. **Make a note of the image size.**

   Because a theme's layout dictates the size of the header, each theme has different size specifications. Usually, you can find this information in the appropriate section of the theme customizer (*Figure 5-12*).

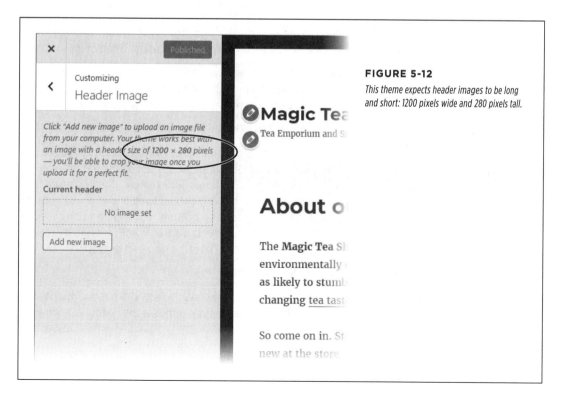

FIGURE 5-12

This theme expects header images to be long and short: 1200 pixels wide and 280 pixels tall.

4. **Prepare your picture.**

   If possible, you should resize or crop your picture to the right size by using an image-editing program before you upload it. That way, you'll get exactly the image you want. If your image doesn't match the dimensions your theme expects, WordPress will ask you to crop it (you'll need to cut part out), or some themes may resize it (which could result in a stretched or fuzzy picture).

5. **Click the "Add new image" button.**

   Or, if you're using Twenty Twenty, you click the "Select logo" button to pick a small (120 × 90 pixels) picture.

   WordPress opens a new window where you can choose the picture you want to use. You can pick a picture you've already added to your site (you won't have any of those yet), or upload something new.

> **TIP** WordPress stores every picture you upload, unless you explicitly remove it. You'll learn more in Chapter 7. But for now, all you need to know is that if you upload a header image, it sticks around—even when you're not using it, and even if you change themes.

6. **Click the Upload Files tab. Then, drag the file that has your header image onto the page, where it says "Drop files to upload."**

   If you don't want to drag and drop, you can hunt down the files instead. Just click the Select Files button to show the standard file selection windows. Then it's up to you to browse to the folder with your picture and select it.

   As soon as you pick a file, WordPress uploads it (*Figure 5-13*). Now the file is on your website (technically, it's in your media library—the repository you'll learn about on page 181). But you still need to stick the picture in your header.

7. **Click the Select and Crop button (in the bottom-right corner).**

   If your picture doesn't fit the required dimensions, WordPress asks you to crop it down (*Figure 5-14*). This works well if your picture is just a little too tall or wide, but it can cause problems in some situations. For example, if your picture isn't wide *enough*, WordPress enlarges the whole thing to fit and then asks you to crop off a significant portion of the top and bottom. You end up with the worst of both worlds: an image of lower quality (because WordPress had to scale it up) and one missing part of the picture (because you had to crop it).

8. **Consider removing the header text.**

   In some themes, the header text meshes neatly with the header image. For example, the old Twenty Thirteen theme superimposes the header text on your header image.

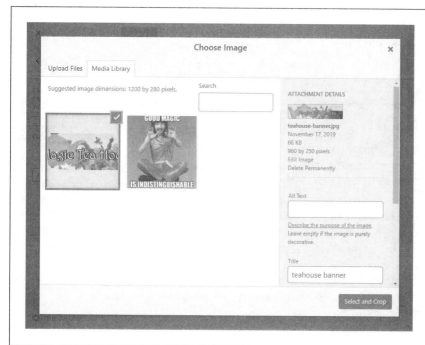

FIGURE 5-13

This website currently has two pictures. On the left is the banner you just uploaded. (A checkmark in the corner shows that it's selected.) Don't worry about the details on the right, like size and title. You'll learn more about the way WordPress treats pictures in Chapter 7.

FIGURE 5-14

To crop a picture, drag the highlight rectangle until you frame the image the way you want it. Here, the picture is a bit too tall—by positioning the highlight rectangle in the middle, WordPress will trim out part of the top and bottom. When you finish, click Crop Image. Or click Skip Cropping if you want to ignore your theme's recommendations and use an oddly sized header that may not fit nicely into your layout.

In other themes, the text and picture are separate. If your header includes text, you will probably want to remove the title from your site (as you saw in *Figure 5-11*). Usually, you do that by going to the Site Identity section and removing the checkmark for the Display Site Title and Tagline setting.

9. **Click Publish.**

As usual, this makes all your changes permanent.

Random Header Pictures

Many themes have the ability to perform a curious header-switching trick. First, they let you upload multiple pictures. Then they randomly choose a different picture for the header on each page.

To try this out, start by uploading more than one picture in the theme customizer. Head back to the Header Image section of the theme customizer, click the "Add new image" button, and go through the sequence of steps you just learned to upload another picture. Then do it again! You'll see each picture appear in the theme customizer, under the heading "Previously uploaded." If you accidentally add one you don't like, hover over it and then click the X icon to remove it from the list.

Here's the magic step: click the "Randomize uploaded headers" button. Now start browsing the posts on your site. As you click from page to page (or if you just hit the Refresh button), you'll see the header change from one picture to another. Eventually, WordPress will cycle through all your images.

Some folks love the changing-picture trick. For example, it's a great way to showcase a number of different and delectable dishes on a food blog. However, most people prefer to pick a single header and stick with it. That gives the site a clearer identity and helps visitors remember your site.

Background Images

Many themes support background images. Usually, you'll find these settings in a theme customizer section named Background or Background Image. Twenty Twenty uses the latter.

To add a background image, you need to upload a picture, just as you did for the header.

Most themes *tile* the background image you choose. That means they repeat the image endlessly, from top to bottom and left to right, filling your visitor's browser window. Twenty Twenty tiles the background image, but leaves a plain background behind the header and footer areas (*Figure 5-15*).

When you're using a tiled background, you need a picture that looks good when it's jammed edge to edge against another copy of itself. Small pictures called *textures* work well for this task, and you can find them online. Often, the effect is distinctly old-fashioned. But you can find some understated choices at *https://tinyurl.com /back-tiles*.

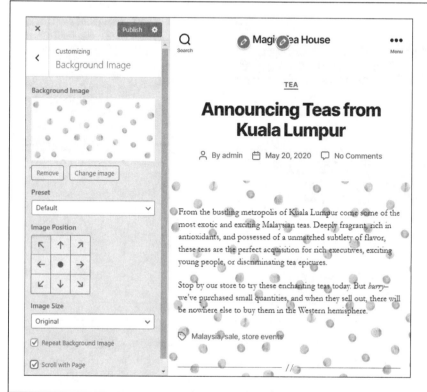

FIGURE 5-15

This Twenty Twenty site uses a background of tiled watercolor dots. On the left, you can see that the theme customizer gives options for aligning the background, tiling it, and choosing whether it scrolls along with the page when the reader moves down.

Themes That Use Post Excerpts

As you become more experienced with themes, you'll notice that many of them change the way your posts look on the home page. Often themes organize the post list into a more compact format, sometimes even tiling posts into a grid with pictures and shortened descriptions (*Figure 5-16*).

FIGURE 5-16

The Oxygen theme uses a magazine-style layout. To fit more articles on the home page, it shows only an excerpt from each one in the list (and it tops each excerpt with an image from the post).

One way themes can shorten the post list is by using a WordPress feature called *excerpts*. With excerpts, the post list shows only the first 50 words or so in a post. That lets it show more posts in less space. (The disadvantage is that you'll now need to click through to the post to read the whole thing.)

Some themes rely heavily on excerpts, while other themes don't use them at all. Twenty Twenty lets you choose. If you want to opt in to excerpts, click the Theme Options section in the theme customizer, look for the "On archive pages, posts show" heading, and choose Summary underneath (*Figure 5-17*).

Excerpts seem straightforward and automatic (and they are). However, the first few sentences of a post aren't always a good reflection of its content. For that reason, you may want to write your own excerpt—in other words, explicitly provide a brief summary of the content in a post. You can do that from the post editor when you're writing or editing a post:

1. **In the sidebar on the right, click the Document tab.**

2. **Scroll down to the Excerpt section, and click it to make it visible.**

3. **Type your excerpt text in the small box.**

4. **Publish or update your post to save the change.**

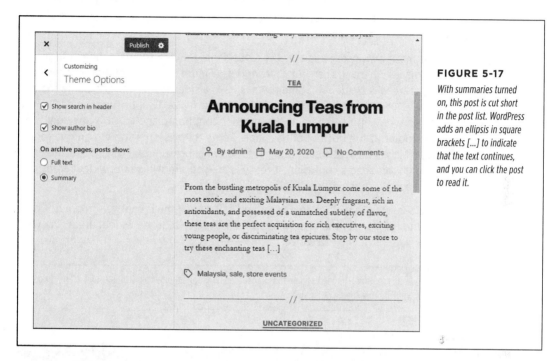

FIGURE 5-17

With summaries turned on, this post is cut short in the post list. WordPress adds an ellipsis in square brackets [...] to indicate that the text continues, and you can click the post to read it.

UP TO SPEED

Writing Good Excerpts

The best thing about excerpts is that they don't need to be directly linked to the text in your post. But don't abuse your freedom—to write a good, genuinely useful excerpt, you need to follow a few rules:

- **Keep it brief.** By keeping your excerpts short (around the 55-word mark, just as WordPress does), you ensure that it fits in a compact layout, and you make it easier to read.

- **Summarize the content of the page.** The goal of an excerpt is to give someone enough information to decide whether they want to click a link to read the full post. An excerpt isn't a place to promote yourself or make flowery comments. Instead, try to clearly and honestly describe what's in the post.

- **Don't repeat the post title.** If you want to make sure every word counts, don't waste time repeating what's clearly visible in the title.

■ Customizing Your Widgets

Widgets are small chunks of useful content. For example, maybe you want to put a short list of links on your site. There's a widget for that! Or maybe you want something a bit fancier, like a calendar that lets readers jump to the posts you published on a particular day. There's a widget for that too. And once you learn how to install plugins (Chapter 9), you can go way off the deep end with specialized widgets for polls, Twitter feeds, ads, and more.

What makes widgets special is their positioning. Every theme defines widget *areas*—places where you can put your widgets (*Figure 5-18*). One of the most common widget areas is a sidebar. Some themes have a sidebar on the right, some themes have one on the left, some have both, and a few (like Twenty Twenty and Twenty Nineteen) have no sidebars at all. Another common place to have a widget area is the footer at the bottom of the page. Some themes offer more than one footer area, which is perfect for creating *fat footers* chock-full of links, ads, or pictures. Although this seems confusing, it really isn't—you simply use the widget areas you need and ignore the others.

Table 5-2 lists the widget areas that you'll find in all the recent year themes. Compared to third-party themes, which can easily have a dozen widget areas, the year themes are rather conservative.

> **NOTE** It's up to the theme to name the widget area. What one theme calls a "Blog Sidebar" might be the same as what another theme calls "Side Panel."

TABLE 5-2 *Widget areas in the WordPress year themes*

THEME	WIDGET AREA NAME	DESCRIPTION
Twenty Twenty	Footer 1	Footer at the bottom.
	Footer 2	A second footer that appears to the right of the first footer (giving a two-column footer effect).
Twenty Nineteen	Footer	Footer at the bottom. This is one or two columns wide, depending on the size of your browser window.
Twenty Seventeen	Blog Sidebar	Sidebar on the right of all posts.
	Footer 1	The left column of the footer.
	Footer 2	The right column of the footer.
Twenty Sixteen	Sidebar	Sidebar on the right, which appears throughout the site.
	Content Bottom 1	The left column of the footer. However, this footer is placed right after the content, not all the way at the bottom of the page.
	Content Bottom 2	The right column of the footer.

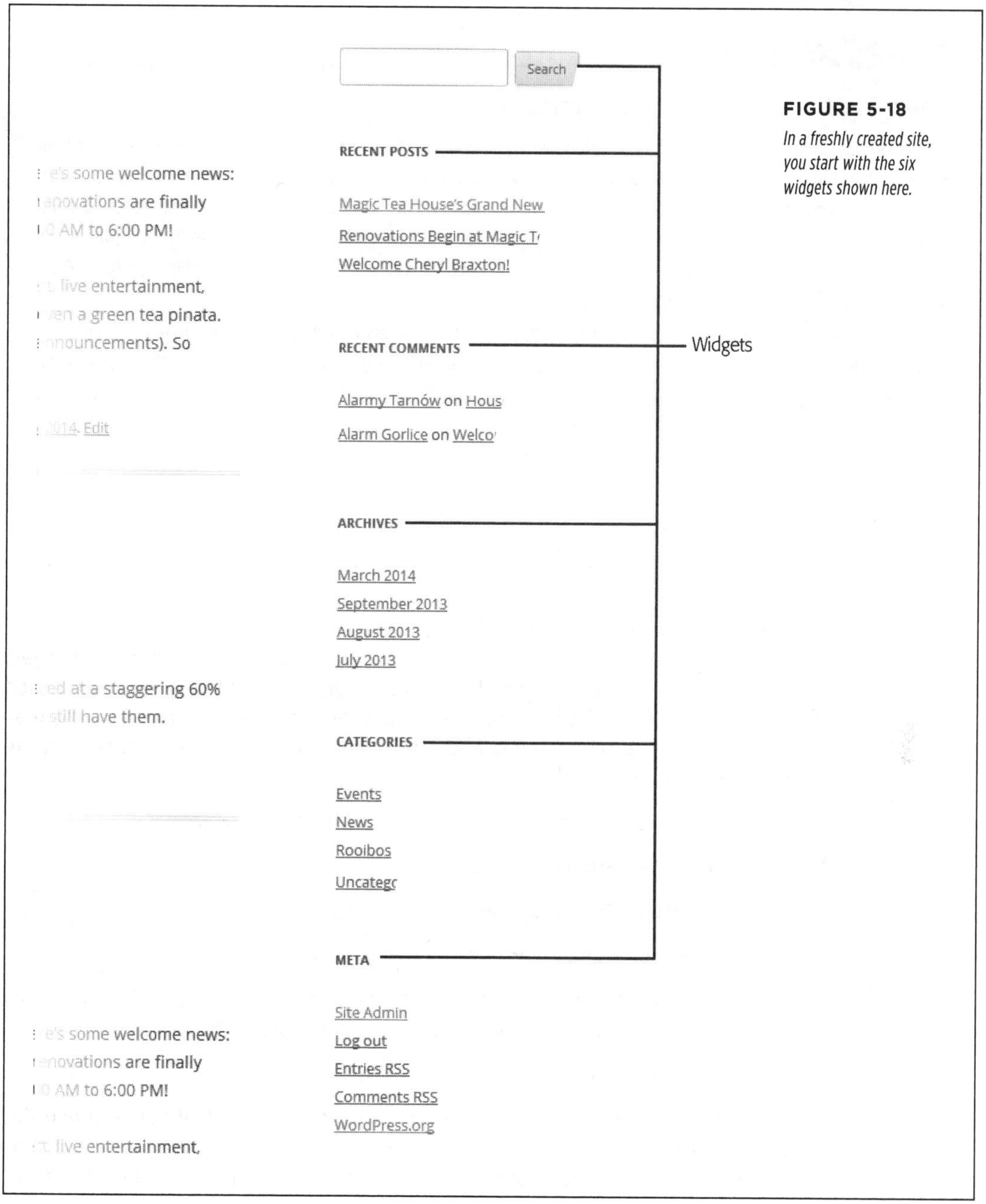

e's some welcome news:

enovations are finally

AM to 6:00 PM!

live entertainment,

en a green tea pinata.

nnouncements). So

14. Edit

ed at a staggering 60%

still have them.

Search

RECENT POSTS

Magic Tea House's Grand New

Renovations Begin at Magic T

Welcome Cheryl Braxton!

RECENT COMMENTS

Alarmy Tarnów on Hous

Alarm Gorlice on Welco

ARCHIVES

March 2014

September 2013

August 2013

July 2013

CATEGORIES

Events

News

Rooibos

Uncatego

META

Site Admin

Log out

Entries RSS

Comments RSS

WordPress.org

Widgets

FIGURE 5-18

*In a freshly created site,
you start with the six
widgets shown here.*

e's some welcome news:

enovations are finally

AM to 6:00 PM!

live entertainment,

No matter where the widget areas are in your theme, it's up to you to choose what widgets go in each area, and to configure the widgets you add. That's the job you'll work on next.

Changing the Layout with Post Templates

Themes are the ultimate enforcers of your website's layout. But sometimes they give you a little wiggle room through a feature called *post templates*.

Here's how it works. When you create a post, WordPress ordinarily uses something called the default template—basically, that's the status quo for your theme. But some themes offer other templates. These templates can change the way layout works for individual posts.

To see your template choices, start editing a post and look at the sidebar on the right side of the post editor. Scroll down to the Post Attributes section and find the Template box. Initially, your post will be set to use the default template, but you may find other options in the Template list.

In Twenty Twenty, for example, you can pick Full Width Template to create a post that breaks out of the standard layout and expands to fill the left and right margin space. (Usually, the effect isn't that attractive.) In other themes, you can use templates to hide the widget-filled sidebar or show different widget areas.

If used carelessly, post templates can break consistency and make your site look sloppy. But they're also the foundation for some features you'll look at later, like the specialized home pages in Chapter 8.

Arranging Your Widgets

There are two ways to configure your widgets. You can use the Appearance→Widgets page, which WordPress has had since forever. Or, you can use the Widgets section of the theme customizer. The theme customizer is a bit more cramped, but it has one definite advantage. As you add the widgets, you can see what they look like in the live preview.

Here's how to add or rearrange widgets in the theme customizer:

1. **Choose Appearance→Customize.**

   This loads the theme customizer.

2. **Choose the Widgets section.**

   You'll see a list of all the widget areas in your theme. The Twenty Twenty theme has just two widget sections, and they're both part of the footer.

3. **Click the widget area you want to change.**

   Now you'll see all the widgets in that area (*Figure 5-19*). If you're trying this out with a new WordPress installation using the Twenty Twenty theme, start out by choosing the Footer 1 area, which represents the left side of the footer.

4. Drag any of the widgets to change the order.

To move a widget, you simply drag one of the widget boxes (in the theme customizer list) to a new spot. For example, click Search and drag it down to the bottom of the list. WordPress will update the preview to put the search controls at the end of the footer. If you're editing a widget area that arranges widgets horizontally (like one in the header), the widget at the top of the list appears first on the left, and subsequent widgets are placed on the right.

You can do more than just move widgets around. The next sections show you how to customize your widgets.

FIGURE 5-19

Initially, Twenty Twenty starts with a bunch of widgets in the Footer 1 widget area (shown here), and no widgets in the Footer 2 widget area.

Changing Widget Settings

Every widget provides a few settings you can adjust. To see them, expand the widget by clicking it in the list. Change the settings you want, and then click Done to make the changes permanent. *Figure 5-20* shows the settings for the Recent Posts and Categories widgets.

Even if a widget provides no other settings, it always includes a Title text box, which you can use to replace the widget's standard headline—for example, "Hot News!" instead of "Recent Posts," or "What People Are Saying" instead of "Recent Comments." If you leave the Title box blank, the widget uses its default title.

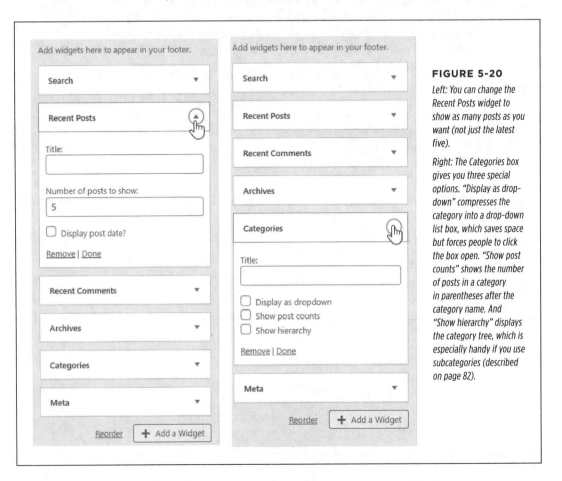

FIGURE 5-20

Left: You can change the Recent Posts widget to show as many posts as you want (not just the latest five).

Right: The Categories box gives you three special options. "Display as dropdown" compresses the category into a drop-down list box, which saves space but forces people to click the box open. "Show post counts" shows the number of posts in a category in parentheses after the category name. And "Show hierarchy" displays the category tree, which is especially handy if you use subcategories (described on page 82).

Adding and Removing Widgets

The next-easiest widget-customization task is *deleting* a widget. To do that, click to expand it. Then click Remove to banish it from your site.

TIP You can practice widget removal with WordPress's most useless widget, the Meta widget. The Meta widget shows administrative links like "Log in" and "Site Admin" (which takes you to the admin area). These links aren't particularly useful, because you already know how to get around your site. Even worse, these links look unprofessional to your readers. The Meta widget is really just a holdover from ancient versions of WordPress.

To add a widget, click the Add a Widget button at the bottom of the widget list in the theme customizer. This pops open a second column (*Figure 5-21*), with a list of all the widgets that are currently installed on your site. (Right now, you have a handful of WordPress classics.) To add one of these widgets, click it.

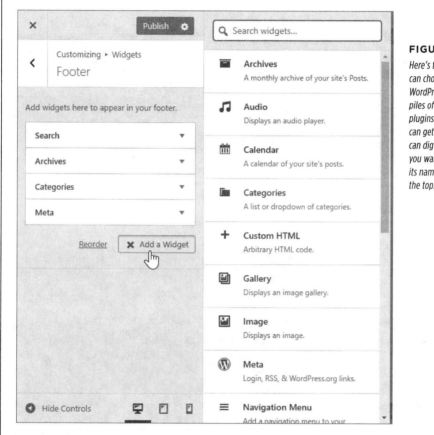

FIGURE 5-21

Here's the list of widgets you can choose from in a fresh WordPress site. If you add piles of new widgets with plugins (Chapter 9), this list can get long. In that case, you can dig up exactly the widget you want by typing part of its name in the search box at the top.

You are now the master of your widgets. But what does each widget actually do? Table 5-3 describes the standard WordPress widgets that you'll get in every WordPress website.

TABLE 5-3 *WordPress widgets*

WIDGET NAME	DESCRIPTION	FOR MORE INFORMATION...
Archives	Shows links that let readers browse a month of posts at a time. You can convert it to a drop-down box and display the number of posts in each month.	—
Audio	Shows a miniature audio player you can use to play an MP3 file.	You'll learn to use audio and video in Chapter 7.
Calendar	Shows a miniature calendar that lets guests find posts on specific dates.	See page 132.
Categories	Shows links that let readers browse all the posts in a category. You can convert it to a drop-down box and display the number of posts in each category.	See page 132.
Custom HTML	Shows a block of HTML. You can put whatever content you want here, which makes it an all-purpose display tool for small bits of information. However, the Text widget has all the same functionality.	See page 134.
Gallery	Creates an image gallery that shows thumbnails for a bunch of pictures, tiled together.	See page 179.
Image	Shows a single image.	—
Meta	Shows administrative links (for example, a "Log in" link that takes you to the admin area). Once you're ready to go live with your site, you should delete the Meta widget.	—
Navigation Menu	Shows a menu of pages or other links that you create using WordPress's menu feature.	Menus are explained in Chapter 8.
Pages	Shows links to the static pages you pick. (Static pages act like ordinary web pages, not posts. You can add them to your website to provide extra information or resources.)	Static pages are explained in Chapter 8.
Recent Comments	Shows the most recent comments left on any of your posts. You can choose how many comments WordPress displays (the standard is five).	Comments are explained in Chapter 10.
Recent Posts	Shows links to your most recent posts. You can choose how many posts WordPress displays (the standard is five).	—

WIDGET NAME	DESCRIPTION	FOR MORE INFORMATION...
RSS	Shows links extracted from an RSS feed (for example, the posts from another person's blog).	—
Search	Shows a box that lets visitors search your posts.	—
Tag Cloud	Shows the tags your blogs use most often, sized according to their popularity. Readers can click a tag to see the posts that use it.	See page 133.
Text	Shows a block of formatted text. You can add lists, bold and italic formatting, and links. You can even use HTML tags for more control.	See page 134.
Video	Shows a miniature video player. You could use this to play a video in a proper web-friendly format, or a video from a hosting service like YouTube.	You'll learn to use audio and video in Chapter 7.

POWER USERS' CLINIC

Moving Widgets from One Area to Another

As you get a bit more ambitious, you may want to try moving widgets from one area to another.

For example, maybe you have a perfectly configured widget in a sidebar that you want to move into your footer. Or vice versa. Unfortunately, you can't do that in the theme customizer. (You'd be stuck deleting the widget you don't want, adding a new copy to the other widget area, and then reconfiguring it.) But you can drag things around in the admin area.

Start by choosing Appearance→Widgets from the menu. There you'll see all the widget areas at once, along with an Available Widgets box that stocks all the widgets you can add. Now find the widget area that has the widget you want to move. If you can't see your widget, just click the title of the widget area to expand the box and show all the widgets inside. Finally, drag your widget over to a different widget area and drop it there.

The Widgets page has another hidden feature, called the Inactive Widgets section. If you have a fully configured widget that you don't want to use right now but want to save for later, put it here. (If you don't see Inactive Widgets, scroll down. It's all the way at the bottom of the page.)

You're already acquainted with the basic set of widgets that every blog begins with: Search, Archives, Recent Posts, Categories, Recent Comments, and Meta. In the following sections, you'll take a closer look at a few widgets you might consider adding to your site. You'll get to many of the more specialized widgets later in this book.

TIP You can add the same widget more than once. For example, you can add two Navigation Menu widgets to your page, give each one a different title, and use each one to show a separate set of links.

The Calendar Widget

The *Calendar widget* gives readers a different way to browse your site—by finding posts published on a specific day (*Figure 5-22*). It's most commonly used in blogs.

FIGURE 5-22

In the month of March, four days have at least one post—the 3rd, 4th, 6th, and 25th. Click the date to see the corresponding posts.

Years ago, the Calendar widget used to be a staple of every blog. These days, it's far less popular. The problem is that unless you blog several times a week, the calendar looks sparse and makes your blog feel half-empty. Also, it emphasizes the current month of posts while neglecting other months. Most readers won't bother clicking their way through month after month to hunt for posts.

You probably won't use the Calendar unless your posts are particularly time-sensitive and you want to emphasize their dates. (For example, the Calendar widget might make sense if you're chronicling a 30-day weight-loss marathon.) If you use the Calendar widget, you probably won't use the similarly date-focused Archives widget, or you'll at least place it far away, in another widget area.

The Categories Widget

The *Categories widget* is one of the widgets that every WordPress site starts out with. You learned about it (and WordPress's post categorization system) in Chapter 4. However, the Categories widget has a few interesting settings that you haven't seen yet. If you expand it, you'll see the following options:

- **Display as dropdown.** This setting saves space in tight layouts. Switch it on, and the category list shrinks to a single box that says "Select Category." The reader can then click this box to show the full category list and pick a category for browsing.

- **Show post counts.** This shows the number of posts in each category after the category name in brackets. For example, if you have 12 posts in a category, you'll see "Green Tea (12)" instead of just "Green Tea."

- **Show hierarchy.** This turns the standard list of categories into a tree of categories. This is a useful way to show how your subcategories are arranged (*Figure 5-23*).

NOTE No matter what setting you tweak, WordPress always orders categories alphabetically. If you want to put a specific category on top, you're better off creating a menu and using the Navigation Menu widget (page 228).

CATEGORIES

Black Tea
Events
Herbal Tea
News
Rooibos
Tea
Uncategorized

CATEGORIES

Events
News
Tea
    Black Tea
    Herbal Tea
    Rooibos
Uncategorized

FIGURE 5-23

Ordinarily, the Categories list ignores subgroups (left). But fear not: with a simple configuration change, you can get a more readable tree (right).

The Tag Cloud Widget

You've already seen how the Categories widget lets visitors browse through the posts in any category. The *Tag Cloud widget* is similar in that it lets readers see posts that use a specific tag.

There's a difference, however. While categories are well-defined and neatly organized, the typical WordPress site uses a jumble of overlapping keywords. Also, the total number of categories you use will probably be small, while the number of tags could be quite large. For these reasons, it makes sense to display tags differently from categories. Categories make sense in a list or tree. Tags work better in a *cloud*, which shows the most popular tags sorted alphabetically and sized proportionately. Tags attached to a lot of posts show up in bigger text, while less-frequently used tags are smaller (*Figure 5-24*).

There's no secret to using the Tag Cloud widget. Just drag it into an area of your theme and see what tags it highlights. The tag cloud might also tell *you* something about your site—for example, which topics keep coming up across all your posts.

NOTE If clouds work so well for tags, it might occur to you that they could also suit categories, especially in sites that have a large number of categories, loosely arranged, and with no subcategories. Happily, a category cloud is easy to create. You need to use a setting in the Tag Cloud widget called Taxonomy. To create your category cloud, change the Taxonomy setting from Tags to Categories.

TAGS

antioxidants children

Christmas Cosmic Harmony

gourmet tea

health Kuala Lumpur

Malaysia music recipes stevia

store tea plant The

Black Teas

FIGURE 5-24

This tag cloud shows that "health" is the most frequently used tag, with "store" close behind. As with categories, clicking a tag shows all the posts that use it.

Taming the Tag Cloud

What do I do if my tag cloud shows too many tags? Or not enough? Or makes the text too big?

The Tag Cloud widget is surprisingly uncustomizable. If you use fewer than 45 tags, it shows every one of them (although it ignores any tags you added to the Posts→Tags list but haven't yet used in a post). If you use more than 45 tags, the Tag Cloud widget shows the 45 most popular.

Occasionally, people want a tag cloud with more tags. But usually they have the opposite concern and want a smaller tag cloud that's slim enough to fit into a sidebar without crowding out other widgets. One solution is to crack open your template files. That's because the behind-the-scenes code (the PHP function that creates the cloud) is actually very flexible. It lets you set upper and lower tag limits, and set upper and lower boundaries for the text size. You can get the full details from WordPress's function reference at *http://tinyurl.com/wptagcloud*. (However, this information won't be much help until you learn how to dig into your WordPress theme files to change your code, a topic explored in Chapter 13.) Another solution is to search for a plugin that lets you pick the tag options and then generates a customized tag cloud. You'll learn how to find and install plugins in Chapter 9.

The Text Widget (and Custom HTML)

The *Text widget* is simple but surprisingly flexible. You can use it anywhere you want to wedge in a bit of fixed content. For example, you can use it in a sidebar, to add a paragraph about yourself or your site. Or you can put it in your footer with some copyright information or a legal disclaimer.

Using the Text widget is easy. First, add it to a widget area (as you do with any other widget). Then, type your text inside. You don't need to stick with plain boring text. The Text widget comes with handy buttons for making bold text, italicized text, numbered lists, bulleted lists, and links. They're all tucked into a tiny toolbar (*Figure 5-25*).

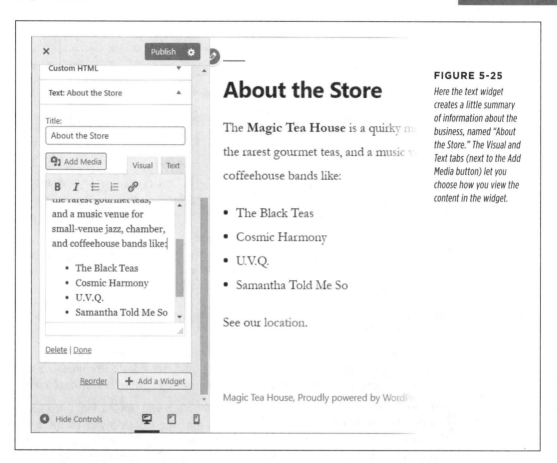

FIGURE 5-25

Here the text widget creates a little summary of information about the business, named "About the Store." The Visual and Text tabs (next to the Add Media button) let you choose how you view the content in the widget.

If you want more control—and you know a bit about HTML—you can write the *exact* markup you want in the Text widget. To do that, click the tiny Text tab in the top-right corner of the expanded widget (*Figure 5-21*). (The Text tab really means HTML text—in other words, you see the raw codes that structure your text. The Visual tab is like a word processor, where you format things and see what they look like.)

When you use the Text tab, you type in your content with the exact HTML tags you want. Here's an example that puts a word in bold type:

```
The following word will be <b>bold</b> on the page.
```

And here's the HTML-formatted text you'd use to replicate the store description in *Figure 5-25*:

```
The <b>Magic Tea House</b> is a quirky mash-up: it's a fine tea importer with
the rarest gourmet teas, and a music venue for small-venue jazz, chamber, and
coffeehouse bands like:
<ul>
<li>The Black Teas</li>
<li>Cosmic Harmony</li>
<li>U.V.Q.</li>
<li>Samantha Told Me So</li>
</ul>
See our <a href='http://tinyurl.com/cyboj83'>location</a>.
```

If you want to edit *only* in HTML view, you can use the Custom HTML widget. This makes sense if you're copying a whole block of HTML from another website or service (for example, a complete sign-up form).

The Custom HTML widget doesn't have a Visual tab, so you don't have the option of seeing a graphical preview. It also adds some HTML editing frills, like line numbers, automatic tag completion, and error checking.

■ Responsive Themes

If you build a website today, it's a safe bet that many of your visitors will browse your pages by using smartphones, tablets, and other mobile devices. Unfortunately, pages that look good in a desktop or laptop browser aren't so swell on smaller screens.

The problem is space—namely, how to adapt your pages when there's not nearly as much room to work with. If you're using a modern, mobile-aware theme, it can adapt itself to work in cramped spaces. It can rearrange your content, simplify your site's layout, and increase the size of your text. This design philosophy is sometimes called *responsive design* because it *responds* to the needs of the viewer.

Themes that aren't responsive don't make any effort to tailor your pages for tiny screens. The result is a postage-stamp-sized page that's difficult to navigate and read. Fortunately, most themes today are responsive, including all the WordPress year themes since Twenty Eleven.

To check what your theme looks like on a mobile device, there's no substitute for pulling out your smartphone and taking a look. However, you can get a quick preview in the theme customizer. At the bottom left of the theme customizer panel are three buttons that represent three screen sizes: standard desktop computer, iPad-style tablet, and smartphone. When you click one of these buttons, the theme customizer adjusts the page size to match the device, and you can see how your layout reflows to fit (*Figure 5-26*).

NOTE If you don't see the view buttons at the bottom of the theme customizer, try making your browser window bigger. They don't appear at small sizes.

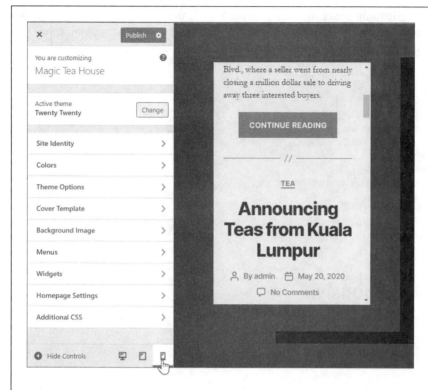

FIGURE 5-26

When you click the smartphone icon in the theme customizer, you get a simulation of what your site will look like on a smartphone. Of course, this is no substitute for checking your design on a real device.

If your theme doesn't recognize mobile devices, all is not lost. You can use a mobile-aware plugin. One is WPtouch (*http://tinyurl.com/wptouch*), which identifies smartphones and other mobile devices, and makes sure they get a simplified theme that better suits their capabilities (and looks pretty slick too). You'll learn more about it in Chapter 14.

■ The Last Word

In this chapter, you took your first look at themes, which are one of WordPress's most important features. Most of the style decisions that control your site—whether it's about layout, typefaces, colors, or images—start with the theme you choose.

You now know how to install themes and change your site from one theme to another. You should also know how to load up the theme customizer and use it for basic customization tasks like changing header images and adding widgets. This is a good foundation to start experimenting with your site.

You'll explore some more advanced theme customization topics later in this book. For example, in Chapter 7, you'll see how themes use featured images to let readers browse your posts. In Chapter 8, you'll think about themes that help you create custom home pages. And in Chapter 13, you'll learn advanced techniques for customizing the styles and code in any theme.

Making Fancier Posts

You know what a basic WordPress post looks like—it starts with a title, followed by one or more paragraphs of text. And there's nothing wrong with that. If you pick a good theme, your WordPress site can look great, even if it holds nothing more than plain-text content.

However, there are plenty of good reasons to make fancier posts with more formatting. For example, long posts are easier to read if you break up your text with subheadings and spacers. Pictures, lists, links, and pull quotes can transform your writing from a drab wall of text to a professional product, whether you're making a chatty blog or a buttoned-down business site. To add details like these, you need to take charge of WordPress's post editor. You'll learn how to do that in this chapter.

■ Simple Text Formatting

Before you plunge deep into the world of WordPress formatting, it's worth taking a quick look at a few simple styling tricks you'll use every day. These are basic formatting touches, like applying bold and italics. WordPress scatters these settings in a few places, so it's easy to lose track of them when you're starting out.

If it's not already obvious, all of these tricks and techniques happen in the post editor. So before you continue to the next section, start with the standard Post→Add New command to create a new post. That way, you have somewhere to test out these techniques.

NOTE The exact appearance of the post editor depends on your theme. That doesn't mean the post editor looks exactly the same as your published page (it won't), but WordPress makes a good effort to get it as similar as possible, and some themes cooperate more than others. For example, if you're using the Twenty Twenty theme, the post editor includes the familiar shaded background and a heavyweight, centered title. If you're using Twenty Nineteen, you get a white background, tighter spacing, and different fonts. And so on.

Applying Bold and Italics

Here's the executive summary: you can **bold** and *italicize* text in WordPress in the same way you do in any word processing program. Just select your text and hit the handy Ctrl-B or Ctrl-I keyboard shortcut (Command-B or Command-I on a Mac). Presto!

But if you want to go a little deeper, you should take a look at WordPress's tiny formatting toolbar, which holds these basic formatting commands. The toolbar appears just above the paragraph you're editing (*Figure 6-1*).

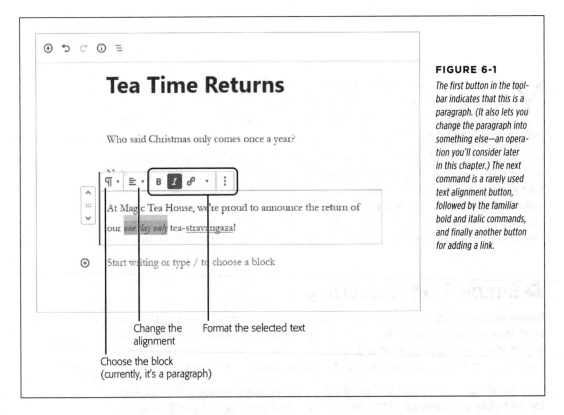

FIGURE 6-1

The first button in the toolbar indicates that this is a paragraph. (It also lets you change the paragraph into something else—an operation you'll consider later in this chapter.) The next command is a rarely used text alignment button, followed by the familiar bold and italic commands, and finally another button for adding a link.

Change the
alignment

Format the selected text

Choose the block
(currently, it's a paragraph)

To change the formatting of your text with the toolbar, begin by selecting your text. (Mouse lovers can click and drag to make a selection. Keyboarders can hold down Shift while using the arrow keys.) Then, click the corresponding formatting button in the toolbar.

The bold and italic buttons are easy to spot. But a few less common choices are tucked away behind the down-pointing triangle icon. For example, you can choose Strikethrough to get crossed-out text ~~like this~~, or Inline Code to get a typewriter-like font that's useful for writing code commands like `timeWasted=timeWasted+1;`.

The toolbar's text alignment setting isn't as useful. It lets you switch between text that's lined up on the left margin, centered, or lined up against the right margin. There's no option for full justification, which fits each line to the full margin space (like what you see in this book). You might use the alignment button to tweak a subheading, but usually you'll stick with the standards enforced by your theme.

UP TO SPEED

The Indispensable Undo Feature

Like any good editor, WordPress's post editor can save you from your own mistakes. If you delete a sentence, rearrange your paragraphs, or make another change and then realize you've gone too far, WordPress's trusty undo feature can help you out. Use it, and the post editor will reverse your last change.

To perform an undo, you click the Undo button in the toolbar at the top of the post editor. (It looks like a backward-curving arrow.) Next to the Undo button is the Redo button, which has a similar arrow that curves forward. If you undo something but then decide you really want it, hit the Redo button to reverse your undo and get things back to the way they were.

You may already know the keyboard shortcuts for these commands, because they're the same in most word processing programs. To undo a change, press Ctrl-Z (Command-Z on a Mac); and to redo a change, press Ctrl-Shift-Z (Command-Shift-Z on a Mac).

WordPress doesn't keep track of just one change. As you work, it keeps a long undo history. That means you can reverse a whole sequence of actions by pressing Undo several times in a row. (But be careful—if you then make a new change, you won't be able to use Redo if you want to restore your changes.)

The undo feature isn't the only way WordPress can save you from an editing disaster. It also keeps old versions of your entire post. Every time you save your draft manually (by clicking Save Draft), or publish your post, or update your published post, WordPress adds a new entry to its revision history. You can peek at the revision history, and even restore an old version of a post if an edit ruined it. Page 327 has more on this feature.

Adding Links

The web wouldn't amount to much without links, those <u>blue underlined</u> bits of text that let you jump from one web page to another. You can easily add links to a post. For example, imagine you have this sentence:

From the bustling metropolis of Kuala Lumpur come some of the most exotic...

To turn "Kuala Lumpur" into a link, first select the text. Then, click the Link button in the toolbar, or just use the handy Ctrl-K keyboard shortcut (Command-K on a Mac). A text box pops up, asking you to fill in the full web address, starting with *http://* (*Figure 6-2*).

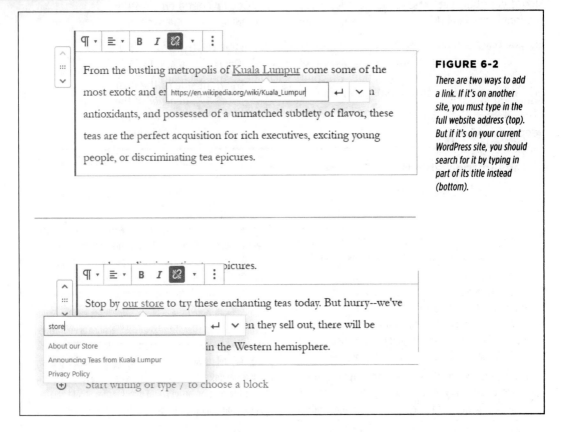

FIGURE 6-2

There are two ways to add a link. If it's on another site, you must type in the full website address (top). But if it's on your current WordPress site, you should search for it by typing in part of its title instead (bottom).

Often you'll want to make a link to one of the posts or pages you've made on your site. In this special case, you don't need to type in the full link. Instead, you can quickly search for the post you need by typing in part of its title. WordPress will then create a link for the post you pick.

The only other detail you can change is the Open in New Tab setting. If you switch this on, the person clicking your link won't leave your site—instead, their browser will open a new tab with the linked page. That sounds convenient, but it's actually a less common design choice, because it can irritate people and cause problems with ad blockers.

> **NOTE** If you want to link to a file—for example, a document that your guests can download or a picture they can view—you need to store that file in WordPress's media library first, and *then* create a link. You'll get the full details on page 205.

To change a link, click it in the post editor (at which point the linked address appears), and then click the tiny pencil icon next to the address. To remove a link and return your text to normal, click anywhere inside the link text and then click the toolbar's Link button.

Formatting Drop Caps, Size, and Background Color

WordPress has a few more formatting frills tucked away into its sidebar. To see these, start by making sure that the sidebar is visible on the right side of the post editor. (If you don't see a sidebar, click the gear icon in the top-right corner of the post editor to make it appear.) Then, click the paragraph you want to format. If you keep your eyes on the sidebar, you'll see a few miscellaneous settings appear in the Block tab (*Figure 6-3*).

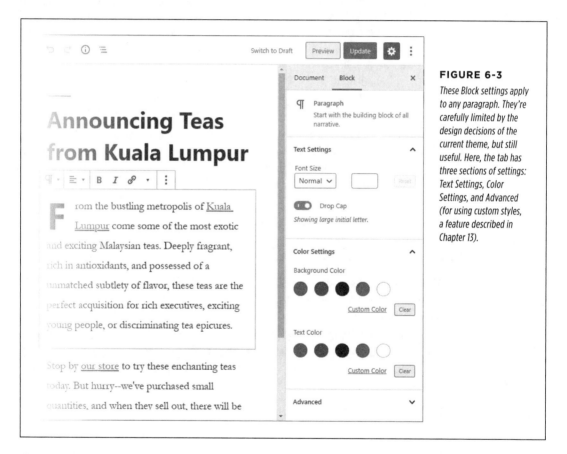

FIGURE 6-3

These Block settings apply to any paragraph. They're carefully limited by the design decisions of the current theme, but still useful. Here, the tab has three sections of settings: Text Settings, Color Settings, and Advanced (for using custom styles, a feature described in Chapter 13).

As you'll see a little later in this chapter, WordPress supports different types of text-based design elements, called *blocks*. The sidebar allows you to format the current block, whatever it is. So far, you've worked only with paragraph blocks, which give you a small number of settings to emphasize text:

- **Change text size**. You can increase or decrease the size of all the text in your paragraph by picking an option in the Font Size box. The idea is to use this feature sparingly to distinguish certain design elements. For example, you might decide to start each post with a single large-text paragraph. You should never use this technique to change *all* your paragraphs. If you want everything to be a bit larger or smaller, you're better off editing your theme to ensure consistency (and to avoid the hassle of endless formatting), as explained in Chapter 13.

> **NOTE** You shouldn't use formatting to try to turn a paragraph into something it isn't (like a subheading). Instead, pick a more suitable block.

- **Change the background color.** Different themes give you different colors, which you pick by clicking the colored circle you like. For example, Twenty Nineteen provides dark blue and dark gray backgrounds, while Twenty Twenty includes a customizable dark red. These colors are chosen to mesh with the style of the overall theme, and WordPress switches your text from black to white based on the background you use, so your text remains visible. Once again, the goal is to highlight specific, occasional design elements. For example, you could add a summary paragraph with a red background to the bottom of every post.

- **Add a drop cap.** That's the stylized jumbo-sized letter that starts a paragraph, and is usually reserved for the first paragraph of your post. To add a drop cap to the beginning of a paragraph, just turn on the Drop Cap setting. But be aware, you won't see the drop cap while you're editing the paragraph. Instead, click a different paragraph to see the result of your change, or just publish the post.

■ Understanding Blocks

To unlock the full capabilities of the WordPress post editor, you need to learn a new concept: *blocks*, the rich, customizable, and rearrangeable components that make up every post.

So far, you've concentrated on typing your content into paragraphs. But in the eyes of WordPress, paragraphs are just one type of block. In fact, WordPress has several dozen block types. You can add a new block by using the Add Block button, which looks like a small plus sign in a circle.

The Add Block button appears in several places:

- **In the toolbar.** If in doubt, you can always find the Add Block button in the toolbar at the top of the post editor. Click it to insert a new block at your current position. This Add Block button is slightly fancier than the others—it gives you a small preview of each block you might want to use when you hover over it in the list.

- **Between blocks.** The Add Block button also appears when you move your mouse around. For example, if you position your mouse in the middle of the space between two existing blocks, the Add Block button appears, allowing you to insert a new block between them.

- **Beside a new block.** If you press Enter to start a new paragraph, the Add Block button appears in the margin on the right. Click it to change your new paragraph block to something else.

Figure 6-4 shows all these variations.

FIGURE 6-4

You can find the Add Block button in three places. It appears (from top to bottom) in the toolbar, between blocks when you move the mouse there, and beside new blank paragraphs. (Of course, you can't use all three buttons at once. We've simply combined them into one picture so you can see them at a glance.)

To take a quick peek at all the blocks WordPress provides, click one of the Add Block buttons. You'll pop open the long, subdivided list shown in *Figure 6-5*. You can scroll through to find what you want, or—if you know the name of the block you want to use, but you don't know exactly where it is—type its name into the search box at the top of the list.

TIP Here's a pro WordPress trick for quickly adding blocks. Start a new blank paragraph where you want to add your block; then type / (the backslash key). WordPress will show a compact list of suggested blocks. Now start typing the *name* of the block you want (like "head" for a heading), and WordPress will update the list with the blocks that match. Click the one you want, or press Enter when it reaches the top of the list.

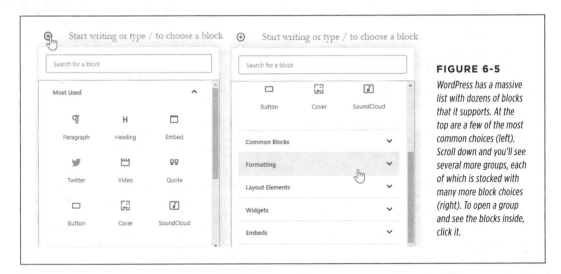

FIGURE 6-5

WordPress has a massive list with dozens of blocks that it supports. At the top are a few of the most common choices (left). Scroll down and you'll see several more groups, each of which is stocked with many more block choices (right). To open a group and see the blocks inside, click it.

It's easy enough to add blocks, but how do you figure out what WordPress offers in its overflowing selection? You'll explore most of WordPress's blocks in this chapter and the next, but Table 6-1 gives you a quick preview of each group.

TABLE 6-1 *The WordPress block groups*

BLOCK GROUP	WHAT IT'S FOR
Most Used	WordPress keeps track of the blocks you use the most, and adds them to this list for easy access.
Common Blocks	Includes some of the most popular and useful blocks, like paragraphs, headings, quotes, and lists, all of which you'll see in this chapter. It also includes pictures and audio, which you'll get to in Chapter 7.
Formatting	Includes some specialized formatting blocks, like big, eye-catching pull quotes, code listings, and tables. You'll learn about all of these blocks later in this chapter.
Layout Elements	Includes blocks that let you split up posts with visual breaks and paginate them, which you'll learn about in this chapter. This group also includes the blocks you use for columns (covered in Chapter 7) and—weirdly—buttons (page 156).
Widgets	Includes blocks that let you take any widget and put it in a post. Widgets were covered in Chapter 5.
Embeds	Includes blocks that lets you put in rich content from another site, like a YouTube video, tweet, Spotify playlist, and so on. You'll learn about embeds in Chapter 7.
Reusable	You can add your own personalized blocks to this section so you can reuse them whenever you need. You'll learn how to do that at the end of this chapter.

■ Using the Essential Blocks

The best way to get familiar with WordPress's sprawling catalog of blocks is to try using some of them. The following sections will lead you through the most popular types, and by the end of this book you'll have seen them all.

Subheadings

Every blog post starts with a heading—the title of your post, which sits above the post content. But if you're writing a long post, it's a good idea to subdivide your writing into smaller units by using *subheadings*.

To create a subheading, click Add Block to show the list of blocks, and pick Heading. (It's stashed in the Common Blocks group, but usually appears in the Most Used group at the top.) Then type away.

Technically, when you add a heading, WordPress uses a level-2 heading (known to HTML nerds as an *<h2>*). This makes sense, because the title of your post is a level-1 heading (or *<h1>*). It's important to use the right headings to structure your posts. Otherwise, you may confuse search engines and cause formatting problems if you switch to another theme.

However, sometimes you want to further subdivide your posts. For example, you might start a new section with a level-2 heading ("Visiting France"), and then decide to make level-3 subheadings inside ("Paris" and "Lyon"). And then, if you're feeling particularly crazy, you might decide to subdivide a level-3 section with level-4 subheadings. Before you know it, you'll have an epically long post filled with fine-grained subdivisions.

To change the heading level, you can use the toolbar or Level setting in the sidebar (*Figure 6-6*). Either way, look for the level (H2) and change it to something else.

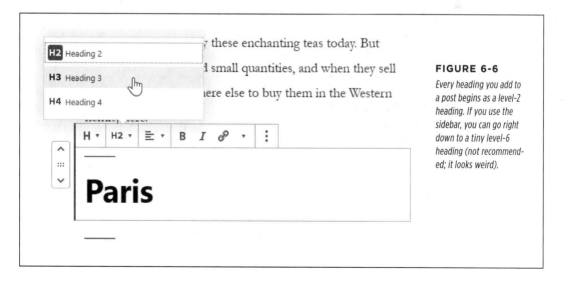

FIGURE 6-6

Every heading you add to a post begins as a level-2 heading. If you use the sidebar, you can go right down to a tiny level-6 heading (not recommended; it looks weird).

Usually, level-2 headings are enough for a post. If you need to get a little more organized, level-3 headings are good. But going deeper rarely makes sense, because the formatting may become inconsistent (for example, your headings may become smaller than your text) and readers probably won't be able to distinguish that many heading levels anyway.

Much as you can format text (just a bit), you can also tweak your headings (just a bit). The size, typeface, and exact appearance of a heading depend on your theme. But sometimes you might want to emphasize a specific heading, or distinguish one type of heading from another. To see what you can do, check out the formatting options in the sidebar on the right. Most themes allow you to change the color of the heading text, but some limit your choices to approved shades that match the overall color scheme. Twenty Twenty is more freewheeling than most—it recommends its usual colors (the dark crimson accent color, along with a few shades of black and beige), but it also lets you pick a different custom color. Twenty Twenty also allows you to change the background color of the heading to really make it stand out.

POWER USERS' CLINIC

Use Anchors to Link to Specific Parts of a Post

You've already learned how to create links to other posts. But here's a nifty trick: creating a link that takes you to a specific *section* of a post. You could use this technique to link from one post to important information in another, or even to build a mini table of contents at the top of a long post that lets readers jump right to the content they want.

The secret to building more specific links is adding an *anchor*—an invisible HTML code at a certain place in your post. WordPress makes it easy to add anchors for any of your headings, which is exactly where you want them.

To add an anchor, click a paragraph, and then expand the Advanced section in the sidebar. Next, type a name for your anchor in the HTML Anchor box. A typical anchor is a unique word from your heading. If you want to use more than one word, separate them with hyphens (-), because spaces aren't allowed. Reasonable anchor names include *paris-travel*, *budget*, and *ravioli-recipe*.

Once you have an anchor, you can link to it by adding a number sign (#) to the end of your link, followed by the anchor name. So if the post is at *http://magicteahouse.net/kuala-lumpur*, and your anchor is named *#jasmine*, your direct link becomes *http://magicteahouse.net/kuala-lumpur#jasmine*. Try it out in a browser, and you should see that it brings you straight to the heading you want.

Separators and Spacers

Subheadings aren't the only way to split up a post. You can also break up posts with lines, a series of dots, or just some extra space in the right place. All of these divisions help to space out long content without dicing it up into separate sections (*Figure 6-7*).

To get a graphical divider, you use the *Separator block*. Click the Add Block button, open the Layout Elements section, and click Separator.

The exact appearance of your separator depends on the theme. (Twenty Twenty, for instance, uses a stylized line with a break in the middle.) Usually, you'll be able to choose between a narrow line, an extra wide line, and a sequence of three dots.

You make your pick from the Styles section in the sidebar. You may also be able to alter the color of the separator in the Color Settings section.

A Nice Cup of Tea

👤 Sarah Samdt   🕐 January 22, 2020   ■ Leave a comment   ✎ Edit

by George Orwell for the Evening Standard, January 12, 1946

The famous English novelist wrote about more than just dystopian fiction.

—— A paragraph with a colored background

—— A wide line separator

I f you look up 'tea' in the first cookery book that comes to hand you will probably find that it is unmentioned; or at most you will find a few lines of sketchy instructions which give no ruling on several of the most important points.

This is curious, not only because tea is one of the main stays of civilization in this country, as well as in Eire, Australia and New Zealand, but because the best manner of making it is the subject of violent disputes.

· · · ———— A dots separator

When I look through my own recipe for the perfect cup of tea, I find

FIGURE 6-7

Here's a formatted post that benefits from two types of spacers.

WordPress also includes a *Spacer block*, which appears in the same Layout Elements section. However, spacers are able to add only extra space, not dividing lines or symbols. Once you've added the spacer, you can drag its bottom edge down (to make it bigger) or up (to make it thinner). The most common reason to use a spacer is to add a bit of breathing room in a dense post with lots of pictures or other types of rich media content. For example, if you have a post with videos (which you'll learn to add in the next chapter), you might not want your heading to show up right underneath the video. By using the spacer, you can fill in a perfectly sized gap without resorting to blank paragraphs.

Lists

Another handy block is *List*, which lets you make a bulleted list (for example, to list recipe ingredients) and a numbered list (say, for recipe steps). Using lists is easy: click the Add Block button and choose List, which you'll find in the Most Used section or the Common Blocks section.

You always start with a bulleted list, but you can switch to a numbered list by using the toolbar buttons (*Figure 6-8*).

TIP WordPress has a handy shortcut for inserting list blocks. To quickly make a bulleted list, just start a new paragraph, and then type an * followed by a space. To make a new numbered list, type *1.* followed by a space.

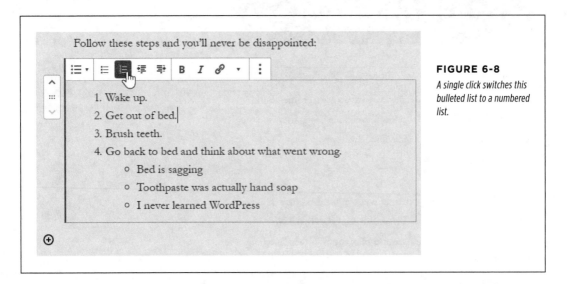

FIGURE 6-8

A single click switches this bulleted list to a numbered list.

You can also use the toolbar to create a *sublist*—a list inside another list. You can see an example in the bulleted list under step 4 in *Figure 6-8*. To create a sublist, add a new item to the list, and then click the Indent button in the toolbar. If you click the Indent button while you're on the first list item, it has no effect. But if you add a second item and then click Indent, you'll start a new indented list underneath the first item.

If you like weird and exotic formatting, you can add another entry and indent it to make a sub-sublist. In fact, there's no limit to how many levels deep you can indent, although ordinary people stop at the first indent.

However, you can't change the visual styling of your lists. If you're creating a numbered list, you can use the Ordered List Settings section of the sidebar to start on a different number (3, 4, 5, 6) or reverse the order (4, 3, 2, 1). But that almost never makes sense.

Fancy Quotes

The first thing you need to know about quotes is that there are actually two quote blocks: Quote and Pullquote. A *Quote block* is the simpler of the two. It simply separates a small section of text where, presumably, you're quoting someone else's words. Quotes are often italicized, indented, and enlarged to distinguish them from the surrounding text, but their exact appearance is determined by the theme you're using (*Figure 6-9*).

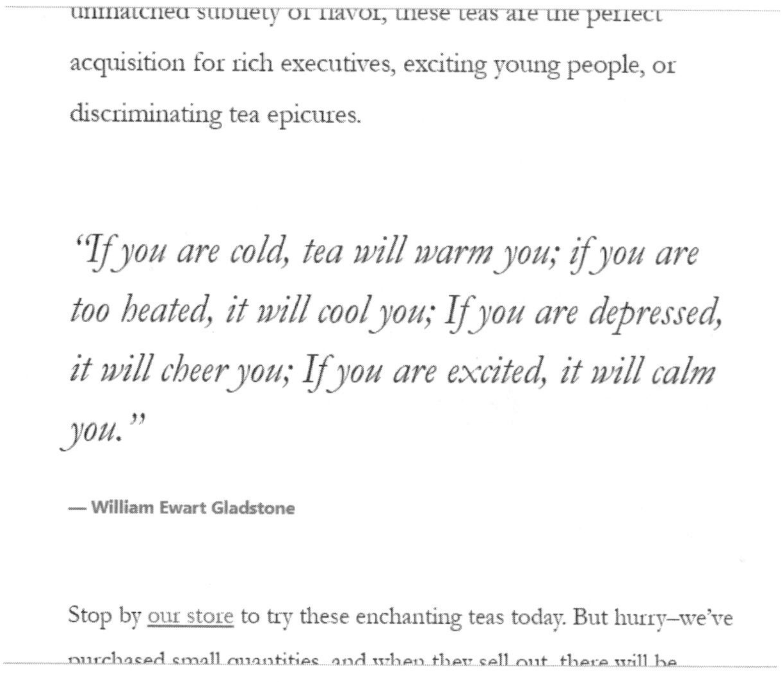

unmatched subtlety of flavor, these teas are the perfect acquisition for rich executives, exciting young people, or discriminating tea epicures.

> *"If you are cold, tea will warm you; if you are too heated, it will cool you; If you are depressed, it will cheer you; If you are excited, it will calm you."*
>
> **— William Ewart Gladstone**

Stop by our store to try these enchanting teas today. But hurry—we've purchased small quantities, and when they sell out, there will be

FIGURE 6-9

Here's a quote between two paragraphs. The Quote block has two details. First is the quote text, which is often italicized. Under that is an optional citation line, which you can use to identify the source of the quote.

To create an ordinary quote, click Add Block and choose Quote. You'll see this option in the Most Used section at the top, but it also appears in the Common Blocks group.

Every quote has two parts. First you type in the quote itself (that's the italicized text in *Figure 6-9*). Then you can enter citation information underneath, if you wish. If you don't enter a citation line, WordPress just skips that part in the final, published post.

TIP   To move from the quote text to the citation line (and to move from the citation line to the next block), you can use the arrow keys or click with the mouse. Hitting Enter won't take you there—it just adds extra blank lines.

Most themes support two quote styles: default and large. You can choose the one you want from the sidebar, in the Styles section. In the case of the Twenty Nineteen theme, for example, the large style uses the italics shown in *Figure 6-9*, while the default style uses ordinary text, slightly indented and with a red line in the margin. In Twenty Twenty, the large style uses large bold text.

Now that you're comfortable with quotes, let's take a look at their fancier sibling, pull quotes. The name *pull quote* comes from the fact that the quote is *pulled* out of your layout and blown up, so it can command attention (*Figure 6-10*). Unlike quotes, pull quotes usually aren't quotations from other people. Instead, a pull quote repeats something that's written in the post itself—something you want to emphasize in jumbo letters. Essentially, you're quoting yourself.

FIGURE 6-10

Here's a pull quote in the Twenty Twenty theme, using the Solid Color style with a red accent background. The quotation mark icon is a style detail that's baked into Twenty Twenty.

To create a pull quote, click Add Block and choose Pullquote from the Formatting group. Modern themes give you some powerful formatting options for pull quotes. In the sidebar, you can use the Styles section to switch your pull quote to white text on a colored background. You can change the color of the background or text in the Color Settings section underneath. Remember, if you don't see these choices, it's simply because your theme doesn't support them (although that's not a problem with any of the WordPress year themes).

Some themes let your pull quote burst out of the usual layout. The secret is changing the alignment setting, which appears in the toolbar while you edit your pull quote. (You might expect to see the alignment setting in the sidebar too, but for some reason it doesn't appear there.) *Figure 6-10* shows a "full alignment" pull quote in Twenty Twenty. You can also use left alignment or right alignment to float your quote to one side of your text. *Figure 6-11* shows the setting you need, and *Figure 6-12* shows its effect.

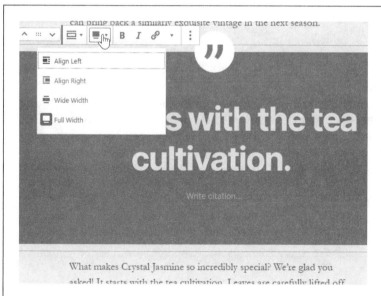

FIGURE 6-11

Use the alignment setting to make a pull quote float to the side or expand into the margins, depending on how the theme uses this setting.

FIGURE 6-12

The Twenty Twenty theme has some of the fanciest pull quote formatting features to date. If you give your quote left alignment, it may display normally (in a narrow browser window), on the left side with text wrapping around it (on medium-sized windows), or completely in the left margin (on wide windows).

FREQUENTLY ASKED QUESTION

Formatting Your Text

How can I adjust typefaces and font sizes?

WordPress's post editor lets you structure your content (for example, put it into lists and headings), add more content (like pictures), and apply certain types of formatting (like boldface and italics). However, plenty of other formatting details aren't under your control. The typeface of your fonts is one of them.

This might seem like an awkward limitation, but it's actually a wise design decision by the people who created WordPress. If WordPress gave you free rein to change fonts, you could easily end up with messy markup and posts that didn't match each other. Even worse, if you switched to a new theme, you'd be stuck with your old fonts, even though they might no longer suit your new look.

Fortunately, there's a more structured way to change the appearance of your text. Once you're certain you have the right theme for your website, you can modify its styles. For example, by changing the style rules, you can change the font, color, and size of your text, and you can make these changes to all your content or to just specific elements (like all level-3 headings inside a post).

Modifying styles is a great way to personalize a theme, and you'll learn how to do that in Chapter 13.

■ Working with More Exotic Blocks

WordPress also has some less common blocks. You might use these blocks for certain types of content or might never need them at all. These include blocks for creating tables (Table), buttons (Button), social media links (Social Icons), and code listings (Code). You'll take a look at these blocks next.

You'll also find a couple of blocks that plug into two often-overlooked WordPress features: the Page Break block and the More block. Both of these blocks are interesting and occasionally useful, but a bit dated. You can read about them in the following sections, or just skip ahead if they don't suit your site.

Tables

Tables are meant for displaying tabular data—for example, a list of countries and their populations, or a record of weight measurements in your 30-day calorie challenge.

> **NOTE** You might think you could use tables to arrange blocks of content into columns, but don't do that. WordPress has much more flexible layout blocks that you'll learn about in Chapter 7.

Here's how to create a table:

1. **Add the Table block (from the Formatting group).**

   When you insert the Table block, it doesn't appear right away. Instead, the post editor shows a box asking you to set the number of columns and rows (*Figure 6-13*, top).

2. Type in the column and row numbers, and click Create Table.

If you want headings, don't count them as one of your rows. But don't overthink this step, because you can easily add rows and columns to your table after you've created it.

Once you create your table, you'll see a basic grid of evenly spaced cells (*Figure 6-13*, bottom). It's a good idea to type in at least one row of information so you can see how the table adjusts itself before you start tweaking it.

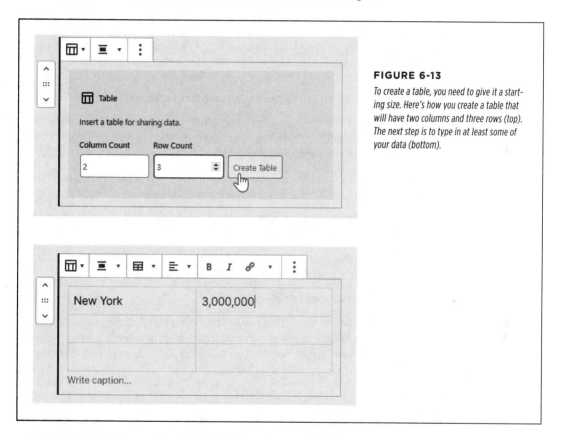

FIGURE 6-13

To create a table, you need to give it a starting size. Here's how you create a table that will have two columns and three rows (top). The next step is to type in at least some of your data (bottom).

3. Optionally switch on "Fixed width table cells" in the sidebar.

Ordinarily, WordPress tries to size the table columns proportionately to fit your content. So if you type a sentence in the first column and a number in the second, WordPress makes the first column larger. (When performing this adjustment, WordPress just looks at the longest values in each column.)

You can override this behavior with the "Fixed width table cells" setting. If you do, the space will be equally divided across all your columns. Currently, Word-Press doesn't let you fine-tune the size of each column by hand.

4. **Optionally, switch on "Header section" (in the sidebar) and add your headings.**

   When you switch on the "Header section" setting, WordPress adds a row at the top of the table where you can put column titles. The titles get bold formatting.

   You can also turn on the "Footer section" setting to get an extra row at the bottom. Sometimes people use this spot to sum up the numbers in the column above.

TIP Tables have some baked-in formatting. In the Twenty Twenty theme, you get a no-frills sans-serif font that's set at a slightly larger size than the body text. If you want to customize the font that's used in your tables, you'll need to modify the styles in your theme, as described in Chapter 13.

---

5. **Optionally, turn on striped rows to make long rows easier to read.**

   If you want this effect, which gives every other row a darker shaded background, click the Styles section in the sidebar and then choose Stripes. Other themes may add more table-formatting choices.

   You can also click the Color Settings section of the sidebar to choose a different background color, but that applies for the whole table, not individual rows.

6. **If you want to change the alignment of a column, click it, and then pick a different alignment option in the toolbar.**

   You can align the contents of a column on the left (the standard choice), on the right, or in the center. Your selection affects the entire column.

7. **Finish filling out your table.**

   Press the Tab key to move from one cell to the next (or just click where you want to go). If you decide to change the structure of your table, click the toolbar's Edit Table button, which has a menu full of options for adding and removing rows and columns.

Buttons

You can think of a *button* as a fancy way to make a link. But instead of having the usual blue underline text, you have a colorful square (you pick the color) that, when clicked, takes the visitor to the new page.

Buttons make sense when a navigation feels like an *action*. For example, if you want to ask readers to sign up for your mailing list, you might have a Subscribe button that links to your sign-up page. Or if you want them to send you some money on PayPal, you might have a Donate button that takes them to a PayPal donation page that you've set up. In both cases, the button is really just a fancy wrapper on an ordinary link. But because the clicker feels like they're about to *do* something, it helps to distinguish this link from ordinary links that take them to different content.

Creating a button is easy:

1. **Add the Buttons block.**

    It's named in plural, because it lets you put more than one button side by side, in a neat row.

2. **Type the caption for your button.**

    For example, "Subscribe" or "Register Now."

3. **Create the link for your button.**

    A button is really just a colored square. Right now, clicking it has no effect. To change that, you need to add a link, just as you did at the beginning of this chapter. Select the button text, click the Link icon in the toolbar, and type in the web address where the button should go.

NOTE When you link a button, the whole rectangle becomes clickable. You don't need to click exactly on the text.

---

4. **Optionally, tweak the appearance of your button.**

    WordPress has plenty of options in the sidebar to satisfy your interior designer (*Figure 6-14*). Some themes have different styles. (For example, Twenty Twenty lets you choose solid buttons or buttons that have an outline only.) You can also change the text color and the background color, and you can even shade the background color into an attractive (and somewhat distracting) blend of colors called a *gradient*. Scroll down a little more and you'll see the Border Radius slider. Pull that to the side to begin rounding off the corners of your button.

5. **Optionally, add another button next to the first one.**

    Sometimes you'll want to put two similar buttons next to each other. For example, you might have a Sign Up button next to a Learn More button. You can add as many buttons as you want in the Buttons block—just click the plus icon that appears immediately to the right of the last button. Each button gets its own text, link, and style settings.

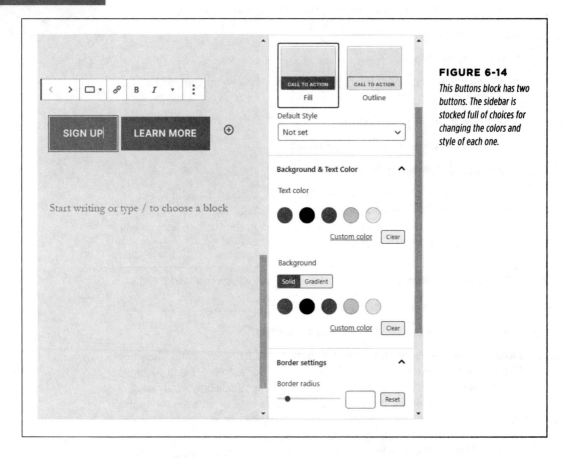

FIGURE 6-14

This Buttons block has two buttons. The sidebar is stocked full of choices for changing the colors and style of each one.

Social Icons

If you're an avid social media user, nothing stops you from dropping a link to your social profiles in a post. For example, you could write "Follow me on Twitter" and turn the word "Twitter" into a link that leads to your Twitter profile. But WordPress also has a fancier way to advertise your social media presence with the Social Icons block.

Essentially, the *Social Icons block* is a group of one of more small buttons (*Figure 6-15*). You choose which services you use, and WordPress creates the corresponding icons for you. For example, if you want a link to your Facebook profile, WordPress asks for your profile page URL and then creates a tiny Facebook icon that leads there.

When you first add the Social Icons block, all it has is a useless link to the WordPress. org site. To get rid of that, click the WordPress icon so the URL box appears, and delete the URL.

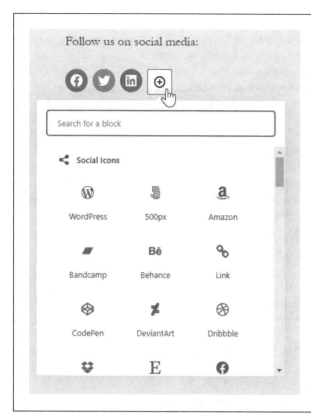

FIGURE 6-15

Really, the Social Icons block is just a way to get nice-looking buttons for your social media profiles. Here, three icons have been added—one each for Facebook, Twitter, and LinkedIn.

To add new icons, click the plus button (shown in *Figure 6-15*). You can add as many or as few as you want. WordPress gives you a huge menu of about 40 websites to choose from. It doesn't matter whether you livestream video games on Twitch, record your musical creations for Bandcamp, or sell crocheted hats on Etsy—there's an icon for you.

After you add a new icon, you need to supply a URL. For example, if you add a Facebook icon, you need to click the icon and type in a URL that points to your Facebook profile. If you add an icon but forget to fill in the URL, it won't appear in your published page. The process is a bit quirky, but you'll get the hang of it.

Using the Social Icons block has two drawbacks. First, visitors might think they're actually *sharing buttons*—buttons that let visitors share your posts with their followings. You'll learn more about sharing buttons in Chapter 12, but for now it's worth noting that they often look the same as these social icons. To avoid confusion, you can add some text above the Social Icons block, with a message like "Follow Us on Social Media."

The other quirk is that you need to add the Social Icons block separately to every post where you want them to appear. If you decide that you want your social media icons to appear on every post, that's a lot of extra work. And if you then decide that you want to change your lineup of links, you need to edit *every* post to change the Social Icons block in each one. To avoid this nightmare, create a reusable block (as described on page 167) if you plan to make frequent use of your social icons. Or consider using the Social Icons widget in the Jetpack plugin (page 256), which lets you put your social icons in a widget area (like the footer). That way, they appear all over your site, but there's still just one widget to configure.

Code Listings

WordPress has a simple *Code block*, which is meant for listings of code in a computer programming language. Code blocks use a *fixed-width* font, so every letter is just as wide. Code blocks also keep your spaces, which makes it easier to indent lines. But code blocks don't have any fancy features like colorization. Some themes, like Twenty Twenty, draw a faint border around the code listing, but that's about all you get.

Interestingly, WordPress has another block that uses a fixed-width typewriter-like font and preserves your spacing exactly the way you type it. That's the *Verse block*, which is meant for poetry—although most people consider it something of a work in progress, because it doesn't look very attractive. And yet *another* similar block, called *Preformatted*, maps to HTML's *<pre>* element. On most themes, it's identical to Verse. But you probably won't use either of them.

Showing Part of a Post with a Teaser

Blocks aren't just about content and fancy typography. Some of them plug into other WordPress features. The *More block* is one such example. But to understand the More block, you need to back up and remind yourself about how post lists work.

As you've seen in previous chapters, WordPress likes to show posts in a reverse-chronological list, with the newest stuff up top. You see this when you browse the posts in a specific category or perform a search. You also see an all-inclusive list of posts on the home page when you first arrive at your site.

When WordPress shows its list of posts, it shows the *entire text* for every post, unless you tell it not to. Here's where things get a little complicated, because you can override WordPress's natural inclination in a couple of ways:

- You can use a theme that does things differently. For example, many themes support a post excerpt feature that shows a summary instead of the full content. You saw this on page 121.

- You can use *teasers*, a WordPress feature that lets you show just the beginning of a post in the WordPress post list. You do this with the help of the More block.

Here's how teasers work. You insert the More block at the spot where you want to divide a post. The content that falls before the tag becomes the teaser, which WordPress displays in the post list (*Figure 6-16*, left). Click the Continue Reading link (or button) and you'll get to the post page, which has the entire post and no trace of the More block (*Figure 6-16*, right). The appearance of the Continue Reading link depends on the theme. For example, in Twenty Twenty, it's actually a button.

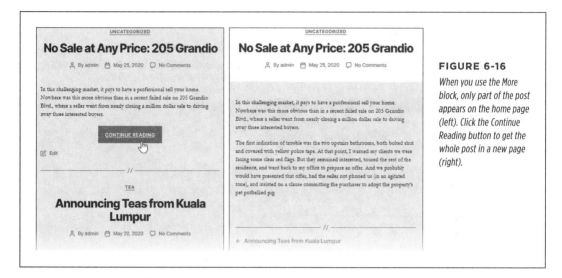

FIGURE 6-16

When you use the More block, only part of the post appears on the home page (left). Click the Continue Reading button to get the whole post in a new page (right).

When you add the More block, you'll see a light-gray dividing line in the post editor. Confusingly, the line has the text "READ MORE," which doesn't match what you see in the link. However, you can get exactly the text that you want by editing the READ MORE label. For example, type in "Tell me more," and that's what you'll see in the teaser link instead of "Continue reading."

Teasers have two possible disadvantages. First, there's the extra link readers need to follow to read a full post. If someone wants to read several posts in a row, this extra step can add up to a lot more clicking. Second, you need to explicitly tell WordPress what part of a post belongs on the home page. It's an easy job, but you need to do it for *each post* you create, which is a pain. So before you adopt teasers all over your site, try out different themes and check out the post excerpt feature (page 122) to see which approach works best for you.

Dividing a Post into Multiple Pages

WordPress has a little-used *Page Break block* that lets you split a single post into as many pages as you want. When someone reads the post, they can see only one page at a time. To go from one page to another, they need to click the navigation links at the bottom of the post. These navigation links look different depending on the theme, but *Figure 6-17* gives you an idea of what you can expect.

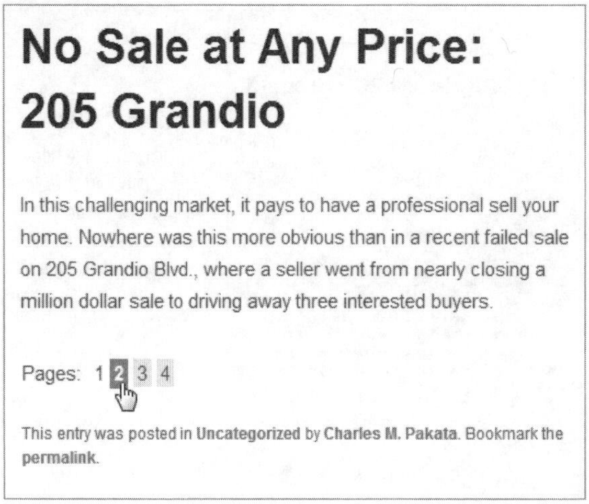

No Sale at Any Price: 205 Grandio

In this challenging market, it pays to have a professional sell your home. Nowhere was this more obvious than in a recent failed sale on 205 Grandio Blvd., where a seller went from nearly closing a million dollar sale to driving away three interested buyers.

Pages: 1 3 4

This entry was posted in **Uncategorized** by **Charles M. Pakata**. Bookmark the **permalink**.

FIGURE 6-17

These page-navigation links let you split a long article into more manageable pieces. But use it sparingly—readers will resent being forced to click without a very good reason.

The page break feature seems pretty nifty at first glance, but most people ultimately decide not to use it. In the distant past, downloading a big page with lots of pictures took time, and even scrolling through a long post was slow. These days, people are accustomed to zooming down through lengthy pages with a quick flick of a finger. If you ask them to keep clicking, they're likely to get bored and go somewhere else.

◼ Managing Your Blocks

Now that you've settled in with the post editor and become familiar with its most useful blocks, it's a good time to learn some tips and tricks to make working with them more efficient. In the following sections, you'll learn a couple of shortcuts for finding and rearranging blocks, a way to peek at the HTML inside a block, and a way to hide blocks you don't want to use.

Jumping from Block to Block

The post editor has an easily overlooked tool—the *Block Navigation button*—that lets you see an outline of all the blocks in your post. You can find this button in the toolbar at the top of the editor, and it looks like a small indented list (*Figure 6-18*). Click it and you'll see an ordered list of all the blocks in your post.

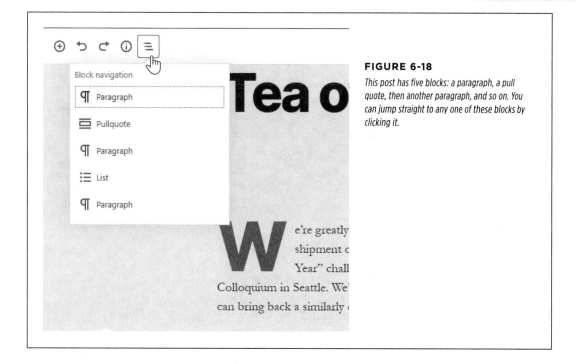

FIGURE 6-18

This post has five blocks: a paragraph, a pull quote, then another paragraph, and so on. You can jump straight to any one of these blocks by clicking it.

TIP Here's a handy way to focus on one block at a time. Click the three dots in the top-right corner of the post editor to see a menu with additional settings, and choose Spotlight Mode. Now, the current block (the one you're positioned on in the post editor) shows up normally, but the other blocks become lighter and slightly transparent. The goal is to make it easier for you to focus on the content you're writing, without getting distracted by everything else. To turn off spotlight mode, pop open the Settings menu and choose Spotlight Mode again.

Moving Blocks

If you want to shuffle things around in your post, you can use the standard cut-and-paste operations. Select your text and use Ctrl-X to cut and Ctrl-V to paste (or Command-X and Command-V on a Mac). But if you want to move a complete block from one spot to another, there's a faster way to work. To try it out, click the block you want to move and look in the left margin. You'll see a tiny strip with two arrows (*Figure 6-19*). You can click these arrows to move your block around.

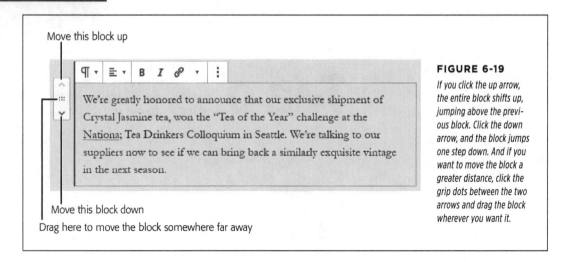

Move this block up

We're greatly honored to announce that our exclusive shipment of Crystal Jasmine tea, won the "Tea of the Year" challenge at the Nationa; Tea Drinkers Colloquium in Seattle. We're talking to our suppliers now to see if we can bring back a similarly exquisite vintage in the next season.

Move this block down

Drag here to move the block somewhere far away

FIGURE 6-19

If you click the up arrow, the entire block shifts up, jumping above the previous block. Click the down arrow, and the block jumps one step down. And if you want to move the block a greater distance, click the grip dots between the two arrows and drag the block wherever you want it.

Seeing the HTML for a Block

So far, you've relied on WordPress's visual designer. That means when you edit your blocks, they look more or less the same way that they'll look when you publish the post. Apply a bit of italic or bold formatting, and that shows up in the editor. Add a link, and you see the familiar underlined blue text. And so on.

But if you have any sort of web design background, you know that pages on the web are built out of HTML, a markup language that uses codes in angle brackets to control spacing and formatting. For example, to show a bold word like **Fire!**, you need to use the HTML markup *Fire!*. The opening ** tag turns on the bold formatting, and the ending ** tag turns it off.

WordPress lets you see the HTML that it's generating for each block. If you're savvy in the ways of the web, you can even decide to edit that markup directly instead of using the visual designer. There are a few reasons you might choose to take this step. Maybe you have an HTML special character to type in (like *©* for the copyright symbol ©). Maybe you're pasting in a block of HTML content from another web page. Or maybe you need to alter a hidden detail in a picture you've added or some special content you've embedded.

Here's how to switch to HTML view:

1. **Click the block in the post editor.**

   You can temporarily turn any block into HTML.

2. **In the toolbar above the block, click the three dots.**

   That's the More Options button (on the right side).

3. Choose Edit as HTML.

Now the HTML markup appears for the block (*Figure 6-20*). Nothing has actually changed about the block—it will still look exactly the same when you publish the post. But now you'll see a different view of it in the post editor.

When you switch a block into HTML mode, it stays that way, even if you click over to another block. However, you can return a block to its normal appearance by clicking the More Options button again and choosing Edit Visually.

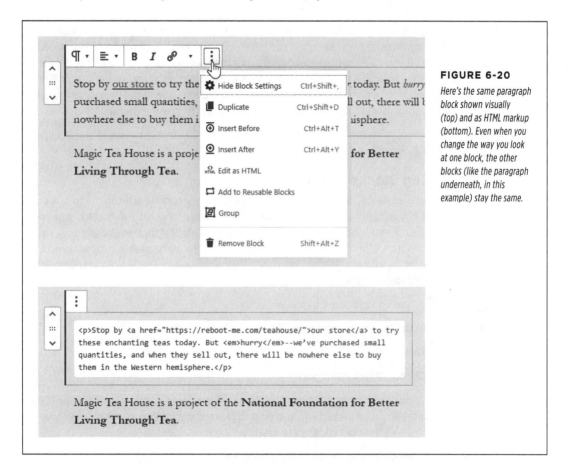

FIGURE 6-20

Here's the same paragraph block shown visually (top) and as HTML markup (bottom). Even when you change the way you look at one block, the other blocks (like the paragraph underneath, in this example) stay the same.

Interestingly, WordPress has a couple of other ways that you can reveal the HTML markup hidden behind your blocks:

- **The Custom HTML block.** When you add the Custom HTML block, it starts out showing a text box with markup, just like the Edit as HTML option shown in *Figure 6-20*. But the Custom HTML block has two extra buttons to its toolbar: HTML and Preview. You can use these buttons to quickly jump back and forth between the HTML markup and the formatted text, which makes the Custom HTML block a tad more convenient than ordinary HTML mode.

- **The code editor.** For a dramatic change, you can look at the HTML for your entire post. To do that, click the three dots in the top-right corner of the post editor, which opens a menu with additional choices, and click Code Editor. To get back to normal, click the three dots again and choose Visual Editor. When you're in the code editor view, absolutely everything in your post is turned into HTML, and you even see the secret WordPress comment codes that surround each block. Don't edit in this view unless you're an expert, because it's easy to break your page or lose some of your content if you accidentally change one of the WordPress codes.

WordPress's HTML features aren't designed for serious editing work. It doesn't have the conveniences you find in a true HTML editor (like, say, Brackets or Visual Studio Code). If you need to make anything other than small changes, you'll probably want to write your HTML first in an HTML editor, and then copy and paste it over. But be warned, you need to know exactly what you're doing or you run the risk of introducing wonky formatting or making it more difficult to change themes.

Most people won't use the HTML editing feature often—or even at all. But if you need it, WordPress has you covered.

Hiding Blocks You Don't Need

Sometimes having a hundred or so blocks kicking around is just too much of a good thing. If you're tired of scrolling through an overly long list every time you add a block, and you suspect there's only a couple of dozen that you're actually using, you can trim WordPress's block list down to more manageable proportions. It's a quick job, and if you decide later that you need one of the blocks you've hidden, it's easy to get it back.

Here's how to hide some blocks:

1. **In the top-right corner of the post editor, click the three dots.**

   This pops open a menu of options.

2. **Choose Block Manager.**

   WordPress shows a list of all the blocks it supports, in their usual groups (Common Blocks, Formatting, and so on). Next to each block is a checkbox and—at first—all the checkboxes are selected.

3. **Click a block to remove the checkmark.**

   WordPress calls this *disabling* a block, but what you're really doing is hiding it.

   When you hide a block, it won't appear in any of the block lists anymore. However, if that block already appears in a post, it stays (and you can continue to edit it as normal).

4. **When you're finished, hit the Esc key to close the Block Manager.**

5. **To restore a block you've hidden, just make a return trip to the Block Manager and switch the checkbox back on.**

■ Creating Personalized Blocks

Here's a neat trick. Let's say you make the perfect block in one post and need to have the same thing in another. For example, maybe you have a little About Me blurb that you want to use all over the place. Or maybe you have a *call to action*—a blurb that asks the reader to follow your social media accounts or sign up for a newsletter, and you want to slap that important message at the bottom of all your posts.

You could do this by editing your first post, copying what you want (Ctrl-C, or Command-C on a Mac), and then editing the second post and pasting the copied stuff (Ctrl-V or Command-V). But an easier option lets you identify a favorite block and use it again and again, whenever you need it. The trick is to make a *reusable block*.

Making a Reusable Block

Here's how to create your own reusable block:

1. **Type in all your content and format it.**

   Don't move on until your block is perfect and ready to be used in other posts.

2. **Click the More Options button (the three dots at the right side of the block toolbar) and choose Add to Reusable Blocks.**

   WordPress asks you to name your block (*Figure 6-21*).

3. **Type in a descriptive name.**

   This is how your block will be labeled in the list of blocks.

4. **Click Save.**

   Now your block is ready to use.

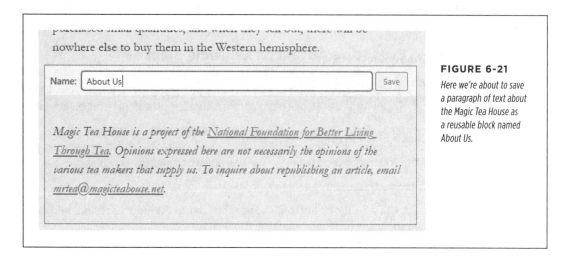

FIGURE 6-21

Here we're about to save a paragraph of text about the Magic Tea House as a reusable block named About Us.

Once you've created a reusable block, you can add it the same way you add any other block. Click the Add Block button, scroll down to the Reusable Blocks section at the bottom, and then give your block a click. You'll get a perfect copy inserted at your current position.

> **NOTE** Reusable blocks copy everything—including your formatting *and* your text. You might think it would be nice to make a custom formatted block for a design element you want to reuse (like a pull quote with a specific color scheme). But custom blocks won't work for this idea, because they'll turn each quote into the same text.

Editing a Reusable Block

It's important to understand that there is only a single version of every reusable block. In other words, if you use a reusable block in several different posts, all those posts will have the exact same content for that section. If you edit a reusable block in one post, WordPress automatically updates every other copy of that block to match.

So how do you edit a reusable block? The most obvious option is to change it right in the post editor. First, click the reusable block (or any of its copies). The block's name appears at the top (as in *Figure 6-21*), with an Edit button. To change your block, click Edit, tweak away, and then click Save. Remember, this changes every copy of the reusable block in *every post* that uses it.

> **TIP** There's another way to edit your reusable blocks. Click the Add Block button, scroll down to the Reusable Blocks section, and click Manage All Reusable Blocks. This takes you to the Blocks page, which lists all your blocks, in much the same way that the Posts page lists all the posts you've written. You can review your blocks here and make your changes.

If you decide you don't want a reusable block any longer, click the More Options button and choose Remove from Reusable Blocks. But don't take this step lightly, because the block will disappear from every page that uses it.

Using Even More Advanced Reusable Blocks

A reusable block is a single block—for example, one paragraph or a single list. But what if you want to create something a bit more elaborate, like a reusable block that includes a heading and a list, or a reusable block that consists of three full paragraphs?

The trick is to use the Group block, which is simply a container that can hold multiple blocks. You make your Group block into a reusable block, and then you can put whatever you want inside, whether it's one block or a dozen.

Here's the exact sequence of steps:

1. **Click the Add Block button, choose the Layout Elements section, and click Group.**

   The Group block starts out as an empty container—it's just an invisible box.

2. **To put something in the Group block, click the Add Block button inside.**

   That's the plus sign that shows up *inside* the group box.

3. **Pick the block you want to add.**

   For example, you can add a heading and then type in its text.

4. **Press Enter to add a new block underneath the first one.**

   Alternatively, you can hover over the bottom edge of the box until a plus sign appears and click that.

5. **Repeat the process in steps 3 and 4 until you have all the blocks you want.**

   When you've added all the content you want inside the Group block, you can move ahead.

6. **Click the faint border around the entire Group block.**

   You're grabbing the container, not the blocks inside.

7. **Turn your group into a reusable block in the usual way.**

   If you've forgotten, you click the More Options dots in the toolbar, choose Add to Reusable Blocks, and give your masterpiece a name.

FREQUENTLY ASKED QUESTION

When You Want to Stop Reusing a Block

Can I change a reusable block without affecting all the other copies?

Every once in a while, you may want to disconnect one of your reusable blocks—in other words, turn it into a normal block. For example, you might decide that you want to create a slightly different version of the About Us reusable block shown in *Figure 6-21*. The problem is that if you make changes, you'll update that block all across your site. To sidestep this problem, you have two choices:

- **Make a new block by giving it a new name.** Click your reusable block, click the Edit button, type in a different name, and click Save. Now you have two reusable blocks (the original one and the new one). They aren't synchronized, so you can change the new block without affecting the old copies.

- **Convert the reusable block to a normal block.** Click your reusable block, click the More Options button, and then choose Convert to Regular Block. All the content is still there, but now it's an ordinary block that's not connected to anything else. (Your reusable block still exists in the block list, however. If you've used it in other places, it stays there too.)

■ The Last Word

In this chapter, you learned about a key WordPress concept: blocks. Now that you're comfortable with the idea of blocks, you don't need to treat each post like a wall of text. Instead, you can assemble them out of ready-made ingredients, like headings, lists, separators, quotes, and tables.

WordPress's block system has many advantages. For one thing, it lets you build your own blocks with important bits of content, and reuse those design elements wherever you need. But perhaps the most important part of the block system is that you get to treat every section of content as something *separate*. If you want to drag a bit of content somewhere else, adjust its formatting, or peek inside at the HTML that makes it, you can do that without affecting the rest of your post. Everything's modular, like Ikea furniture.

In the next chapter, you'll build on this model with blocks that hold rich content. You'll see how to add pictures, videos, and information that's grabbed from other sites.

Adding Pictures, Videos, and More

You spent the preceding chapter filling your posts with headings, lists, and other types of fancy formatting. This is great if you're mainly concerned with making blocks of text look good. But if you want to stand out in the crowded online world, full of websites jockeying for attention, you need more. You need to add *rich media*, like pictures, video, and interactive widgets.

WordPress has you covered with a wide variety of blocks that are tailored to different types of content. In this chapter, you'll learn to use them. You'll start by adding ordinary pictures, along with a few finer details (like captions, columns, and galleries). Then you'll learn how to show content from other websites, like YouTube videos, SoundCloud playlists, or even a Kickstarter project. As you'll see, WordPress makes it easy.

■ Adding a Basic Picture

On the web, text and pictures are usually stored in separate files, and it's not much different in the world of WordPress. To put a picture into a WordPress post, you need to upload the picture file to your website. You can do that using the familiar post editor. Here's how:

1. **Go to the post editor.**

   It doesn't matter how you get here. You can create a new post, or you can edit something you wrote previously. Go to the spot in your post where you want to insert your picture.

2. **Click the Add Block button and choose Image.**

   When you add an Image block, it starts out as an empty placeholder box. There's nothing in it but a few buttons (*Figure 7-1*). Here's what these buttons do:

   Upload grabs a picture from your computer. This is the approach you're going to try first.

   Media Library uses a picture that's already on your site (because you've uploaded it previously).

   Insert from URL grabs a picture that's already online. For example, maybe it's on a different site you manage, or on a free picture-sharing service like *http:// pixabay.com*. To use it, all you need is to supply the website address that points to the picture.

TIP The Insert from URL feature is risky, because it doesn't copy the picture to your site. It's always possible that the picture you're using gets moved, renamed, or removed, in which case it will disappear from your page too. To be on the safe side, download the picture to your computer first and then use the Upload button instead. And—of course—don't use other people's pictures unless they give you permission.

FIGURE 7-1

There are three ways to add a picture, as represented by these three buttons in the Image block placeholder.

3. **Click Upload.**

   WordPress pops open a standard file-browsing window.

4. **Find the picture you want, select it, and click Open.**

Make sure you pick a type of picture that web browsers understand. Common standards include JPEG (*.jpg*), PNG (*.png*), and GIF (*.gif*) files. Don't use TIFF files (*.tif*) or uncompressed bitmaps (*.bmp*).

WordPress takes a few seconds uploading your picture to your website. Then it inserts the picture into the Image block (*Figure 7-2*).

> **TIP** You can use drag-and-drop as a shortcut to upload your pictures. Instead of clicking the Upload button, start by finding your picture file on your computer. (If it's not on your desktop, use Windows Explorer on Windows, or Finder on the Mac.) Then click the image file, drag it over web browser window, and release it onto the Image block.

FIGURE 7-2

Does your website need this picture of a frog lounging in a bathtub? Why yes, yes it does.

5. **Optionally, resize your picture.**

WordPress automatically makes your picture full size (or as big as it can be while still fitting into your theme's layout). But if you want to shrink your picture, just drag the blue sizing bar on the right or the bottom edge of your picture (*Figure 7-3*). WordPress sizes pictures proportionately, which means it keeps the height and width proportions the same so nothing gets distorted.

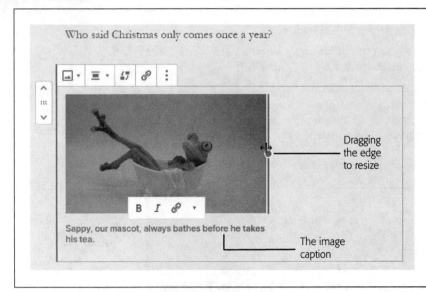

Who said Christmas only comes once a year?

Dragging
the edge
to resize

Sappy, our mascot, always bathes before he takes
his tea.

The image
caption

FIGURE 7-3

*Here's the sizing bar on
the right, which you can
drag to expand or shrink
your picture. There's a
similar sizing bar you can
grab hold of on the bottom
edge.*

6. **Optionally, add a caption.**

   Click the space just beneath the image (where it says "Write caption") and
   type what you want. WordPress makes sure your picture and your caption stick
   together, and the caption doesn't become wider than the picture (*Figure 7-3*).

7. **Optionally, change the alignment.**

   The standard alignment fits your picture neatly into the text column. But you
   have other interesting options. To see them, click Change Alignment in the
   toolbar and choose one of the options:

   Align Left puts your picture on the left and lets your text flow down the right,
   for a more magazine-like layout. Shrink your picture to fit more text.

   Align Right is just like Align Left, except it puts your picture on the right side
   and lets text flow on the left.

   Align Center puts your picture an even distance away from both sides. (You
   might not notice the effect unless you shrink your picture to make some extra
   room.) There is no text flow on the sides.

   Wide Width expands your picture to be bigger than the column that contains
   it (*Figure 7-4*), like the jumbo-sized pull quotes you saw in Chapter 6. Not all
   themes support this alignment option.

   Full Width expands your picture all the way to the sides of the browser window,
   for a truly eye-catching effect. Not all themes support this option either.

NOTE There's one option that, oddly enough, you won't see in the list—standard alignment. If you try out one of the other options and want to get back to WordPress's normal picture placement, click the alignment option you just selected in the menu *a second time*. That removes the alignment setting, returning you to normal.

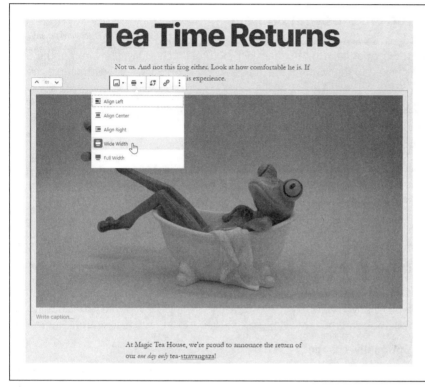

FIGURE 7-4

Wide Width gives you extra space to spotlight truly important or attractive pictures.

8. **Optionally, turn your picture into a link.**

   You can make it so that your picture is a clickable link that takes the reader to another page. To do that, click the "Insert link" button in the toolbar and type in an address, just as you did with the ordinary types of links you learned to create in Chapter 6.

   The Image block also gives you an extra linking option—you can link your picture to the picture *file*. To do this, click the "Insert link" button and then choose Media File at the top of the list. You might use this approach if you have a very large picture. Visitors will see a smaller version in your page (because the size is limited based on your layout), but they can take a closer look at the full image by clicking it.

9. **Publish your post.**

   That's it! As usual, you can see the result of your changes by previewing or publishing your post after you're finished tweaking your picture.

NOTE If you delete an Image block from a post, the picture file still exists on your WordPress site. This might be what you want (for example, it lets you use the picture in another post), or it might be a problem (if you're worrying about someone stumbling across an embarrassing incident you made the mistake of photographing). To wipe a picture off your site, you need to use the media library, as described on page 183.

Pairing Pictures with Content

As you just saw, WordPress lets you put a caption under your pictures. WordPress sizes the caption text to fit under your picture. (Technically, the width of the caption is limited to the width of the image.) Long captions wrap over multiple lines.

There's another way to keep text connected to a picture. You can create a paragraph of text that goes *beside* your picture, as we do with the figures in this book. This layout option looks similar to the Align Left setting, except you control exactly what text goes next to your image.

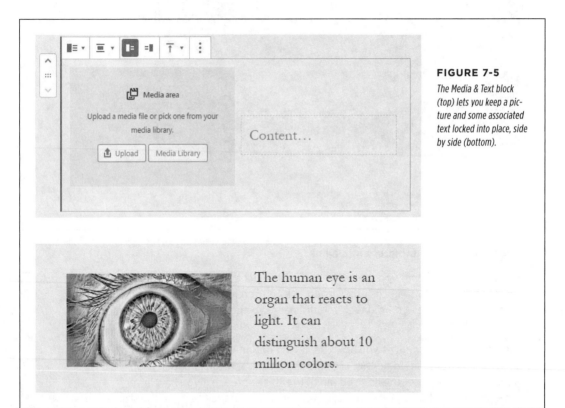

FIGURE 7-5

The Media & Text block (top) lets you keep a picture and some associated text locked into place, side by side (bottom).

To create this effect, you don't use the Image block. Instead, add the block named Media & Text. It creates two side-by-side columns, with space reserved for a picture on the left and text on the right (as in *Figure 7-5*). You can add your picture in the normal way (click Upload on the left) and add your text (type it in on the right).

To get the best results out of the Media & Text block, here are a few things you should know:

- **Horizontal order.** You can swap the two columns so the text is on the left and the picture on the right. To do that, choose "Show media on right" in the toolbar.

- **Image size.** The Media & Text block divides its space evenly. That means your picture gets exactly half that available space. But you can enlarge or shrink your picture by grabbing onto the blue resize bar (which appears on the edge of the picture) and dragging it to either side.

- **Vertical alignment.** If one column is taller than the other, the Media & Text block centers the small column between its top and bottom edges. (For example, pair a tall picture with a short bit of text, and the text floats at the side of the picture, somewhere in the middle.) You can choose a different option from the "Change vertical alignment" list in the toolbar.

- **Text size.** The Media & Text block sizes your text to be slightly bigger than the text in your post. Maybe that's what you want, but if it isn't, you can always select the text and adjust its font size in the toolbar.

- **Going extra big.** Depending on your theme, you may be able to break out of the column and make your Media & Text block supersized, just as you can with an Image block. Look for the "Change alignment" list in the toolbar with options for Wide Width and Full Width.

GEM IN THE ROUGH

Column Layouts

The Image block and the Media & Text block give you most of the layout options you'll want for showing pictures in your posts. However, if you want to get more advanced (and more complicated), you can turn to another block—the Columns block.

The *Columns block* allows you to add two to six linked columns. (When you first add it, the Columns block lets you pick from a few simple two-column or three-column arrangements, but you can go up to the full six by using the settings in the sidebar.) You can adjust the size of each column by using percentages—for example, so one column gets 50% of the space and the other two get 25% each.

You can put whatever you want in each column. For example, you can use the Columns block to show a wide picture with two narrow bits of text on either side. Or you can use it to evenly tile three pictures side by side. Or you can even add one Columns block inside another, to subdivide your page into increasingly infinitesimal regions.

Although the Columns block is a nifty idea, it's not practical on most sites. It's too easy to jumble up your layout or create narrow columns that don't accommodate their content on different screens. After all, one of the great benefits of WordPress design is that it imposes consistency on your posts. So any tool that allows you to break out of those conventions comes with the cost of making your site less unified.

The real reason for the Columns block is to standardize different types of specialized layout, so they can be used to build new types of reusable blocks. Right now, you might use the Columns block for an unusual design element on a very special page. But it's more likely that you'll eventually use a theme or plugin that includes a specialized multicolumned block for you to drop into your pages.

Adding an Inline Image

WordPress has another way to insert images that's often overlooked. It's called the *inline image*, and it's the very poor cousin of the Image block.

An inline image is a picture placed inside a paragraph of text. In fact, it's treated as though it were just another character. Usually, inline images are used for tiny pictures, like icons, emoticons, and symbols. They're meant to appear alongside your words in a sentence. If you use a large picture for an inline image, WordPress enlarges the whole line to fit it, which leaves a big gap of empty space above the rest of the text (*Figure 7-6*).

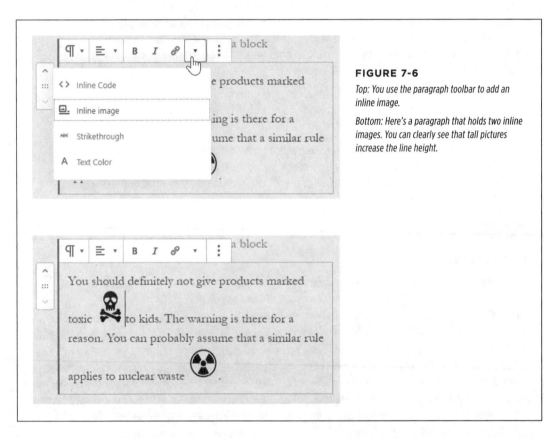

FIGURE 7-6

Top: You use the paragraph toolbar to add an inline image.

Bottom: Here's a paragraph that holds two inline images. You can clearly see that tall pictures increase the line height.

To add an inline image, first add an ordinary paragraph. When you're at the place where you want the picture to appear, click the down-pointing triangle icon in the small toolbar (the one that appears just above the paragraph). You'll see a short list of options including Text Color, Strikethrough, and—the one you want—Inline image. Choose that, and you can upload a picture in the same way you do when you're using the Image block.

Because inline images are inside your text, they're easy to work with. They flow with your text, and you can backspace them right out of existence. But you won't get any of the layout capabilities that the Image block has, like the ability to wrap text along the side of a floating picture. And if you want to change the size of the image, there's no dragging or resize bars. Instead, you need to click the inline image so the Width box appears, and type in a new number. (For example, if your picture has a width of 100, try typing 50 to make it smaller.)

Making a Gallery of Pictures

Pictures are great on their own. And there's no limit to the number of pictures you can include in a single post—you can pile in dozens and arrange your text around them in the best way possible.

But sometimes you might want to group your pictures together in a more organized, browsable way. For example, maybe you're writing a travelogue of your trip through Nepal, or an introduction to a new line of products your company is about to release. In this case, it makes sense to group your pictures into a neat gallery. Each picture is turned into a smaller thumbnail, and all the thumbnails are packed together so the reader can scan through them easily (*Figure 7-7*). To take a closer look at a pic, all you need to do is click one of the thumbnails.

FIGURE 7-7

WordPress's gallery feature creates smaller versions of your pictures and stacks them together, one after the other, taking as many rows as it needs. If you write a caption, it's shown directly on the picture, as with the first image in this gallery.

Here's how to create a gallery like this:

1. **Add the Gallery block.**

   The Gallery block works much like the Image block, except it allows you to add as many pictures as you want.

2. **Upload the pictures you want in your gallery.**

   The fastest approach is to drag your files over from your computer and drop them on the Gallery box. The other option is to click Upload, browse for the pics, and then select all the ones you want. Hold down the Ctrl key (on Windows) or Command key (on a Mac) to pick more than one picture file at a time.

If you've already uploaded the pictures you want, you can click Media Library to find them. You'll learn more about the media library on page 181.

3. **Choose the number of columns you want in your gallery.**

A new gallery is three columns wide. That means WordPress puts pictures one after another, three to a line. The only exception is the last line, which might have fewer than three images left over.

You can set your gallery to show as few as one image per line or as many as five. Look in the sidebar on the right, click the Block tab, and adjust the Columns slider to get the number you want.

4. **Adjust your image cropping.**

Ordinarily, the Gallery block crops each thumbnail down to a perfect square, trimming off the sides if needed. This makes all the pictures neat and even. But if your picture dimensions vary dramatically, square cropping might lose important information. You can turn off the "Crop images" setting to see what your thumbnails look like when they aren't cropped. Now you'll get a perfectly shaped, sized-down version of each picture, but some extra space will remain around the irregular edges (*Figure 7-8*).

Proportional spacing creates islands of empty space

Sick Lego man

FIGURE 7-8

WordPress has proportionally sized these thumbnails. If a picture is tall and skinny, it gets to be as tall as it wants. If a picture is short and wide, WordPress gives it all the width it needs.

5. **Drag your pictures around to set their order.**

You can put the images in your gallery in any order. Just click one and pull it to a new place. If you want to remove one of the images, click the X icon in the corner.

6. **Optionally, caption your pictures.**

You can add a caption to some or all of your pictures. Just click to select the picture and start typing. Your text is superimposed on top of the picture in white lettering, as shown in *Figure 7-8*.

7. **In the sidebar, in the Link To list, choose Media File.**

The Link To setting determines what happens when someone clicks a thumbnail in the gallery. If you choose Media, you'll see the full-sized image on its own. If you choose Attachment Page, you'll see a WordPress-generated page that features the image, along with extra details that you can set in the media library (page 184) and optional comments. Or you can turn off links altogether (choose None).

NOTE If you use the Jetpack plugin running on your site (page 256), WordPress ignores the Link To setting. Instead, it shows a photo carousel when a visitor clicks a picture.

■ Working with the Media Library

When you use the Image block, WordPress takes your pictures and stores them on your site. To do that, it uses a special folder that starts with *wp-content/uploads*.

For example, if you upload a picture named *face_photo.jpg* to the Magic Tea House site in January 2020, WordPress stores it at *http://magicteahouse.net/wp-content/uploads/2020/01/face_photo.jpg*. Upload a picture with the same name in the same month, and WordPress adds a number to the end of the filename, as in *face_photo-3.jpg*.

If you want to check the exact name of an uploaded picture, there's an easy way to do it. Usually, you don't need to know the name of this file, because WordPress handles it for you. But if you're curious, click the Image block so the mini toolbar shows up, and then choose More Options→Edit As HTML. You'll see a snippet of HTML markup, with the full address of the uploaded copy of your picture:

```
<figure class="wp-block-image size-large">
<img src=
  "https://www.magicteahouse.net/wp-content/uploads/2020/01/face_photo-3.jpg"
  alt="" class="wp-image-182"/>
</figure>
```

Choose More Options→Edit Visually to stop looking at the HTML and see the picture again.

WordPress also creates large, medium, and thumbnail-sized copies of your picture with names like *face_photo_300x200.jpg*, and stores them in the same folder. That way, WordPress doesn't waste bandwidth by sending a full-sized picture if a post needs to show just a tiny thumbnail. And if you change a picture (using the basic cropping, resizing, and rotating tools described on page 185), WordPress stores even more versions of the same picture.

NOTE In addition to storing the original file you upload, WordPress creates three extra versions of every picture: a 150 × 150 pixel thumbnail, a 300 × 300 medium-size image, and a 1024 × 1024 large-size pic. You can change these defaults in the Settings→Media section of the admin area, under the "Image sizes" heading. However, the changes affect only new pictures, not the ones you've already uploaded.

WordPress calls this repository, which holds all your pictures and files, the *media library*. To see the contents of your site's media library, choose Media→Library from the admin menu. You'll see a list of all the files you've uploaded, knowingly or unknowingly while you wrote your posts (*Figure 7-9*).

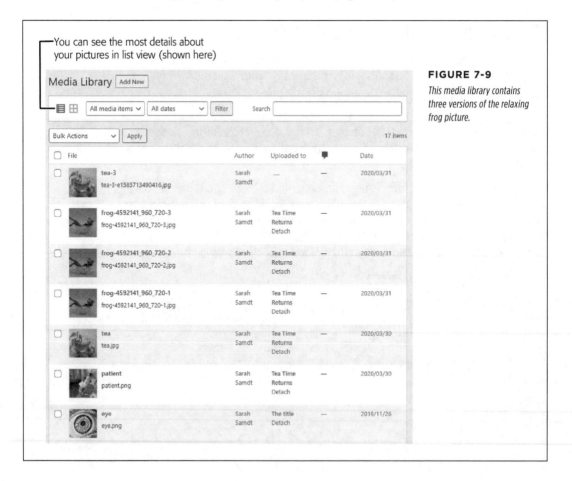

You can see the most details about your pictures in list view (shown here)

FIGURE 7-9

This media library contains three versions of the relaxing frog picture.

You might find more files in the media library than you expect. If you add a picture to an Image block but then delete the Image block, the picture stays in your media library. The media library also accumulates pictures you use for other purposes—say, header images, images in sidebar widgets, and so on. And it holds other types of files you might decide to upload for your visitors to download (page 205).

There are a few reasons you might want to use the media library:

- To remove files you don't need anymore, and save space on your website.

- To upload files you want to use in the future, even though you haven't started writing the post where you're going to use them.

- To change some of the information WordPress stores about a file, or make basic edits to your pictures.

Deleting Pictures from the Media Library

You might choose to delete a media file as part of your basic website housekeeping. After all, why keep a file to clutter up your site if you're not using it?

However, you need to proceed carefully when deleting items from the media library. You don't want to remove a picture that you're using in a post. (Do that, and your picture will quietly disappear from the post, with no warning.)

WordPress gives you limited help in figuring out which pictures are being used. In the media library list (*Figure 7-9*), you'll see a column named "Uploaded to." This tells you the name of the post where you added the picture. In theory, you should be able to look for pictures that aren't linked to a post, and remove those. But it isn't that simple. If you delete an Image block, WordPress doesn't update the "Uploaded to" column. In *Figure 7-9*, for instance, you'll see three versions of the relaxing frog picture. They were all uploaded to the same post (named "Tea Time Returns"), and then all but one was removed. However, there's no way to tell which one is still there unless you visit the Tea Time Returns post.

Not only will unused pictures still look like their linked-to specific posts, but a picture that isn't linked to anything (has a blank "Uploaded to" value) might actually show up somewhere on your site. This can happen in various ways—for example, you might upload a picture to the media library and insert it into a post later on. And don't ask what happens if you put the same picture in more than one post (WordPress won't even try to record that fact). In short, the "Uploaded to" column is a clue that can help you start figuring out whether you're using a picture, but it's not the final word.

If you do decide to delete a media file, it's easy to do. Hover over it in the list, and then click the Delete Permanently link that appears underneath.

> **TIP** The best advice is to remove pictures you don't need immediately (for example, as soon as you delete the corresponding Image block). Otherwise, you'll be forced to do what most WordPress site owners do—keep the picture in your media library, just in case.

Editing a Picture in the Media Library

WordPress lets you tweak some details about your pictures in the media library. This feature is occasionally useful—for example, if you want to preconfigure some information before you use a picture in a post, or if you want to make very simple picture edits.

To edit a picture, hover over it in the list and click Edit. This takes you to the Edit Media page (*Figure 7-10*), which tells you basic information about your picture (where it's stored, how big it is, when it was uploaded) and gives you the chance to change some details.

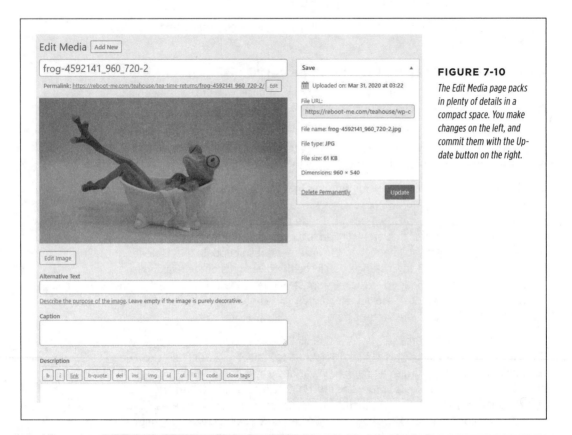

FIGURE 7-10

The Edit Media page packs in plenty of details in a compact space. You make changes on the left, and commit them with the Update button on the right.

Here's a rundown of the information you can change:

- **Alternative Text** is the text sent to assistive devices (like screen readers) to help people with disabilities interpret pictures they can't see. Search engines also pay attention to the alternative text. You should use it to describe the content of a picture (if it's important) or leave it blank (if the picture is just decoration). You can also adjust this information in the block editor when you're writing your post—just select the picture and look in the sidebar.

- **Caption** is the text caption that appears under the picture in the Image block. If you fill it in here, it won't change your Image block. But if you fill it in here and *then* add your picture from the media library, this text is automatically copied into your caption.

- **Description** is a longer, more detailed explanation of the picture. You can use it for your records, or you might show it in a post—if you have a theme that supports it (most don't). The description also gets added to your galleries if you use the Jetpack plugin, as described in Chapter 9.

Remember to click the Update button to save any changes you make.

Optionally, you can edit your picture by clicking the Edit Image button, which appears under the picture. This gives you a bunch of basic editing controls, like cropping (*Figure 7-11*).

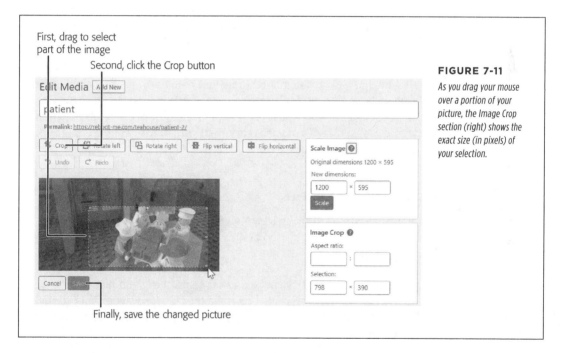

FIGURE 7-11

As you drag your mouse over a portion of your picture, the Image Crop section (right) shows the exact size (in pixels) of your selection.

The most powerful way to adjust your pictures is to use an image-editing program on your computer. But you can make a few useful changes right on your website by using WordPress. They include the following:

- **Cropping.** Drag a rectangle around the region you want to keep and then click the Crop button.

- **Rotating.** You can turn your picture 90 degrees or flip it vertically or horizontally by clicking one of the other buttons above the picture.

- **Resizing.** You can scale your picture down, reducing its size. To do so, type in either a new width or a new height in the Scale Image section on the right, using pixels as the unit of measure. WordPress adjusts the other dimension proportionally, ensuring that you don't distort the picture. Then click Scale to apply the change.

When you perform any of these operations, click Save to make your change permanent.

> **NOTE** When you edit a picture, WordPress actually creates a new file (and a new item in your media library). If you look at the picture's URL, you'll notice that WordPress appends a number to the end of the filename, so *patient.png* becomes *patient-e1585970187424.png*, for example. This sleight of hand occurs for two reasons. First, it makes sure any posts that are using the old picture won't change. Second, it lets you get your original picture back later if you ever need it.

Adding Pictures to the Media Library

You might choose to add an image to your library to prepare for future posts. Maybe you have a company logo, or some other stock photos you plan to reuse on multiple posts. Maybe you have a batch of pictures that detail a home renovation project, and you want to make sure they're on your site when you need them—even though you don't plan to start writing just yet. In all these cases, it makes sense to upload your pictures straight to the media library. That way, they're available when you're writing your posts in the block editor.

Here's how to do it:

1. **Choose Media→Add New.**

   This takes you to the Upload New Media page. You can also get here by clicking the Add New button at the top of the media library page.

2. **Add your files.**

   You upload media files by dragging them onto the "Drop files here" box. Or click the Select Files button and browse to the files you want on your computer. You can pick as many as you want by holding Ctrl (on a Windows computer) or Command (on a Mac).

3. **Optionally, edit the information for each picture.**

   You can edit the descriptive information that's associated with the picture (like the alternative text and the caption), or manipulate the image itself (say, cropping it or flipping it), as you've already seen. Just click the Edit link next to the picture you want to change to get started.

When you want to insert the picture into a post, just add your Image block. Then click the Media Library button to browse the pictures you've uploaded and pick the one you want (*Figure 7-12*).

Select or Upload Media

Upload Files   Media Library

Filter Media   Search

March 2020

All dates
April 2020
March 2020
November 2019

ATTACHMENT DETAILS

frog-4592141_960_720-2.jpg
March 31, 2020
61 KB
960 by 540 pixels
Edit Image
Delete Permanently

Alt Text

Describe the purpose of the
image. Leave empty if the image
is purely decorative.

Title   frog-4592141_960_720-2

Caption

Description

Copy Link   https://reboot-me.com/tea

Select

FIGURE 7-12

Because the media library holds every image that's used in every post (and probably more files besides that), it gets very big. Knowing when you uploaded your picture is helpful, because you can home in on just the pictures from that month. Alternatively, you can use the search box to hunt based on the picture's filename or its description.

FREQUENTLY ASKED QUESTION

Replacing Pictures

If I use a picture in different posts, is there a way to change it?

WordPress doesn't let you replace a picture that's in the media library. Even if you upload a picture that has the same name as one already in your library, WordPress puts it in a different subfolder or gives it a slightly different filename. The same thing happens if you edit a picture that's already in the media library.

WordPress works this way because—everything else being equal—its developers don't want to make it possible for users' small changes to make big mistakes. If it were possible to

replace pictures, it would also be possible for a website creator to replace a picture in one post without realizing that it affects other posts, possibly in unexpected or layout-breaking ways. This system prevents some seriously frustrating problems, but it also means that there's no way to update the picture in a post without editing the post.

However, you can work around this limitation by using a plugin, which you'll learn to do in Chapter 9. Once you understand how plugins work, you can use Enable Media Replace (*https://tinyurl.com/media-replace*) to switch an old image for a newer version, right under WordPress's nose.

■ Adding Featured Images

So far, you've spent your time adding pictures to posts. However, different WordPress themes have different, innovative ways to use pictures. The most common example is with *featured images*.

A featured image *represents* a post, but it isn't part of the post content. Instead, its role varies depending on the theme. Some themes ignore featured images altogether. Others show the featured image at the top of the corresponding post, near the title area (and sometimes even *above* the title area). Some themes get fancy and add an effect, like superimposing the post title over the featured image (*Figure 7-13*), or tinting the picture a matching color (like Twenty Nineteen). Twenty Twenty is more reserved—it simply positions the image at the top of the post, lined up on the dividing line that separates the header area from the shaded post content underneath.

FIGURE 7-13

The simplest way a theme can use a featured image is to place it at the top of the post page (as in Twenty Seventeen, top) or just before the post content (Twenty Twenty, bottom).

More ambitious themes often include a way to use featured images on your site's home page. For example, a theme might show a list of posts, with the featured images shown next to each one.

Assigning a Featured Image to a Post

Each post can have just one featured image. You assign the featured image while you're editing the post. Here's what to do:

1. **Go to the post editor.**

   You can create a new post or edit an existing one.

2. **In the sidebar on the right, click the Document tab.**

   Remember, if you don't see the sidebar you need, click the gear icon in the top-right corner of the post editor.

3. **Scroll down until you see the Featured Image section, and click to expand it.**

   If your post has a featured image, you'll see it in this section. Otherwise, all you'll see is a single "Set featured image" button (*Figure 7-14*).

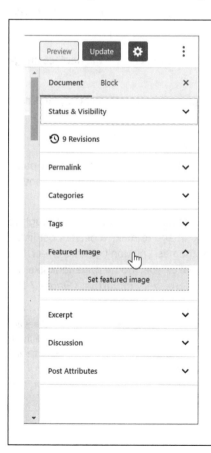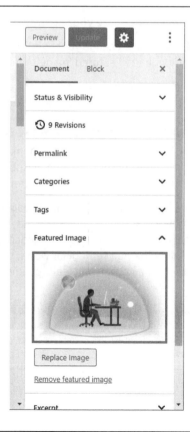

FIGURE 7-14

At first, the Featured Image section is empty (left). But if your post has a featured image, it appears here (right).

4. **Click "Set featured image."**

This opens a new window where you can pick your picture.

5. **Select the image you want.**

Use the Media Library tab to choose a file that you've already uploaded and is already in your media library. Or use the Upload Files tab to select a new file from your computer.

NOTE When you insert a standard image into a post, you get the chance to size it. But when you use a featured image, you don't have this control. If you upload a big picture, it's possible that your theme will automatically crop it or squash it. To prevent this, you can scale down the picture before you upload it (using your favorite image editor) or crop and size it in the media library (page 185). Some themes work best with featured images that have specific dimensions, and you may be able to find those details in the information pages for your theme. Otherwise, you'll need some trial-and-error experimentation.

---

6. **Click the "Set featured image button" in the bottom-right corner to make your selection official.**

Now you'll see the picture appear in the Featured Image section of the sidebar.

7. **Publish or update your post.**

If you want to change your featured image later, you can use the Replace Image button in the Featured Image section.

If you decide to remove your featured image and leave nothing in its place, click the "Remove featured image" link instead.

Themes That Make the Most of Pictures

As you saw in *Figure 7-13*, featured images get a special place on the post page, even if the difference is sometimes subtle. But featured images also play another role—they let some themes build better home pages.

You'll learn all about custom home pages in Chapter 8, but for now it's enough to understand that some themes use the featured image to make a better post list. If you're using one of WordPress's standard year themes, you don't get this effect. But if you use a theme like Regular News, which makes your home page look like the front page of a newspaper, featured images become very important (*Figure 7-15*).

NOTE Featured posts are interesting because they rely on the interplay between WordPress features and theme features. WordPress defines the concept of the feature (in this case, featured images), and the theme decides how to implement that concept, opening a wide, uncharted territory of possibilities. The same idea underpins many other WordPress features, like post excerpts (page 121) and menus.

FIGURE 7-15

Many themes use featured images to create a more attractive post list. If a post doesn't have a featured image, the image square next to it is left blank.

Featured images are also important if you're creating a picture-centric site, like a photoblog that showcases your nature photography. You could use the techniques you learned earlier in this chapter to add images and galleries, but the results won't be ideal. Standard WordPress themes split up your pictures, making it impossible for visitors to browse your portfolio from beginning to end.

The solution is to find a photoblog or *portfolio* theme. There are plenty in the Word-Press theme gallery, and you can dig up good candidates by searching for words like "photo," "photoblog," "picture," and "portfolio."

Every photoblog and portfolio theme works a bit differently, but most share some key features. First, the home page likely displays the pictures in your posts rather than the posts themselves. *Figure 7-16* shows an example from a theme called Foto.

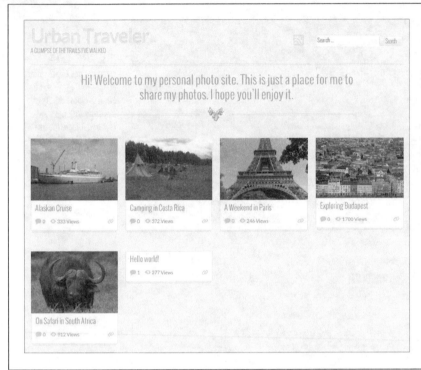

FIGURE 7-16

Instead of an ordinary list of posts for its home page, the Foto theme creates a tiled display of images without any post excerpt or text. It also adds a neat animated effect: when you move your mouse over a picture, the image expands slightly. You can then click the enlarged tile to read the full post.

Embedding a Video

Now that you've jazzed up your site with pictures, it's time for something even more ambitious: video.

There are two reasons you might put a video playback window in a post. First, you may want to use someone else's video to add a little something extra to your content. For example, if you comment on a local protest concert, your post will be more interesting if you include a clip of the event. Similarly, if you review a new movie, you might include its trailer. If you talk about a trip to Egypt, you might want to take visitors inside the pyramids. In all these examples, the video adds a bit of context to your post.

The other possibility is that the video may actually *be* your content. For example, you might use video to show your band's latest live performance, a bike-repair tutorial you filmed in your garage, or a blistering web rant about the ever-dwindling size of a Pringles can. In all these cases, you create the video and then use your WordPress site to share it with a larger audience.

TIP Depending on the type of site you create, your written content doesn't need to be *about* the video—instead, the video could add supplementary information or a bit of visual distraction. For example, you may talk about your favorite coffee blend and add a video that shows the grueling coffee harvest in Indonesia. Used carefully and sparingly, video accompaniment can enhance your posts in the same way a whimsical picture cribbed from a free photo service can.

To present someone else's video on your site, find the movie on a video-sharing site and copy the address. Most of the time, that video-sharing site will be YouTube, the wildly popular hub that rarely drops out of the top-three world's most visited websites.

And if you want to show your *own* videos, you'll probably *still* turn to a video-hosting service like YouTube because the alternative—uploading your video files straight to your web server—has significant drawbacks. Here are some of the reasons you should strongly consider a video-hosting service:

- **Bandwidth.** Video files are large—vastly bigger than any other sort of content you can put on a site, even truckloads of pictures. Even if you have room for your video files on your web server, you might not have the bandwidth allotment you need to play back the videos for all your visitors, especially if your website picks up some buzz. The result could be extra charges or even a crashed website that refuses to respond to anyone.

NOTE *Bandwidth allotment* refers to the amount of web traffic your site host allows. Hosts may limit bandwidth so that an extremely busy site—one with lots of visitors stopping by or downloading files—doesn't affect the performance of other sites the service hosts.

- **Encoding.** Usually, the kind of file you create when you record a video differs from the kind you need when you want to share it online. When recording, you need a high-quality format that stands up well to editing. But when viewing a video over the web, you need a heavily compressed, streamlined format that ensures smooth, stutter-free playback. Sadly, the process of converting your video files to a web-friendly format is time-consuming and often requires technical knowledge.

- **Compatibility.** In today's world, no single web-friendly video format accommodates the variety of web browsers, devices (computers, tablets, and mobile phones), and web connections (fast and slow) out there. Video services like YouTube solve this problem by encoding the same video file multiple times, so that there's a version that works well for everyone. You can do the same on your own, but without a pricey professional tool, you're in for hours of tedium.

For all these reasons, it rarely makes sense to go it alone with video files, even if you produce them yourself. Instead, pick a good video service and park your files there. In the following section, you'll start with YouTube.

NOTE If you do decide to host your own video on your website, the process is similar to using pictures. You add the Video block and upload your video to your site's media library. Of course, this assumes that your WordPress site accepts large uploads, you have the space you need, and your video is encoded in a web-friendly video format.

UP TO SPEED

The Dangers of Using Other People's Video

The risk of embedding other people's videos is that the video-hosting service may take down the videos, often because of copyright issues, and they'll disappear from your site too. This is a particularly acute danger for videos that include content owned by someone other than the uploader. Examples include scenes from television shows and fan-made music videos.

Usually, WordPress authors don't worry about this problem—if a video goes dead, they edit their posts after the fact. To avoid potential problems from the get-go, stay away from clips that are obviously cribbed from someone else's content, especially if they're recent. For example, a video that shows a segment from last night's *Saturday Night Live* broadcast is clearly at risk of being taken down. A decades-old bootleg recording of a Grateful Dead concert is probably safe.

Showing a YouTube Video

Hosting a YouTube video in WordPress is ridiculously easy. All you need is the video's web address.

To get it, start by visiting the video page on YouTube. (If you're one of the six people who haven't yet visited YouTube, start at *www.youtube.com*.) If you haven't already found the video you want, you can spend time searching around. When you find the right video, click the Share link under the video window (*Figure 7-17 , top*).

The URL will look something like this:

```
http://youtu.be/0xKKr0Qrcjg
```

The first part, *youtu.be*, is a more compact form of the video's full web address (which starts with *www.youtube.com/watch?v=*). The string of letters and numbers after it uniquely identifies the video you want.

If you want to queue up your video at a specific point, find the right spot and pause the playback. Then, when you click the Share button, tick the "Start at" box (*Figure 7-17*, bottom). YouTube adds an extra *&t* to the end of the URL with a number of seconds. For example, this video starts one minute in:

```
http://youtu.be/0xKKr0Qrcjg&t=60
```

Once you have the right video link, you can use it in a post on your WordPress site. In the post editor, begin by adding a YouTube block. Then paste the video's URL into the YouTube block, and the video will appear immediately (*Figure 7-18*).

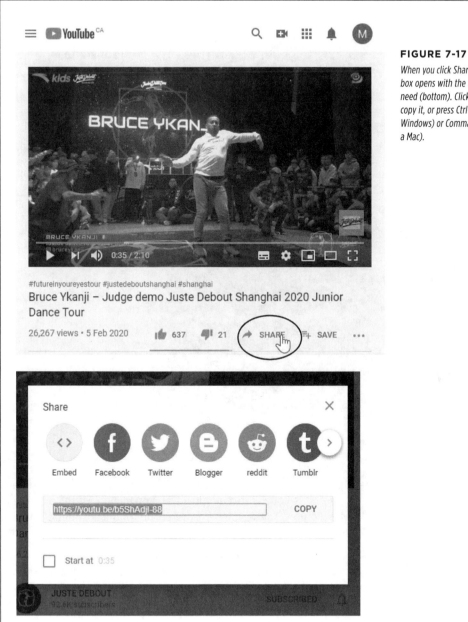

FIGURE 7-17

When you click Share (top), a box opens with the URL you need (bottom). Click Copy to copy it, or press Ctrl-C (on Windows) or Command-C (on a Mac).

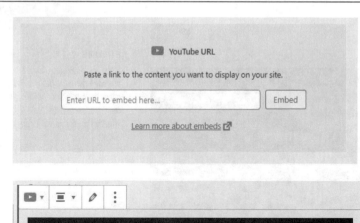

FIGURE 7-18

A fresh new YouTube block just sits empty, waiting for a video (top). Once you add a URL, you'll see a thumbnail of your video (bottom). You won't see other information from the YouTube page, such as its description and comments.

The YouTube block is a pretty basic ingredient. Currently, it doesn't include any YouTube-specific settings you can tweak. However, you can adjust the alignment, just as you do with pictures, so the video box bursts out of the confines of the text column that holds it.

> **TIP** Here's a shortcut for adding YouTube videos—and other types of embeddable content that WordPress recognizes. Copy the video URL and paste it on a new, blank line in your post. The block editor will recognize your URL and turn it into a YouTube block automatically. This trick doesn't work if you type in the URL by hand, or if you paste the link into another paragraph (in which case the block editor turns it into a link instead).

Uploading Your Videos to YouTube

The process of playing your own videos through YouTube is essentially the same as the one for showcasing someone else's. The only difference is that you begin by signing up with You-Tube (if you don't already have an account) and then uploading your video. The steps are fairly straightforward—click the icon of a video camera with a plus sign in it, in the top-right corner of the YouTube home page, choose "Upload video," and follow the instructions YouTube gives you. Make sure you designate your video as *public*, which means that it appears in YouTube's search results. If you don't use this setting, you won't be able to embed your video in a WordPress post. Other settings (for example, whether you allow comments and ratings) have no effect on whether you can embed the video.

For best results, YouTube recommends that you upload the highest-quality video file you have, even if that file is ridiculously big. YouTube will encode it in a more compressed, web-friendly format, while preserving as much of the quality as possible. And if the quality of your video is good enough, YouTube will offer high-speed viewers the option to watch it in high-definition format.

Uploading videos takes a while, because the files are huge and transferring all that data takes time, even on the fastest network connection. YouTube also needs to process your video, although its industrial-strength servers can do most of that as you upload it. When you get the video live on YouTube, click Share to get the URL you need for your WordPress post, which you can paste alongside your content, exactly as you did before.

Showing Videos from Other Video Services

Although YouTube is the most popular video service, it's not the only game in town. WordPress has similar blocks for videos for all of the following services:

- Animoto
- Facebook
- TikTok
- WordPress.tv
- Dailymotion
- Flickr
- VideoPress
- Hulu
- TED
- Vimeo

All these blocks are in the Embeds section, because they represent content that WordPress can borrow from another website and *embed* into one of your posts, as though it belongs there naturally.

The list of eligible sites is ever-growing, so don't be surprised to find additional services listed when you check WordPress's official list at *https://wordpress.org /support/article/embeds*. However, the fastest way to check that a video site works is to try pasting the URL into the block editor. If WordPress recognizes the site, it will automatically convert it to the right type of block. Or you can add the all-purpose Embed block, which does the same thing—it takes a URL you specify and tries to embed the content from that site.

TIP If you want to embed a video service that isn't supported, you can usually find a plugin to do the job. For example, the Jetpack plugin that's introduced in Chapter 9 adds support for embedding Twitch streams.

Premium Video Hosting

As you learned earlier, hosting videos on your own website is generally a bad idea, unless you're a web development god with a clear plan. But at times, a free video-hosting site might not suit your needs. Here are some of the problems you might encounter hosting your files on a free service:

- **Privacy.** Free video services make your videos visible to the entire world. If you have a video that contains sensitive material, or one you want to sell to a group of subscribers, you won't want it on a free video service, where it's exposed to prying eyes.

- **Ads.** Video-sharing services may try to profit from your free account by including ads. They use two basic strategies: playing a television-commercial-style ad before your video starts playing, or superimposing an irritating banner over your video while it plays. YouTube's advertising policy is good—it won't show ads unless you give it permission to (usually in a misguided attempt to make a buck) or you post someone else's copyrighted content and *they* ask YouTube to slap on an ad (in which case they collect the money).

- **Content restrictions.** Free video services won't allow certain types of content. YouTube's content policy is fairly liberal, allowing everything short of hate speech, pornography, and bomb-making tutorials. That said, videos are sometimes removed for disputable reasons, and popular videos are occasionally poached by traffic-seeking video thieves, who post their own copies and try to trick YouTube into removing the original, legitimate videos.

To escape these restrictions, you need to pay for a video-hosting package. For example, the popular video site Vimeo (*http://vimeo.com*) offers free basic hosting and a more flexible premium service that costs $60 per year. The latter offers unlimited bandwidth, generous storage space, no length restriction, and the option to limit your videos to specific people, who must sign in to see them.

WordPress also offers its own video-hosting service called VideoPress, although it's more closely tied to the WordPress.com hosting service. Either way, you can embed your premium videos just as easily as YouTube videos, using the Vimeo and VideoPress blocks.

■ Playing Audio Files

Sometimes you'll want to let readers play an audio clip (or several) without using a full-blown video window. An obvious example is if you're a music artist promoting your work. However, audio files are equally well suited to the spoken word, whether that's an interview, talk show, sermon, audio book, or motivational lecture. Audio files are particularly useful if you want to join the web's thriving community of *podcasters*—providing sites with downloadable, long audio files that users can listen to on the go (for example, on their iPods or smartphones).

Adding a Basic Audio Player

The simplest approach to hosting audio is to upload the file to your website. You can then provide a tiny audio player that lets users listen to the audio without leaving their browser (*Figure 7-19*). Unlike with video files, audio files are small enough that hosting them yourself is no great burden. (Their small size is also the reason that piracy decimated the music industry years before it became a problem for television and movies.)

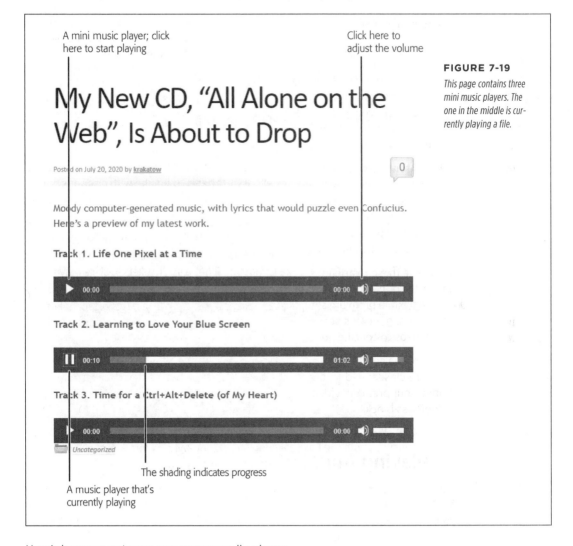

FIGURE 7-19

This page contains three mini music players. The one in the middle is currently playing a file.

Here's how you get your own snazzy audio player:

1. **Add the Audio block.**

   This adds a placeholder square that looks almost the same as an Image block.

2. **Click the Upload button to find the audio file on your computer.**

   Of course, if you've already uploaded the audio file to the media library, then the Media Library button is your ticket.

3. **Pick the MP3 file you want to upload.**

   Although you can upload other formats, the MP3 format has the best browser support. Currently, the only browser that won't play an MP3 is the mobile version of Opera, which you'll rarely encounter in the wild.

4. **Publish or update your page.**

   You're rewarded with a tiny music player. You can repeat this process to add as many music players as you want to the same page.

Using a Music-Sharing Service

If you're serious about sharing a set of audio tracks—for example, you're in a band and trying to popularize your work—your best bet is to sign up with a serious music-sharing service.

The first advantage is that a good music service increases the exposure of your audio tracks. Casual music browsers may stumble across your work, similar artists may link to it, and just about anyone can add a comment or click Like, which boosts your buzz. The second advantage is that you'll get a sleek jukebox-style player that can seamlessly play a whole *list* of songs.

WordPress has built-in blocks for the following music-sharing services:

- SoundCloud
- Spotify
- Mixcloud
- ReverbNation

Once again, all these blocks are in the Embeds section, but the easiest way to add one is to just to paste the link to a song, artist, or playlist. WordPress will recognize the website automatically and choose the right block for you.

> **TIP** If you want to embed Bandcamp albums, the Jetpack plugin (covered in Chapter 9) will help you out. And if you have another audio service you want to use, there's almost certainly a free plugin that can do the job.

Figure 7-20 shows a site that uses SoundCloud.

SoundCloud offers a fair deal: no charge for the first three hours of audio you upload, and a pro plan with a modest monthly fee and unlimited storage. To sign up with

SoundCloud, start at *http://soundcloud.com*, click Sign Up, and follow the instructions. Once you upload some audio files, you can start embedding them in your
WordPress posts. SoundCloud gives you the option of a single-track music player,
or you can assemble a group of tracks into a playlist.

To embed your SoundCloud content in a WordPress post, browse to the song, artist,
or playlist you want to use, and then copy the URL out of the address bar. Here's the
URL that creates the music player in *Figure 7-20*:

```
https://soundcloud.com/antoine-dufour-1
```

FIGURE 7-20

*When you embed an artist's
page from SoundCloud, you
get a picture, a list of tracks,
and the distinctive orange
waveform that shows your
progress during playback.*

Podcasting

Podcasting is a specialized way to present audio. It gives readers the choice of lis-
tening to audio files the normal way (in their browsers by clicking the Play button)
or downloading the audio files so they can put them on a mobile device.

The central idea with podcasting is that you prepare content that busy people will
listen to on the go. Usually, the content is long—30 minutes to an hour is common
for podcasts. If your audio file is only a couple of minutes long, it's not worth a visitor's trouble to download it and transfer it to a mobile device.

Podcast creators also tend to organize podcasts in groups—for example, they make
each audio file an episode in a series, and release them at regular intervals (say,
weekly). Good examples of podcasts include a web talk show with commentary and
interviews, a motivational lecture, or a spoken chapter from an audio book.

You don't need any special technique to upload a podcast audio file. You can simply create a post and add an ordinary Audio block. But there's a slight difference in the way you *present* podcasts. For readers to find your podcasts quickly and download new episodes easily, you need to separate these audio files from the rest of your site.

To do that, you need to create a dedicated category for posts that contain podcasts. You can give this category a name like "Podcasts" or "Lectures" or "Audio Book." Then, when you create a new post that has an audio file in it and you want to include that audio as part of your podcast, assign the post your podcast-specific category (*Figure 7-21*). Depending on the structure of your site (and the way you let viewers browse it), you may decide to set two categories—one to identify the type of post (say, "Sports"), and one to flag the post as a podcast ("Podcasts").

FIGURE 7-21

This post has the two key ingredients that make it a podcast—embedded audio and a podcast-specific category.

Here's the embedded audio file.
A simple link is also acceptable.

This category is for podcasts only.
This allows you to submit
your podcast to iTunes.

The reason you group audio files into a single podcast category is so you can generate a feed for that category. A *feed* is a sort of computer-readable index to your content. In the case of podcasting, your feed tells other programs and websites where to get the podcast files on your site. It also lets you notify visitors when you publish a new file—say, if they subscribe to your podcast through the Apple Podcasts service.

You can get the feed for a category by using a URL with this syntax:

```
http://[Site]/category/[CategoryName]/feed
```

So if you have a site named *http://dimagiosworkouts.com*, you can get a feed for the Podcasts category like this:

```
http://dimagiosworkouts.com/category/podcasts/feed
```

This is a valuable link—it's the piece of information you need to supply to register your podcast so other people can discover it.

> **NOTE** The most popular tools for podcast-lovers are Apple Podcasts (formerly iTunes), Google Podcasts, and Spotify. If your site offers podcasts and you'd like to use them to attract new visitors (and why wouldn't you?), you should submit your podcast information to these services. To register with Apple Podcasts, start by reviewing Apple's instructions at *http://tinyurl.com/podcastspecs*. To submit to Google, check out *https://tinyurl.com/googl-pod*. And to show yourself on Spotify, visit *http://podcasters.spotify.com*.

PLUGIN POWER

Better Podcasting with a Plugin

Podcasting with WordPress is easy. All you need is the right type of audio file (MP3) and a post category just for podcasts.

However, if you're a power podcaster—meaning you plan to invest serious effort in making podcasting a part of your web presence—it's worth considering a plugin that can make your life easier. Two popular podcasting plugins are PowerPress (*http://tinyurl.com/wp-podcast*) and Seriously Simple Podcasting (*https://tinyurl.com/simple-pod*).

Both of these plugins make it easy to create a more detailed podcast feed, complete with cover art and a description. They also integrate with popular podcast services like Apple Podcasts and Spotify, so you can submit your podcast to their catalogs with a couple of clicks. If you're interested in learning more, wait until you get the plugin lowdown in Chapter 9, and then give one of these plugins a whirl on your site.

■ Embedding Other Types of Content

By now, you've seen several examples that use WordPress's embedding abilities to take content from other sites and slide it into one of your posts. Technically, WordPress works this embedding magic by following a decade-old standard called Embed.

The most common type of embeddable content for most sites is video. But as you've probably already guessed by looking at the Embeds group of blocks, WordPress supports an even broader range of embeddable content. Here are some examples:

- Twitter tweets
- Instagram posts
- Crowdsignal surveys
- Imgur memes
- Issuu publications

- Kickstarter projects
- Meetup events
- Reddit posts
- Scribd documents
- SlideShare presentations

To get the latest list of what's available, check *https://wordpress.org/support /article/embeds.*

In each case, the recipe for using embedded content is the same. You browse to the content you want to include in your post, copy the URL address from your browser, and paste it into the post editor. WordPress then hooks you up with the correct block (*Figure 7-22*).

FIGURE 7-22

This post has a taste of the web with no less than three blocks of embedded content. From top to bottom, there's a post from Reddit, a Meetup event, and a poll created on Crowdsignal.

Be careful not to overuse embedded content. The first problem is that your theme doesn't have the power to format the content you embed—instead, it stays with the distinctive formatting of its source site. This can lead to a jumbled, inconsistent effect when paired with your content. The second problem is that embedded content tends to steal your visitors' focus. Many types of embedded content will even pull readers out of your site altogether. For example, click the Reddit post or the "Check out this Meetup" button in *Figure 7-22*, and you'll be sent to the original site that has the full content. (The embedded content is effectively just a preview.) The Crowdsignal poll works better, because it puts all the important details on your post, meaning there's no need for someone to take a trip to another site.

Adding a Download Link

Throughout this chapter, you've concentrated on ways to feature rich content in WordPress, from pictures to audio, video, and embeds. But sometimes you're not after a fancy way to showcase your content. You just want to let someone else download it.

For example, maybe you want to let someone download an old-fashioned price list from your site in the Microsoft Excel spreadsheet format. Or maybe you want to give your visitor the ability to snag a PDF, a Word document, a compressed ZIP file with your complete collection of Pokémon icons, or even a picture or video file. You aren't interested in showing the data in your post. You just want a link that your visitors can use to download what they want.

To make this happen, you need to put the file you want to share in the media library and then add a link to it in your post. WordPress can automate the whole process with the File block. Here's how it works:

1. **Add the File block.**

   This adds a placeholder square that looks similar to the Audio and Image blocks you've already seen.

2. **Click the Upload button to find the file on your computer.**

   Of course, if you've already uploaded the file you want to share to the media library, you can use the Media Library button instead.

3. **Pick the file you want to upload.**

   WordPress doesn't need to be able to understand your file, so you can upload absolutely anything.

   Once the file is uploaded, WordPress adds a link and a download button (*Figure 7-23*). Depending on the type of file, these two pieces work in slightly different ways. If you click the Download button, your browser starts downloading the file. If you click the link, your browser tries to *open* the file. If it's a type of file that it understands—like a picture or a PDF—the browser shows it to you. If it's a type of file that the browser doesn't understand—like an Excel document or a ZIP file—the browser starts a download, just as if you clicked the Download button.

4. **Optionally, you can change the link text by using the HTML view.**

   The File block creates a link that uses the exact name of your file, without the file extension. So *SalesSpreadsheet.xls* becomes *SalesSpreadsheet*. But if your filename is something like *tomato-soup-2108418-200.png*, the link text isn't going to be pretty.

The File block doesn't let you rename anything, but you can fix up your text in HTML view. First, click the More Options button in the toolbar (the three dots) and choose Edit As HTML. This turns the File block into a mess of HTML markup. Now, look for the link text, which will appear right before the closing *</a>* tag. You can rename that text to whatever you want. When you're finished, click More Options and choose Edit Visually to return to the normal view.

5. **Optionally, hide the Download button.**

   The Download button can be a nice convenience, but it also clutters up your post, particularly if you have more than one File block. To hide it, turn off the "Show download button" setting in the sidebar.

6. **Publish or update your page.**

   You can repeat this process to add as many download links as you need. Just make a separate File block for each one.

FIGURE 7-23

Here's an example with two File blocks. The first is a picture; the second is a PDF.

■ The Last Word

In this chapter, you saw how to add many types of rich content. First, you looked at pictures, galleries, and the media library. After that, you branched out to consider featured images, and took a quick side trip to see how themes can use your pictures to build photoblogs and fancy home pages—a topic you'll consider again in the next chapter.

Pictures aren't only fancy ingredients you can put in a post. WordPress has extensive support for embedding different types of media, and you saw how easy it is to turn a plain link into a video window or audio player. You can even branch out to embedding content from a variety of specialized sites. The same link-pasting magic works on URLs that point to content on a host of different sites. The trick is to keep this content from overwhelming your own.

Creating Pages and Menus

n previous chapters, you focused most of your attention on WordPress *posts*—the blocks of dated, categorized content at the heart of most WordPress blogs. But WordPress has another, complementary way to showcase content, called *pages*.

Like posts, pages are built out of blocks and styled by your theme. Unlike posts, pages aren't dated, categorized, or tagged. One way to understand the role of WordPress pages is to think of them as ordinary web pages, like the kind you might compose in an HTML editor.

In this chapter, you'll learn to use pages to supplement your blog. Then you'll learn to manage page navigation with *menus*, so your visitors can find the content they want. After that, you'll use pages to build a custom home page that replaces the standard WordPress list of posts. You'll even see how you can build an entire site out of pages, with no posts required.

Understanding Page Basics

You're likely to use pages for two reasons. First, even in a traditional blog, you may want to keep some content around permanently, rather than throw it into your ever-advancing sequence of posts. For example, let's say you make a personal blog and want to include a page called About Me with biographical information. It doesn't make sense to tie this page to a specific date, like a post. Instead, you want it to be easily accessible all the time, as a page. Similarly, if you're making a WordPress site for a business, you might use a page to provide contact information, a map, or a list of frequently asked questions.

Another reason to use pages is to build simple sites that don't feel like blogs. For example, if you create a site for your small business, you might use pages to display the core content of your site (information about your company, your policies, the brands you carry, and so on), while adding a blog-powered section of posts for news and promotions. You can even make a fine-tuned home page to greet your visitors, instead of using the default reverse-chronological list of posts. But before we get to that, it's time to make your first page.

Creating a Page

Although pages behave differently from posts, the process of creating and managing them is similar. Just as you work with posts in the Posts menu, you work with pages in the Pages menu.

Here's how to create a new page:

1. **Choose Pages→Add New.**

   This loads up the page editor (*Figure 8-1*). Surprise—it's almost identical to the editor you use to write posts.

FIGURE 8-1

The page editor is essentially the same as the post editor, with all the same features for creating blocks and formatting your content. The only exception is that pages aren't classified with categories and tags, so you won't see that information in the sidebar.

2. **Give the page a title and some content.**

   For example, why not create an About the Practice page like the one in *Figure 8-1?*

3. **Click Publish to call up the sidebar with publication options, and click Publish again to make the page go live on your site.**

   As with posts, you can save a page as a draft (page 68), schedule it for future publication (page 72), or make it private (page 335).

   When you publish a page, a "Page published" message appears at the bottom of the editor, confirming that the page is up and open to the public. Now is a good time to click the "View Page" link to take a look at the live page (*Figure 8-2*). Remember, if you hold down the Ctrl key while you click (or the Command key on a Mac), you can open a new tab to see the page, and keep your editor open for the next step.

FIGURE 8-2

A page in WordPress looks suspiciously like a post. The only obvious difference is that you won't get comments (unless you specifically switch them on, as described on page 277).

4. **Optionally, change the permalink (the web address) for your page.**

Page addresses are important. If you want to direct someone to your newly created About Me page, it's easier to tell them to check out *http://thoughtsofalawyer.net/about* than *http://thoughtsofalawyer.net/about-the-family-law-practice*. Fortunately, making the change is easy. In fact, changing the address of a web page is exactly the same as changing the address of a post, with one difference—unlike posts, pages don't use your permalink setting (page 92). That means the address for a web page never includes date or category information.

To change your address, click the title of your page in the editor. The permalink will appear just above the title. (If you don't see it, you haven't published or previewed your page yet.) Click Edit, clean up the address to something nice and crisp, and then click Save.

FREQUENTLY ASKED QUESTION

Understanding Pages

Why do some people call pages "static pages"?

Although WordPress calls this feature *pages*, many webheads find that confusing. After all, isn't a page anything you view on the web with a browser? And don't posts appear in web pages?

For these reasons, WordPress experts—and WordPress itself, sometimes—often use a different term. They call WordPress pages *static pages*. Sadly, this term is almost as confusing. It stems from the old days of the web, when designers distinguished between dynamic pages that could do incredible feats with the help of code, and static pages that showed fixed, unchanging content. That fits with the way most people use WordPress pages—they create them, fill them with content, and then publish them.

However, WordPress pages aren't really static; they do change. Flip your blog over to a different theme, and all your posts

and pages update seamlessly to match the new style. That's because WordPress stores all the content for your pages—as well as the content for the rest of your site, including posts—in its database.

If you're still confused, here's the bottom line: a WordPress site can hold both posts and pages, and you create, format, and manage them in much the same way. The key difference is that WordPress automatically dates, orders, and groups posts. WordPress also puts them on the home page and assumes that people will want to read them from newest to oldest. From WordPress's point of view, posts are the lead actors on your site, while pages are supporting characters. But you're not bound by that narrow definition of a site, as you'll see on page 228.

To review a list of the pages on your site, choose Pages→All Pages. You'll see a familiar table of pages, which works the same way as WordPress's list of posts and media files. Point to a page title, and you have the choice to view the page, edit it, or delete it (see *Figure 8-3*). And if you're working with piles of pages, you can use the same bulk actions you use with posts to delete or change a whole group of pages in one step.

FIGURE 8-3

Here's a WordPress site with three pages. We just created the About the Practice page. WordPress adds the other two pages to every new site. Privacy Policy is a draft page. (If you need a privacy policy on your site, you might decide to edit and publish it.) The other page is named, less imaginatively, Sample Page. Now is a good time to delete Sample Page. Just point to the title and then click Trash.

Linking to Pages

You can probably think of a couple of pages you could add to improve your site. If nothing else, you could add an About Me page containing your biographical information. But a key question remains. Once you make your pages, how do guests visit them?

Unlike posts, pages don't have categories and tags, so you can't find them by browsing or searching with the search box. And they aren't shown on your home page in a reverse-chronological list. To bring someone to a page, you need a link that takes them there.

One simple approach is to add an ordinary link to a post in the block editor (*Figure 8-4*). Ordinary links are fine for small jobs, but they don't work well if you need a way to let your readers jump around your entire site. Odds are you have many pages and so you need many links—ideally, arranged in an attractive way and made available to your visitors, no matter where they are on your site. The WordPress solution is *menus*.

Most themes have designated menu areas, where you can put neatly arranged, multilevel menus that link to all the pages you want. All you need to do is group your pages into a menu and then decide where to place it. You'll take a closer look in the next section.

Attorney **Sandra Chapelle** has practiced law in Illinois since 1998. Ms. Chapelle is a member of the Illinois Bar, and earned her B.A. at Michigan State University. She has a general law practice that handles both civil and criminal cases, with a special focus on complex and highly expensive litigation. Ms. Chapelle has a deeply felt need to help others, and sincerely hopes that your legal problems can help fund her next trip to the Mediterranean.

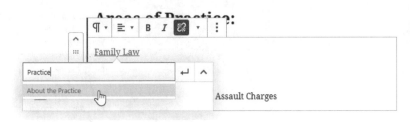

FIGURE 8-4

To add a link, select the text you want to linkify and click the Link button. There's no need to remember the page URL. Instead, just start typing the page title, and WordPress will find the page you want (in this case, that's About the Practice). Click it in the list.

◼ Showing Pages with a Menu

All WordPress themes support menus. But there's a twist: the theme is in charge of the menu. The theme decides how your menu works, where it goes, and what it looks like.

Many themes put menu links in a traditional, horizontal menu bar. In the Twenty Twenty theme, for example, the menu sits at the very top of the page, between the site title and search button (*Figure 8-5*). Some themes arrange the menu vertically and place it on the side, like WordPress's admin menu. And some even support multiple menus, which they put in different places.

Before you start creating and customizing menus, there's something else you need to know. Some themes have an *automatic menu* feature. That means they always show a menu, provided you have at least one published page in your site.

Other themes don't show anything unless you create a *custom menu* and specifically designate the pages that go in it. Custom menus are powerful tools. In fact, even the themes that have the automatic menu feature *also* support custom menus, so you have the choice of which approach you want to use.

In the following sections, you'll start by taking a quick look at the automatic menu feature in Twenty Twenty. Next, you'll learn how to build custom menus with any theme, which is the more professional approach.

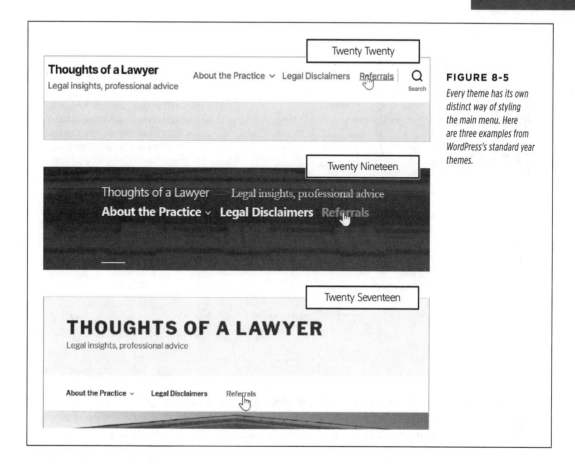

FIGURE 8-5

*Every theme has its own
distinct way of styling
the main menu. Here
are three examples from
WordPress's standard year
themes.*

Using an Automatic Menu

The Twenty Twenty theme supports the automatic menu feature. (Twenty Nineteen, by comparison, does not.) When you first create a new site with Twenty Twenty, the automatic menu has just one item: a Sample Page link that opens the WordPress sample page (unless you took our earlier advice and deleted it).

So what makes an automatic menu *automatic*? Every time you add a new page to your site, WordPress adds a matching link in the menu. For example, if you created the About the Practice page shown earlier, it appears in the menu immediately, with no extra steps required. WordPress orders your menu alphabetically by page name, which means About the Practice will show up first on the left, followed by Sample Page. Unfortunately, alphabetical order won't always get you the order you want.

To control the exact order of your pages, you use the Order setting. To see it, edit your page, click the Document tab in the sidebar, and expand the Page Attributes block (*Figure 8-6*).

FIGURE 8-6
The Page Attributes section holds the key to arranging automatic menus. You change the order of pages with the Order number, and you group pages into submenus with the Parent Page setting.

Ordinarily, WordPress sets the order number of every page to 0. Technically, that means that each page is tied for first position, and so WordPress arranges them alphabetically. But if you want to change the order (say you want "Our Story" followed by "Our Location" followed by "Contact Us"), you assign your own order numbers (say, 1, 2, and 3). The exact number doesn't matter—WordPress simply makes sure that the lowest numbers come first.

TIP If you rearrange a bunch of pages, you need to change all their order values. The easiest way to do that is to go to the Pages list (choose Pages→All Pages), point to a post, and click the Quick Edit link. This way, you can quickly modify some page information, including the order, without opening the whole page for editing.

Automatic Submenus with Child Pages

There's another way to arrange pages: you can group them together into *submenus.* *Figure 8-7* shows an example.

FIGURE 8-7

This site groups some of its pages into submenus. At first, all you see in the menu bar is the top level of headings. When you click a heading (like About the Practice), you see the child pages.

Creating a submenu is easy, once you learn how. First, you need to understand the difference between parent pages and child pages. A *parent page* is a page that corresponds to top-level menu headings (like About the Practice). A *child page* is a page that corresponds to a submenu item (like Family Law). Initially, all pages are parents. You tell WordPress that a given page is a child (and therefore belongs in a submenu) by using the Parent Page setting in the Pages Attributes box, which you can see back in *Figure 8-6*.

For example, let's say you want to put the Family Law page in the About the Practice submenu. Here's what you need to do:

1. **Edit the Family Law page.**

2. **Click the Parent Page box, which says, "(no parent)."**

3. **WordPress shows a list of all the pages on your site. Pick "About the Practice."**

4. **Publish your changes.**

You need to repeat this process for each page that you want to put in the submenu.

WordPress orders every submenu separately. That means you use order numbers to control the order of the menu headings *and* to control the order of the items inside each submenu. At first, this can make life a bit confusing. Just remember that you can treat each submenu as a separate ordering job. For example, you can use the page order to adjust the position of the Assault Charges, Drug Offenses, Family Law, and Personal Injury Defense pages with respect to one another. However, WordPress won't compare the order values of the Family Law and Legal Disclaimers pages, because they are in separate submenus.

To better understand the effect of ordering and grouping, check out the following table, which shows the Parent Page and Order settings you need for all the pages that make up the About the Practice submenu:

PAGE	PARENT PAGE	ORDER
About the Practice	-	0
Assault Charges	About the Practice	3
Drug Offenses	About the Practice	2
Family Law	About the Practice	0
Personal Injury Defense	About the Practice	1
Legal Disclaimers	-	1
Referrals	-	0

NOTE Child pages get slightly different permalinks that include the name of the child page itself and the name of the parent page. For example, if the child page Family Law has the parent About the Practice, the full permalink might be something like *http://thoughtsofalawyer.net/about-the-practice/family-law*.

Building a Custom Menu

WordPress's ordering and grouping features give you enough flexibility to create a good-looking, well-ordered automatic menu. However, most WordPress developers eventually decide to build a custom menu, for the following reasons:

- **To get more types of menu items.** An automatic menu includes links to your pages, and that's it. But a custom menu can include other types of links—for example, ones that lead to a particular post, a whole category of posts, or even another website.

- **To hide pages.** An automatic menu always includes *all* your pages. This might not be a problem for a relatively new WordPress site, but as your site grows, you'll probably add more and more pages and use them for different types of information. Eventually, you'll create pages that you don't want to include in your main menu (for example, you might want to add a page that readers can visit only by clicking a link in a post). The only way to hide a page from a menu is to abandon the automatic menu and build a custom menu.

- **To have multiple menus.** Some themes support more than one home-page menu. However, a site can have only one automatic menu. To take advantage of the multiple-menu feature, you need to create additional menus as custom menus.

- **Because sometimes automatic menus are hard.** To get an automatic menu to look the way you want it to, you need to think carefully about the order and parent settings. If you have dozens of pages, this sort of planning can twist your brain into a pretzel. If you build a custom menu, you can drag and drop your way to a good-looking menu. It still takes time and work, but it requires a lot less planning and thinking.

WordPress gives you two ways to build custom menus. One option is to go to the Appearance→Menus page in the admin area, and assemble one there. This is the old-fashioned way of doing things, because you can't see what your menu looks like until you're done. The other, more popular way to create a menu is in the theme customizer. Here's how to do that:

1. **In the admin area, choose Appearance→Customize.**

   This launches the familiar theme customizer.

2. **Click the Menus section.**

   If you had a menu, you'd see it now. But instead, you'll start with a Create New Menu button (*Figure 8-8*).

3. **Click Create New Menu.**

   Now you're going to create your menu in the sidebar (*Figure 8-9*).

4. **Type in a name for your menu in the Menu Name box.**

   The name uniquely identifies the menu. Normally, you name the menu after its function (Main Menu, Navigation Menu, Page Menu, and so on). You shouldn't name it based on its position (as in Side Menu or Top Menu Bar), because that detail may change if you switch themes.

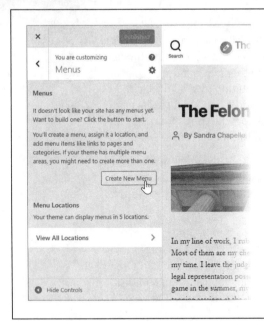

FIGURE 8-8

The Menus section lets you see all the menus you've created (whether you're currently using them or not) and all the places in your theme where you can show a menu. But your first step is to create a new menu of your own.

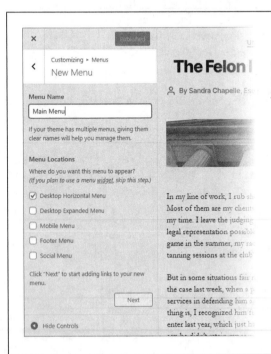

FIGURE 8-9

This menu will be named Main Menu and placed in the Desktop Horizontal Menu location, as defined by Twenty Twenty.

5. **Choose where to show your menu by turning on one of the Menu Locations checkboxes.**

   Every theme offers its own specific choices for menu placement. Some themes offer just one location. (In that case, you still need to make sure the checkbox is checked in order to show your menu.) The Twenty Twenty theme is a bit of an outlier, with five theme areas. You'll learn about them all on page 225, but for now just choose Desktop Horizontal Menu, which is the standard menu choice.

6. **Click Next.**

   This creates a new, empty menu. Now it's time to add your first menu item.

7. **Click the Add Items button.**

   WordPress opens a new side panel with options for creating a menu item (*Figure 8-10*). Right now, we'll focus on the Pages section, which lists all the pages you've created.

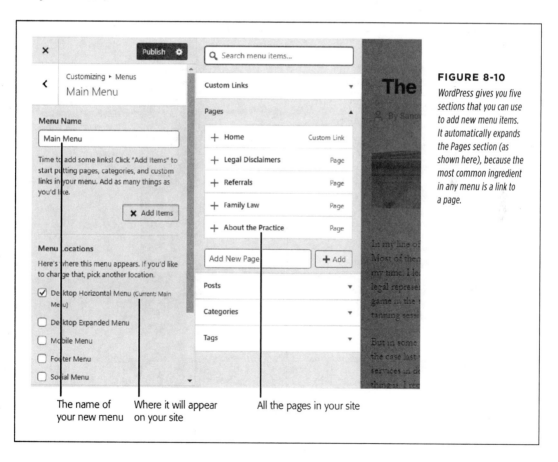

FIGURE 8-10

WordPress gives you five sections that you can use to add new menu items. It automatically expands the Pages section (as shown here), because the most common ingredient in any menu is a link to a page.

The name of your new menu   Where it will appear on your site   All the pages in your site

NOTE Technically, Home isn't a page, which is why you'll see "Custom Link" written next to it. That's because Home simply takes the visitor to the entry point of your site (like *http://magicteahouse.net*). But most menus need a link that takes you back to the beginning, and because it's common to think of the home page as an actual page, WordPress includes it in the page list.

8. **Click the plus (+) button next to each page you want to add.**

   For example, if you want to add the Home page Referrals page to your menu, click each one. They'll move to the left, under your menu name (*Figure 8-11*).

   If you accidentally add a page you don't want (or you add the same page twice), click the X icon to remove it from your menu.

   If you have a huge list of pages, you might not want to search through the list to find the ones you want. In this situation, you can use the search box above the list of pages to hunt for a specific page by name. When you find the one you want, just click it to add it to your menu.

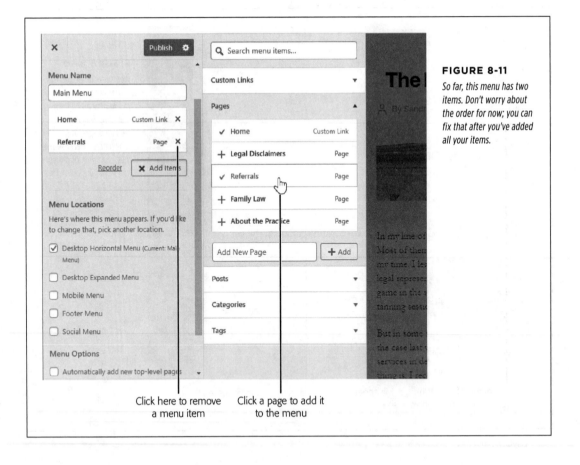

FIGURE 8-11

So far, this menu has two items. Don't worry about the order for now; you can fix that after you've added all your items.

Click here to remove a menu item

Click a page to add it to the menu

NOTE WordPress adds items to the menu in the order you click them. So the items you add first appear on the left of a horizontal menu, or at the top of a vertical menu. However, don't worry about the order yet, because you'll learn how to move everything around in a moment.

9. **Optionally, you can rename your menu items.**

When you add a menu item, WordPress gives it the same name as your page title. This makes sense if you have a short, no-frills title like "Family Law" but isn't as suitable for something like "Inside Oklahoma's Premier Family Law Clinic." In the latter case, you'll want shorter menu text.

To change the menu text, find your item in the list of all the items you've added (on the left, under the menu name). Click it to open a box with extra menu item settings (*Figure 8-12*). For pages, the only extra setting you get is Navigation Label, which sets the text you see in the menu.

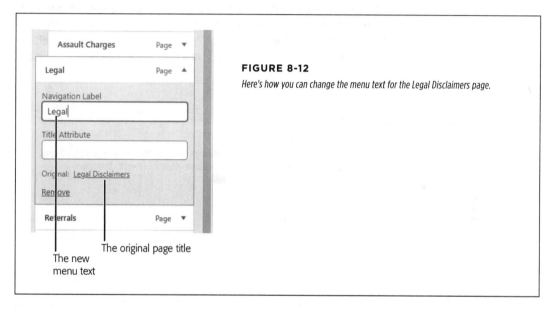

FIGURE 8-12

Here's how you can change the menu text for the Legal Disclaimers page.

10. **Now you need to arrange your menu items. Drag them around to position them and group them into submenus.**

Unlike automatic menus, custom menus don't pay attention to your page order or parent settings. This is good for flexibility (because it means you can arrange the same commands in different menus in different ways), but it also means you need to do a little more work when you create the menu.

Fortunately, arranging menu items is easy. To move an item from one place to another, simply drag it up or down the list. WordPress displays items in top-to-bottom order, so if you use a horizontal menu (as the Twenty Twenty theme does), the topmost item is on the left, followed by the next menu item, and so on.

Creating submenus is just as convenient, once you know the trick. First, arrange your menu items so that the child items (the items you want to appear in the submenu) appear immediately after the parent menu item. Then, drag the child menu item slightly to the right, so that it looks like it's indented one level (*Figure 8-13*).

NOTE You can easily create multilayered menus (menus with submenus inside submenus). All you do is keep dragging items a bit more to the right. However, most well-designed WordPress sites stop at one level of submenus. Otherwise, guests may find it awkward to dig through all the layers without accidentally moving the mouse off the menu.

11. **Optionally, you can turn on the "Automatically add new top-level pages" setting.**

If you do, every time you create a new page, WordPress automatically tacks it on to the end of your custom menu. This is similar to the way an automatic menu works (although you can edit your custom menu at any time to move newly added items to a better place). Most WordPress experts avoid this setting, because they prefer to control what gets into their menu and where it goes.

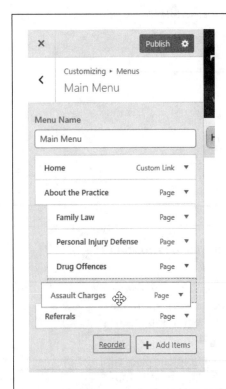

FIGURE 8-13

By dragging the Assault Charges page slightly to the right, it becomes a submenu item under the About the Practice page, along with several other pages.

12. **Click the Publish button at the top of the theme customizer.**

Now that your changes are official, you can visit your site and try out your new menu.

You can repeat this process to create more menus (which you can show in different places, as you'll see in *Figure 8-15*). If you uncheck all the Menu Locations checkboxes, you'll stop using your menu, but WordPress will keep it in case you want to use it later. To ditch a menu completely, go to the Menus section of the theme customizer, pick your menu, and then click the Delete Menu link at the bottom of the list of menu items.

Adding Different Types of Links to a Menu

Pages aren't the only thing you can add to a menu. When you click the Add Items button, WordPress shows five groups, one for each type of link.

WordPress opens the Pages section first, because that's the most common type of link. But you can click any of the other four headings to add a different type of link (*Figure 8-14*):

- **Custom Links.** A custom link is an address that points to any location on the web, either on your site or another site. For example, you could create a custom link that goes to a product page on Amazon, a Wikipedia article, a friend's blog, or something else. When you add a link, you supply the web address and the text that appears in the menu.

- **Posts.** This section shows all the posts you've added to your site. You can add one of them to your menu just as easily as you add a page.

- **Categories.** This section shows all the categories you've created to organize your posts. A category link goes to the corresponding category page, which shows all the posts in that category, from newest to oldest (page 96).

- **Tags.** This section shows all the tags you've added to classify your posts. A tag link goes to the corresponding tag-browsing page, which has all the posts that use that tag (page 97).

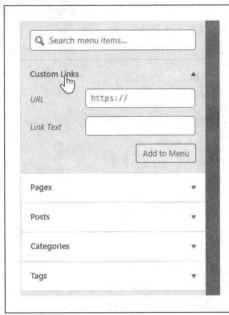

FIGURE 8-14

Decide which section you want to use, and click to expand it. You can use the Custom Links section to add a link to your menu that points anywhere you want.

Creating a Menu Item That Does Nothing

Can I make a submenu heading that visitors can't click?

When you make a menu that has more than one level, every level is clickable. This can be confusing.

For example, look at the menu modelled in *Figure 8-13* (which is similar to the one shown in *Figure 8-7*). If you want to see what's in the About the Practice menu, you need to hover over that heading. Then you'll see its contents (Family Law, Personal Injury Defense, and so on). But if you accidentally click the About the Practice heading, you'll go to the About the Practice page instead.

This behavior can be confusing, especially because traditional desktop programs don't work this way. Some WordPress themes, like Twenty Twenty, try to make this difference clear by adding a little drop-down arrow that lets you open a submenu without clicking the heading. But other themes just assume you'll know to hover over the heading rather than click it if you want to see what's inside.

To avoid this problem, you can make a heading that doesn't do anything when it's clicked. To do that, add a new menu item from the Custom Link section. Set the Link Text to the text you want to see (like "About the Practice") and set the URL to # (the number sign character). To web browsers, the # symbol represents the current page, so if you click the menu item, you won't go anywhere. In fact, you won't even see the page flicker, which is exactly what you want.

Understanding the Twenty Twenty Menu Areas

As you learned, it's the theme that decides where to show a menu. Some themes offer just one place. Other themes have two or more (*Figure 8-15*).

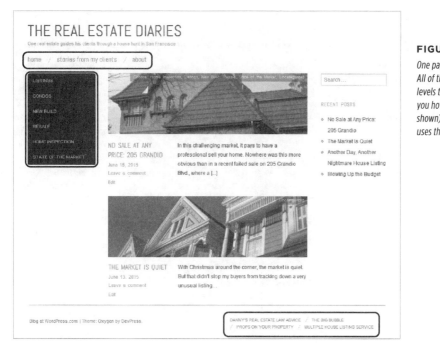

FIGURE 8-15

One page, three menus. All of them have multiple levels that pop open when you hover over them (not shown). This example site uses the Oxygen theme.

You don't *need* to use all the menus a theme provides. But if you want to, you have a choice. You can set the same menu to appear in more than one menu area. Or, more usefully, you can create different menus for each area. You can create as many menus as you need by following the steps starting on page 217.

When themes offer several placement choices, your options are usually self-evident. You can guess that the Main Menu area goes at the top of the page, Footer Menu shows up in the footer, and so on. But the Twenty Twenty theme is a bit confusing. It offers no fewer than *five* menu-placement options. Some can be used at the same time, while others can't.

■ THE DESKTOP HORIZONTAL MENU

The simplest approach is the one you've been using so far. You create a single menu and set it to be the Desktop Horizontal Menu. The Desktop Horizontal Menu tries to put your top-level menu items horizontally across the top of your page (on the right side), provided there's room. If there isn't enough room to fit all your top-level headings, the menu is tucked away behind a three-dot icon (*Figure 8-16, middle*). Click that to pop open the whole menu.

NOTE   Twenty Twenty uses some slightly confusing names. Even though it call its standard menu, it still works for phones and mobile devices. On mobile screens, the page is smaller and the horizontal menu changes to the three-dot menu icon (Figure 8-16, bottom).

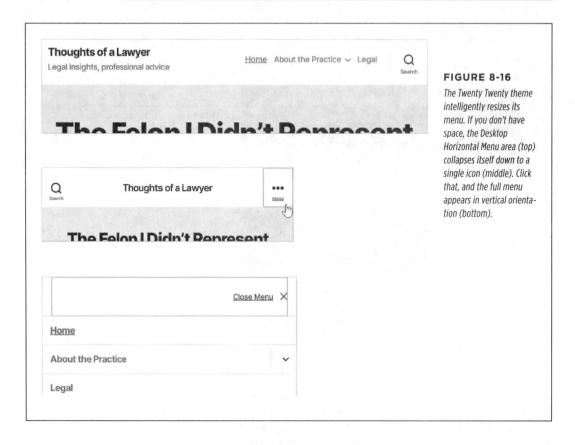

FIGURE 8-16

The Twenty Twenty theme intelligently resizes its menu. If you don't have space, the Desktop Horizontal Menu area (top) collapses itself down to a single icon (middle). Click that, and the full menu appears in vertical orientation (bottom).

■ THE DESKTOP EXPANDED MENU

The Desktop Expanded Menu is an alternative to the Desktop Horizontal Menu for people who always want to show the compact version of the main menu. If you use the Desktop Expanded Menu, you'll see the three-dot menu icon no matter how big the web browser window is. (It's called *expanded* because you need to expand the menu—by clicking the three-dot menu icon—to see it.)

Like the Desktop Horizontal Menu, the Desktop Expanded Menu works perfectly well on mobile devices.

NOTE   Don't accidentally use the Desktop Expanded Menu and the Desktop Horizontal Menu at the same time. If you do, you'll end up with two menus side by side, which is confusing!

■ THE MOBILE MENU

The Mobile Menu is an optional enhancement to Twenty Twenty's menu system. It's meant to be used in conjunction with either the Desktop Horizontal Menu or the Desktop Expanded Menu. Here's how it works: when you resize your window extra small, the mobile menu kicks into action, replacing the small version of the desktop menu. In other words, you use the Mobile Menu to tailor your menu for mobile devices. For example, maybe you want to simplify the menu by giving it fewer levels, removing items that don't apply, or shortening the menu item text. If you don't want to make any customizations, you don't need to set the Mobile Menu, because the two desktop options already collapse down to a perfectly suitable mobile version.

> **TIP** You don't need to use a mobile device to see the mobile menu. If you've put a menu in the Mobile Menu area, WordPress uses it automatically when you shrink the browser window down small enough. That means you can test it on your desktop computer.

■ OTHER TWENTY TWENTY MENU AREAS

The Footer Menu area is at the bottom of your page, just above the Powered by WordPress line. If your window is wide enough, the Footer Menu appears as a horizontal strip, just like a lower-down version of the Desktop Horizontal Menu. But if the window is narrow (for example, on a mobile device), it turns into a vertical menu that stacks the menu items from top to bottom.

The Social Menu also appears at the bottom of the page. But it's an unusual menu area, because it uses Twenty Twenty's social icon feature. You shouldn't put ordinary links to pages, posts, and categories in your social menu. Instead, it's meant for a menu made up of custom links to your social profiles (for example, on sites like Twitter, Facebook, and Instagram).

When Twenty Twenty shows the Social Menu, it doesn't show any menu text. Instead, it simply shows the icon that matches the social media service. So if you add a link to your Twitter feed, the Social Menu shows a Twitter icon (*Figure 8-17*).

FIGURE 8-17

This social menu includes two custom links. The first points to a Twitter profile (note the bird icon), and the second points to a Facebook profile. You'll get many more details on Twitter and Facebook integration in Chapter 10.

Showing a Menu in a Widget

There's another way to show a custom menu: in the Navigation Menu widget. That lets you put your menu anywhere you can place a widget. So if your theme has only one menu area but you want to put your menu somewhere else—say, a footer or a sidebar—you can do it, provided there's a suitable widget area for the location.

When you add the Navigation Menu widget, you give it a title and choose the menu you want to show. (For a refresher on adding widgets to your site, flip back to page 126.) However, there's one caveat. Often, your menu won't look the same in a widget as it does in one of the menu areas. That's because the Navigation Menu shows an *expanded* view of your menu that shows all the menu items at once. Submenus are indented. That means you can easily see that the Family Law and Referrals pages in our earlier example belong to the About the Practice submenu. But it also means you need more space (or you need to use a smaller menu) so that everything fits nicely in your widget.

Incidentally, if you just want to show a list of pages in a widget, you don't need to create a menu. Instead, you can add the slightly old-fashioned Pages widget, which grabs all the pages you have on your site, except those you explicitly exclude.

■ Changing Your Home Page

Right now, your WordPress site has a home page dominated by a familiar feature: the reverse-chronological list of your posts. Visitors can use your site's navigation menu to travel somewhere else, but they always begin with your posts.

This setup is perfectly reasonable—after all, your posts typically contain the newest, most relevant content, so it's a good idea to showcase them up front. However, this design doesn't fit all sites. If the list of posts is less important on your site, or if you want to show some sort of welcome message, or if you just want to direct traffic (in other words, give readers the option of reading posts or going elsewhere on your site), it makes sense to start by showing a page instead of a post.

In the following sections, you'll find out how. First, you'll use one of your custom-created pages as your site's home page, all in the interest of building a brochure site. Next, you'll see how to get the best of both worlds: a fixed home page with the content you want *and* a list of posts, tucked away in another place on your site.

Creating a Brochure Site

The simplest way to change your home page is to ditch the post system altogether, using pages instead of posts throughout your site. The resulting all-pages site is sometimes called a *brochure site*, because it resembles the sort of informational pamphlet you might pick up in a store (*Figure 8-18*).

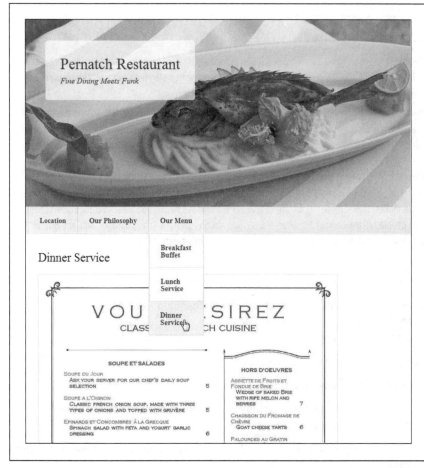

FIGURE 8-18

This restaurant website is a collection of WordPress pages, including those labeled Location, Our Philosophy, and Our Menu. Unlike posts, pages aren't related in any obvious way, nor are they dated, categorized, or tagged.

To create a brochure site, you follow some simple principles:

- Build a site that consists entirely of pages, each one handcrafted by you.

- Add these pages to a custom menu, and arrange the pages the way you want.

- When visitors arrive at your site, the first thing they see is one of your pages and a navigation menu.

You already know how to perform each of these tasks, except the last one (changing the home page). That's what you'll learn next.

Should You Build a Brochure Site?

A brochure site may make sense if you're building a small site with very simple content. The restaurant site in *Figure 8-18* is one example.

But if you're trying to decide between a brochure site and a post-based site, consider two questions. First, would your site be more attractive to readers if you included posts? Even the bare-bones restaurant site might be more interesting with posts that chronicle restaurant news, menu experiments, and special events. Not only that, but having posts that are frequent, dated, and personal makes the site more vibrant. In addition, if you want to get people talking on your site—for example, posting comments about recent meals or sending in requests and off-the-wall recipe ideas—you'll have more luck if you include posts. Think of it this way: a brochure site feels like a statement, while a blog feels like a constantly unfolding conversation.

Then again, you may decide that a brochure site is exactly what you want. Maybe you don't have time to spend updating and maintaining a site, so you simply want a place to publish some basic information on the web and leave it at that. You can still take advantage of several of WordPress's best features, like themes, which ensure that your pages look consistent. You'll also get WordPress's help if you want to track visitors (page 381), add sharing buttons (page 355), or add any one of a number of features described in this book.

The key step in building a brochure site is changing your home page, replacing the traditional list of posts with a page of your own devising. So your first step is to create a substitute home page, using the familiar command Pages→Add New.

Once you create your new, replacement home page, follow these steps:

1. **Choose Settings→Reading.**

2. **In the "Your homepage displays" setting, choose "A static page" (rather than "Your latest posts").**

3. **In the Homepage list underneath, pick the page you want as your new home page.**

4. **Click Save Changes at the bottom of the page.**

Now, when you go to your site's home page, WordPress automatically serves up the page you picked (although the URL won't change in the browser's address bar—it's still the home page of your site). Similarly, when you click Home in the menu, you return to your custom home page.

Creating a Custom Posts Page

What happens if you want to keep your site focused on posts, but you still want to greet visitors with a custom home page? WordPress makes this design almost as easy. But there's a trick—because your posts won't live in their normal spot on your home page, you need to put them in a different place. That "different place" is an address of your choosing.

For example, let's say you own the website *http://magicteahouse.net*. In the past, your list of posts was on the home page. But now, you want *http://magicteahouse .net* to show a welcome page you've created. You decide that you can put your list of posts at *http://magicteahouse.net/posts*.

Here's where things get slightly bizarre. To tell WordPress you want the post list to be at a URL named *posts*, you need to create yet another page. This page is just a placeholder—its sole purpose is to provide the web address for the posts page. You don't actually need to put any content on this page, because WordPress automatically creates the list of posts.

Here's the whole process:

1. **Decide on a URL for the posts section of your site.**

   For example, if your home page is at *http://magicteahouse.net*, you might put the posts at *http://magicteahouse.net/posts* or *http://magicteahouse.net/blog*.

2. **Create a new page and give a title that matches the URL you want.**

   For example, if you want the post list to be at *posts*, you create a page with the title Posts. WordPress sets the page permalink to match.

TIP Remember, once you preview or publish your page, you can see the page permalink and edit it. Just click the page title in the block editor, and the permalink appears above. If you need to make a change, click the Edit button to type in a different permalink.

---

3. **Optionally, add some content to your page.**

   You don't need any content (and you probably don't want any either). But if you do add an introductory paragraph or two, WordPress displays it just above the list of posts.

4. **Publish your page.**

   Your placeholder page is ready. Now all you need to do is change your site settings.

5. **Choose Settings→Reading.**

This brings you back to the setting page that you used in the previous example, when you were creating a brochure site.

6. **If you haven't already set a custom home page, do that now.**

In the "Your homepage displays" setting, choose "A static page" (rather than "Your latest posts"). In the Homepage list underneath, pick the page you want to use for your new home page.

7. **In the "Posts page" list, pick the page you created in step 2.**

For example, if your page is named Posts, choose it here. This step tells WordPress to start using your placeholder page to show the list of posts.

8. **Click Save Changes at the bottom of the page.**

Now visitors can see your old home page—the list of posts—using the URL for the placeholder page you created in step 2. So if you created a page named Posts, when you request that page (say, *http://magicteahouse.net/posts*), you see your list of posts. But if someone requests the home page (*http://magicteahouse.net*), they'll see the custom home page instead.

9. **Optionally, edit your menu and add a new menu item for your new posts page.**

Even though you created a posts page, that doesn't mean your visitors know about it. They need a way to get there, and the best option is a link in your site's menu. Creating that is easy—you simply add a new menu item that points to your placeholder page (*Figure 8-19*).

TIP In some cases, you may decide that the single reverse-chronological stream of posts isn't that useful for your website. Maybe it just crams way too many types of posts in one big undifferentiated list. Instead, you might decide to use separate category links so visitors browse the posts one category at a time.

If you're going to make a custom home page for your site, it had better be useful. That means you shouldn't just show an ordinary message; you should also provide at least one path (preferably several) that can take the viewer to the rest of your site. These paths can include links in your home-page content; a menu; a footer; widgets with links to pages, categories, or recent posts; and so on. All of these techniques are shown in *Figure 8-19*.

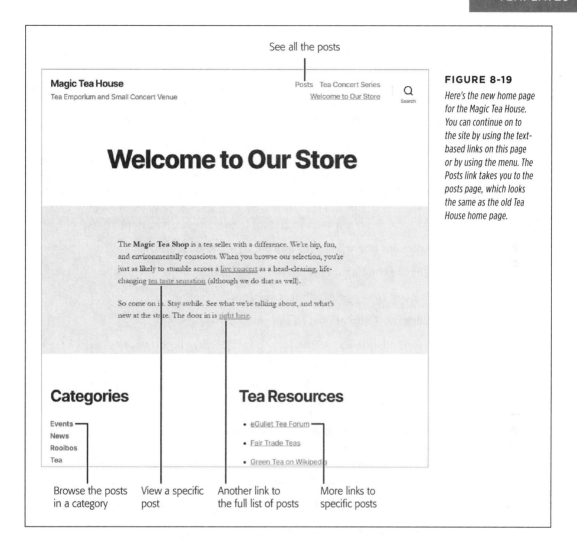

See all the posts

Magic Tea House
Tea Emporium and Small Concert Venue

Posts   Tea Concert Series
Welcome to Our Store

Q
Search

Welcome to Our Store

The **Magic Tea Shop** is a tea seller with a difference. We're hip, fun, and environmentally conscious. When you browse our selection, you're just as likely to stumble across a live concert as a head-clearing, life-changing tea taste sensation (although we do that as well).

So come on in. Stay awhile. See what we're talking about, and what's new at the store. The door in is right here.

Categories

Events
News
Rooibos
Tea

Tea Resources

- eGullet Tea Forum
- Fair Trade Teas
- Green Tea on Wikipedia

Browse the posts
in a category

View a specific
post

Another link to
the full list of posts

More links to
specific posts

FIGURE 8-19

Here's the new home page for the Magic Tea House. You can continue on to the site by using the text-based links on this page or by using the menu. The Posts link takes you to the posts page, which looks the same as the old Tea House home page.

◼ Creating Better Home Pages with Templates

You now know how to take an ordinary page and put it front and center at the entrance to your site. But some sites have much fancier types of home pages. Sometimes the home page doesn't resemble an ordinary page at all. It may have special features (like a carousel of featured posts), different visual styling (like animated pictures and headings), and a customized post list. The trick is that all these features depend on your theme. Some themes let you fancy up your home pages in certain ways, and other themes don't offer much of anything beyond the WordPress basics.

So how do themes give you the ability to opt in to fancy home pages? The secret is a feature called *page templates*. You may remember the template feature, which lets posts choose different layout options (page 126). For example, Twenty Twenty has a rarely used Full Width template, which lets a post expand to fill the margin area. Other themes use templates that let you choose whether your posts show a sidebar next to their content.

The page template feature is identical to the post template feature. It lets you choose from different layouts for your pages. There are a number of reasons that you might want an alternate layout with a different page template, but fancy home pages are one of the most common.

Like post templates, page templates are an optional part of a WordPress theme. Your theme may include multiple templates or none at all. If your theme provides a page template that's designed for better-looking home pages, it probably has a name like Homepage or Front Page template. To check what templates your theme has, you need to create a page and try to use one of them. Here's how:

1. **Edit your home page (or any page) in the post editor.**

2. **Choose the Document tab in the sidebar.**

3. **Expand the Page Attributes section.**

4. **Click the Template list to see your options.**

If you expand the Page Attributes section but there is no Template list, your theme doesn't include any templates at all.

Now, to understand how page templates work to make a better home page, you need to consider a few themes that put them to use.

Page Templates in the Year Themes

The WordPress year themes use page templates inconsistently. Twenty Nineteen and Twenty Seventeen don't have any templates at all. But Twenty Twenty adds a new Cover template, which lets you create a fancy graphical home page (*Figure 8-20*).

Using the Cover template is easy. Here's a complete walk-through of what to do:

1. **Create (or edit) an ordinary page.**

   In *Figure 8-20*, it's a page with the title "Tea Is Magic."

2. **Change the template to Cover.**

   To do that, look in the sidebar, choose the Document tab, and expand the Page Attributes section. In the Template list, choose Cover.

3. Set a featured image.

The top part of the cover page is an image. Twenty Twenty uses the featured image for your page and tints it with the current accent color (page 113).

Featured images were discussed in detail in Chapter 7. But if you're a bit hazy, here's a quick refresher: choose the Document tab in the sidebar, scroll to the Featured Image section, and click "Set featured image" to pick the picture you want.

FIGURE 8-20

With the Cover template, your page has two parts. First, visitors see your cover image filling the entire browser window (top). They can either click the down-pointing arrow or scroll down to see the actual page content (bottom).

4. **Add the content for your page (*Figure 8-21*).**

   All the page content appears "under the fold." To see it, the visitor needs to scroll down past the picture.

5. **Publish or update your page.**

   If the page isn't already set as your website home page, remember to visit Settings→Reading and change your home-page settings.

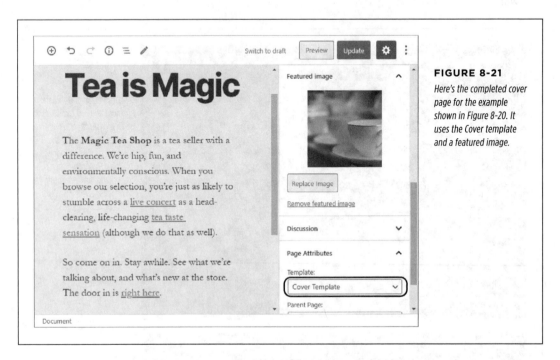

FIGURE 8-21

Here's the completed cover page for the example shown in Figure 8-20. It uses the Cover template and a featured image.

Other year themes use page templates, but they're much older. For example, Twenty Twelve has a Front Page template that gives your home page a different widget area. That way, you can add widgets to your home-page footer that won't appear on other posts and pages in your site.

The Twenty Eleven theme takes the concept a bit further with its Showcase template that combines page content, a group of featured posts, and a list of recent posts (*Figure 8-22*).

You probably won't use the older year themes these days, because they're getting a little creaky. However, they provide good examples of how different themes can give different options for home-page customization. Plenty of modern themes in WordPress's theme repository use page templates to make even fancier home pages. In the next section, you'll take a closer look at one example.

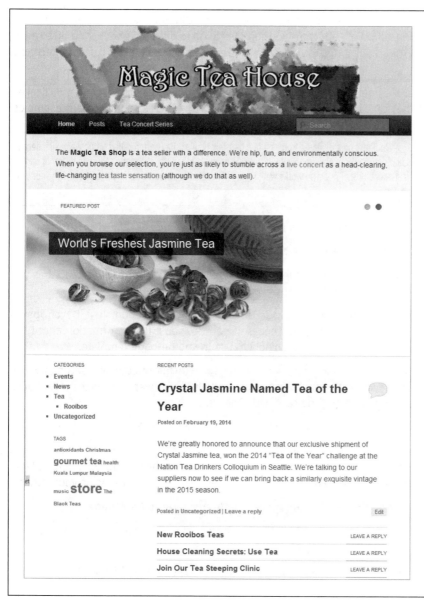

FIGURE 8-22

*The showcase page in
the older Twenty Eleven
theme fuses together your
page content, a gallery
of featured posts, a list
of recent posts, and a
showcase sidebar.*

Page Templates in the OnePress Theme

You've probably seen modern websites that stuff their home page full of sections. As you scroll down, you'll see animated pictures, fancy types of bulleted points, profiles of people, a featured list of articles, and more. Different sections provide different jumping-off points—buttons you can click to go to different pages. And often these elements are *animated*. As you scroll, the different sections swoop or shift into view.

You can create this type of home page for your own site. All you need is a good theme, the right page template, and the theme customizer. In this section, we'll take a look the popular OnePress theme (*Figure 8-23*), which is just one of many examples.

NOTE OnePress gets its name from the fact that this type of site is often called a *one-page site*, because it's possible to put a complete simple site in this one page. You then use the menu to link to different parts of this mega page. But OnePress also works well if you just want a fancy jumbo home page along with your usual posts and pages.

Here's how to create a home page like this:

1. **Add and activate the OnePress theme.**

    You can find OnePress in the WordPress theme repository. You add it as you would any theme—check Chapter 5 if you need a refresher.

    The OnePress theme adds an admin page you can visit by choosing Appearance→OnePress Theme. If you do, you'll find information about the theme, and instructions on how to use the home-page template.

    Many themes have their own admin page with information and settings. Some also need you to install and activate a plugin that powers certain theme features, although this isn't the case with OnePress.

2. **Create your home page (choose Pages→Add New).**

    You aren't going to put any content in your home page. In fact, OnePress will ignore its title and any blocks you put inside. This home page is simply a place-holder for OnePress's real home page, which it will create based on your settings.

3. **Change the page template of your home page to Frontpage.**

    This is the magic link that tells OnePress to turn this blank page into a customized home page.

4. **Give your home page a name that makes sense (like "Home").**

    This information won't appear anywhere on your site, but it will help you remember which page to set for your home page.

FIGURE 8-23

Here's a bird's-eye view of a three-section home page created with the OnePress theme. What you can't see here are the animated effects that highlight the different sections as you scroll.

HOME   ABOUT H2H   BOOKS   LABS   NEWS   PROSETECH

BE A HERO.

Read the book. Do the exercises.
*The unique format of **Halfwit to Hero** books guarantees that you'll pick up real skills, not papercuts.*

SEE THE BOOKS   HOW IT STARTED

WHAT MAKES US

DIFFERENT

Hands-on Learning	Expert Guidance	Laser Focus	Fun in the Trenches
Every *Halfwit to Hero* book has a trusty companion—a tutorial website with hands-on, step-by-step exercises for each chapter.	Follow expert explainer Matthew MacDonald, the author of dozens of bestselling tech books and a three-time Microsoft MVP.	Don't have time for fluff? *Halfwit to Hero* books focus on the practical skills you need, not the sales pitch.	We keep it light, with a fun, breezy style that'll have you flipping pages without falling asleep.

THE LATEST

NEWS

HALFWIT TO HERO / NEWS

Take "Excel Formulas" for a test drive

Curious how Halfwit to Tutorials work? Need to learn Excel, but not sure if you should invest hours watching Excel videos or chewing over Excel theory in a reference book? ...

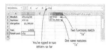

HALFWIT TO HERO / NEWS

"Excel Formulas" is in the pipeline!

The first H2H book is on its way! Excel Formulas: Halfwit to Hero is a compact guide to the essential art of formula writing, whether you're using Excel to build ...

SEE ALL THE NEWS

5. **Choose Settings→Reading, and set your home page.**

   As you did before, you change the "Your homepage displays" setting to "A static page" and then pick your home page in the list below.

   If you try visiting your site now, you'll see a generic introductory page about One-Press. To customize it with your content, you need to use the theme customizer.

6. **Choose Appearance→Customize to launch the theme customizer.**

   In the theme customizer, you'll see a layout that's very different from usual. You'll see a long list of custom sections that belong to OnePress (*Figure 8-24*).

FIGURE 8-24

The OnePress theme has a separate section in the theme customizer for every section you can add to your home page.

NOTE If you don't see the extra sections, you probably haven't set the "Your homepage displays" setting to the right home page. OnePress gives you these options only if you're working with its specialized home-page template.

7. **Choose which sections you want to appear.**

   For example, if you want to see the Hero section (which is the name for the picture with superimposed text at the top of the page), click Section: Hero. Then click Hero Settings, and make it appear (or disappear) by setting the "Hide this section" checkbox.

8. **Configure the content of each section you want.**

   For example, if you want to fill in the text in the Hero section, click Section: Hero, then Hero Content Layout. Click around the theme designer, and you'll eventually find the settings that let you change the buttons, colors, background picture, and animated effect.

TIP Often you can home in on exactly the section you want to change by using the preview in the theme customizer. Just look for a pencil icon next to the section you want to change, and click it to jump to the corresponding settings in the sidebar.

9. **Publish your site and enjoy.**

   You'll need to invest some time clicking around the theme customizer before you learn your way around the OnePress theme. Not everything is customizable. And sometimes in themes like OnePress, you'll find additional features that are available only if you upgrade to a more feature-filled professional version, which isn't free. (The professional version of OnePress costs about $70, which is a typical price. It's a fair cost if you decide to commit to the theme for a long time, but it's less reasonable when you're trying out a theme and still not sure if it's the best option for your needs.)

Unfortunately, if you switch to another fancy home-page theme, you'll need to learn its slightly different arrangement of settings. You'll also need to fill in your home-page content all over again. That's the only problem with fancy home pages—because they aren't standardized, there's no common model that all of them can share. But if you're looking for a polished, professional effect, it's usually worth the trade-off.

■ The Last Word

Up until this chapter, posts were the star of the show. You spent all your time creating, categorizing, and sorting them. But in this chapter, you learned about their similar but different partner: *pages*.

Pages are a fundamental ingredient in almost every WordPress site. In a traditional blog, you'll spend most of your time writing posts, and you'll probably have just a couple of pages for permanent, undated content (like an About Me write-up). If you're building a site to promote a business, you might fall into the opposite extreme, and create a site with a small section of dated posts (for example, for news), while putting all your most important content in a series of pages, anchored by a menu.

In other words, it's up to you to determine the exact balance of posts and pages—but you'll almost certainly put both to good use.

Getting New Features with Plugins

W ordPress offers an impressive set of built-in features. In the previous chapters, you used them to write posts and pages, and to glam up your site. But serious WordPress fans have a way to get even more from the program—or, technically, about 60,000 ways to get more, because that's how many free WordPress plugins you can use to supercharge your site.

In this chapter, you'll learn how to search for interesting plugins, decide whether a plugin is a good fit, and see how to install it on your site. This gets you ready for the rest of the book, where you'll use plenty of plugins to solve stubborn problems, fill gaps, and add frills.

You'll also try out Jetpack, a feature-packed plugin developed by Automattic, the same people who built WordPress. Jetpack is an unusual creation—instead of offering a few neat features organized around a specific theme, it gives you a grab bag of unrelated enhancements. The original idea was for Jetpack to give WordPress website builders (that's you) the same features that people get when they sign up for a free blog at WordPress.com. But Jetpack is useful for anyone. It's a starting point for many of the features you'll see later in this book, like Facebook and Twitter comments (Chapter 10), site statistics (Chapter 12), and even backups (Chapter 14).

How Plugins Work

Technically, a *plugin* is a small program written in the same programming language that runs the entire WordPress system, PHP.

Plugins work by inserting themselves into various WordPress operations. For example, before WordPress displays a post, it checks to see if you've installed any plugins related to displaying posts. If you have, WordPress calls your plugin into action. This sort of check is called a *hook*, and WordPress has a long list of hooks that can launch different plugins at different times. A WordPress page can also use a special code, called a *template tag*, to ask a plugin to insert something in a specific place. But either way, the interaction between a WordPress site and its plugins happens behind the scenes, without your intervention.

NOTE If you want to learn more about the way plugins work, you can check out the WordPress guidelines for plugin authors *http://tinyurl.com/write-plugin*. But a hefty disclaimer applies—while we encourage *learning* about plugin building, we don't suggest you try to make one of your own for a real public website. There are too many little ways to cause big problems that will affect the security and performance of your site.

Real Talk About Plugin Security

There's no way to get around this: as you add features with plugins, you also add potential security vulnerabilities.

As you already know, the core WordPress engine is a massive open source project. During the development cycle, its code is put in front of hundreds of working programmers. By comparison, many plugins have just a single developer—one who may be busy juggling other projects and responsibilities. Out-of-date plugins are a particular problem. If a developer abandons your favorite plugin, the old version lingers on, potentially putting your site in harm's way. And the risk remains even if you deactivate a plugin, as long as you've left it installed on your site.

But don't panic yet. You can protect yourself by practicing good plugin hygiene. Here's what you need to do:

- Avoid installing plugins from plugin creators that don't have an established reputation.

- Avoid installing plugins that haven't been updated recently (at least once in the previous year).

- When you have a choice, go with the most popular plugin (the one that's been installed on the largest number of sites).

- When a new version of a plugin you use is available, update it as soon as possible.

- Uninstall any plugins you don't use.

The bottom line is this: to avoid security problems, stick with well-used and well-loved plugins—those that have stood the test of time, are in widespread use, and are still regularly revised. As you'll see on page 250, you can get all the essential information about a plugin before you install it, so you can make good decisions and keep your site safe.

Plugin Management

You manage plugins in the WordPress admin area. Once you've logged in, choose Plugins→Installed Plugins. You'll see a list of all the plugins that are currently installed on your website (*Figure 9-1*).

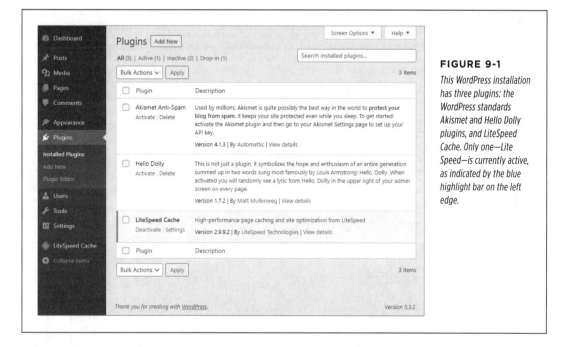

FIGURE 9-1

This WordPress installation has three plugins: the WordPress standards Akismet and Hello Dolly plugins, and LiteSpeed Cache. Only one—Lite Speed—is currently active, as indicated by the blue highlight bar on the left edge.

WordPress installations always start with two plugins: Akismet and Hello Dolly. They're installed but not *active*. In other words, they're sitting on your website, ready to be called into service but not actually doing anything yet.

Some web hosts may top you up with a few extra plugins. For example, the Softaculous installer might have offered to add the Limit Login Attempts Reloaded plugin (page 27) for extra security. Or, your web host might add a caching plugin like LightSpeed to prevent poor performance. (You'll learn more about caching plugins in Chapter 14.) But to get the really good stuff, you need to install plugins on your own.

Activating (and Deactivating) a Plugin

Plugins are like themes—you can install as many as you want, but they have no effect until you turn them on. Unlike themes, you can activate more than one plugin at a time. To turn on a plugin, click the Activate link under its name. When a plugin is active, WordPress shows a blue highlight bar next to it and changes the Activate link into a Deactivate link.

To practice activating and deactivating a plugin without risking anything on your site, you can try the harmless (but also useless) Hello Dolly sample plugin. When you switch it on, it adds a random lyric from Louis Armstrong's song "Hello, Dolly" to the top corner of every page in the admin area. (*Figure 9-2* shows what that looks like.)

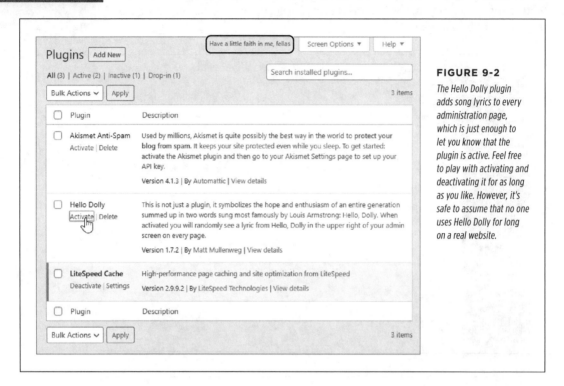

FIGURE 9-2

The Hello Dolly plugin adds song lyrics to every administration page, which is just enough to let you know that the plugin is active. Feel free to play with activating and deactivating it for as long as you like. However, it's safe to assume that no one uses Hello Dolly for long on a real website.

You can't always tell what a newly activated plugin is up to. Often there'll be evidence that a plugin is active somewhere in the admin area, but every plugin works differently. Some provide a new page of options in the menu; others add new widgets to the theme customizer or new blocks to the post editor. And some (like Hello Dolly) simply start doing their work quietly in the background.

Confusingly, different plugins put their settings in different places. Here are the three places to look:

- **The admin menu.** Some plugins add an entirely new item to the menu. For example, if you install Jetpack, you'll see a Jetpack menu item near the top of the menu, just above Posts (*Figure 9-3*). This is the most common place for plugin settings, especially if the plugin offers plenty of options. But there's no surefire way to know where your plugin's link will appear. Some plugins put it near the top of the menu (as does Jetpack), while others put it at the bottom (LiteSpeed Cache).

- **The Settings menu heading.** Some plugins add a new menu under the Settings heading in the admin menu. For example, to configure the Limit Login Attempts Reloaded plugin, you choose Settings→Limit Login Attempts. Plugins are most likely to use this approach when they have a small group of settings that fit on a single page. You may even find a rogue plugin that stashes itself under the Tools heading.

- **The Plugins page.** As you know, WordPress lists every plugin you install on the Plugins page. Underneath each plugin are links that let you activate, deactivate, edit, and delete the plugin. Some plugins add an additional Settings link that you can click to review and edit the plugin's options. Depending on the plugin, this link may provide the only way to configure the plugin, or it may duplicate an option in the admin menu. For example, clicking the Settings link under the LiteSpeed Cache plugin has the same effect as clicking LiteSpeed Cache in the menu (*Figure 9-3*).

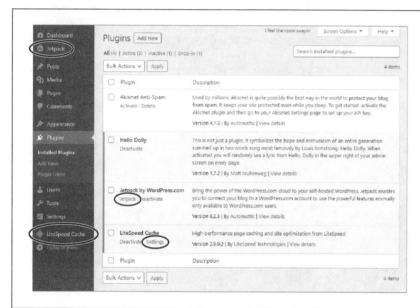

FIGURE 9-3

This site has three active plugins. The Hello Dolly plugin has no additional settings, but both the Jetpack and LiteSpeed Cache plugins do, and you can get to them through the admin menu (circled on the left). LiteSpeed Cache also includes a Settings link on the Plugins page, and Jetpack includes a similar link named, unhelpfully, "Jetpack."

This variation in plugin settings is part of a common theme. Namely, different plugins often work in ways that are similar but not exactly the same. For that reason, you need to learn the way each plugin works individually.

Finding New Plugins

The process of installing a plugin is simply the process of copying the plugin files to your website's plugin folder. WordPress names the plugin folder */wp-content /plugins*. So if you put WordPress on your site at *http://pancakesforever.com*, the plugin folder is *http://pancakesforever.com/wp-content/plugins*. Of course, you don't need to worry about the exact location, because WordPress manages the plugin folder for you.

To install a plugin, start by choosing Plugins→Add New, which takes you to the Add Plugins page (*Figure 9-4*). At first, the Add Plugins page shows you a list of a few recommended plugins, like the Akismet spam fighter (which you already have) and the Classic Editor for people used to the old-fashioned way of writing posts that existed before WordPress 5.

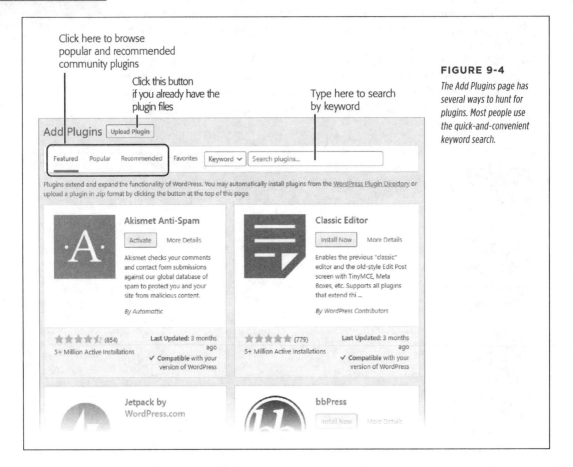

Click here to browse popular and recommended community plugins

Click this button if you already have the plugin files

Type here to search by keyword

FIGURE 9-4

The Add Plugins page has several ways to hunt for plugins. Most people use the quick-and-convenient keyword search.

WordPress gives you several options to find new plugins:

- **Search for it.** This is the most common way to hunt for a plugin. If you can think of a keyword for the sort of WordPress plugin you want (for example, "statistics" or "Twitter"), you can probably find it through a search. Type one or more keywords in the search box; then press Enter. WordPress will scan its massive plugin directory for matches.

- **Browse for it.** If you're not sure what type of plugin you want, you can browse WordPress's collection of plugins. Click the Featured, Popular, or Recommended tab to browse a curated list of useful plugins—and maybe even discover a new one you don't know about.

- **Get a favorite.** Next to the Featured, Popular, and Recommended tabs is another tab, called Favorites. Initially, you won't have any favorites. But if you want to use this feature, here's how it works. First, you visit WordPress's plugin directory at *http://wordpress.org/plugins*. You'll find the same plugins here as you do in the admin area, but you can browse them without logging in to a live WordPress website. If you find something you might want, you can flag it as a favorite. You can then review your favorites list in the admin area. The catch is that this feature works only if you go through the effort of signing up for a free WordPress.org account.

- **Upload it.** Do you already have the plugin files on hand? If you already down-loaded a plugin to your computer, you can use the upload command to transfer it to your website. To do that, click the Upload Plugin button (at the top of the page, next to the Add Plugins title). Then, browse to your plugin ZIP file, select it, and click Install Now.

> **NOTE** You might use plugin uploading if you acquire a plugin from somewhere other than WordPress's plugin directory (for example, if you bought a commercial plugin from a third party).

The quickest way to find a plugin is to perform a search. Type in some keywords (like "contact form") and press Enter. WordPress scans its staggering collection of more than 60,000 plugins and shows a list of matches (*Figure 9-5*).

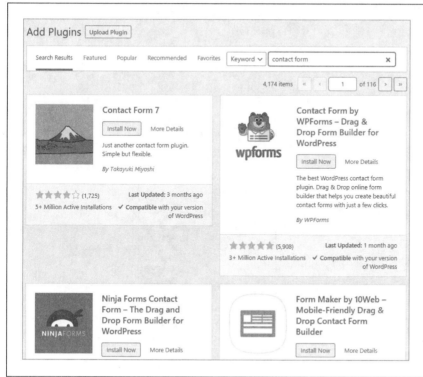

FIGURE 9-5

WordPress found over four thousand plugins that use the words "contact form." But it puts the most popular ones that best match your search at the top of the list (like Contact Form 7 and Ninja Forms).

The next step is to determine if you actually want to install the plugin you found. Or, if you've found several plugins, you need to decide which one you'll try first.

Before you decide to install a plugin, you need to spend a little time investigating it. To take a closer look, click the More Details link (which appears near the Install Now button). When you click More Details, WordPress pops opens a new window with extra information about the plugin. This information includes a detailed list of the plugin's features, the current version number, the person who created it, its compatibility with different versions of WordPress, and the plugin's rating (*Figure 9-6*).

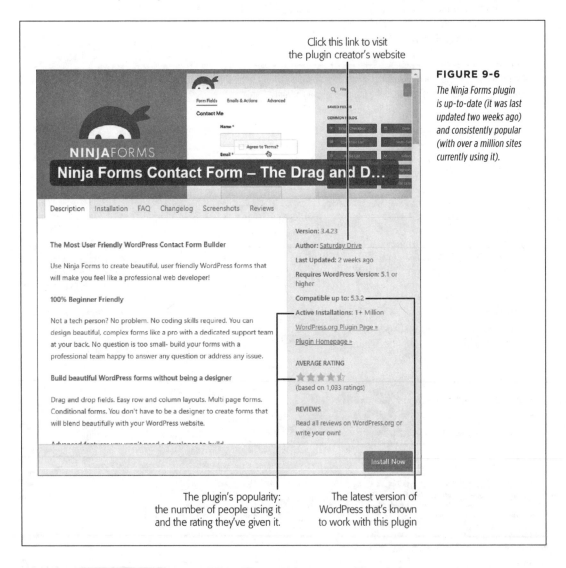

Click this link to visit
the plugin creator's website

FIGURE 9-6

The Ninja Forms plugin is up-to-date (it was last updated two weeks ago) and consistently popular (with over a million sites currently using it).

The plugin's popularity:
the number of people using it
and the rating they've given it.

The latest version of
WordPress that's known
to work with this plugin

So how do you decide if you've found the right plugin? You'll probably need to put it on your site, try it out, and see how you like it. But before you get to that point, a bit of quality control can help you avoid plugin lemons. Here's what to check on the details page:

- **Does it have the features you want?** By now, you've probably read the plugin description, which usually includes a detailed feature list. If you're still not satisfied, you can click one of the other tabs in the details page to get more information: Installation (to review extra installation steps for complex plugins), FAQ (to read the answers to common questions), Changelog (to see what changes or fixes have recently been made), and Screenshots (to get a feel for what the plugin looks like when it's installed).

- **When was it last updated?** Although old plugins often keep working, it's best to stick with plugins that have been updated within the last year (at least). Regular maintenance also increases the chances that a plugin is getting new features and fixes, which are two more attributes that make it a good candidate for your site.

- **What WordPress version does it support?** The Requires WordPress Version detail tells you the minimum version of WordPress you need to use this plugin. Because you're a security-conscious webmaster who always makes sure your site runs the latest WordPress updates, this part isn't so important. But the "Compatible up to" detail is more significant, because it tells you the *latest* version of WordPress that the plugin supports. You should avoid old plugins made for old versions of WordPress (*Figure 9-7*).

- **How many people are using it?** The number of active installations tells you how many sites are using the plugin right now. The more people using the plugin, the less likely it has hidden bugs.

- **Do people like it?** The star rating shows you an average of all the reviews WordPressers have left. Scroll down a bit and you'll see the breakdown—how many five-star reviews, how many four-star reviews, and so on. Click one of these links (for example, "1 star") to read the actual reviews people have left, which will help you weed out plugins with obvious problems.

- **Are support questions getting answered?** If you run into problems with a plugin, you can ask for help on the WordPress.org site. Of course, you won't have any questions right now—you haven't even tried the plugin—but you might be curious to see what other people have asked, and if their troubles were resolved. To take a peek, click the WordPress.org Plugin Page link. Once you get there, scroll down until you see the Support tab, and click that. You'll see a list of posts from plugin users and their replies.

TIP Finding a good plugin requires a bit of detective work. If you find a few plugins that seem to offer the feature you need, compare them to see which has the highest number of downloads, the best star rating, and the most recent updates.

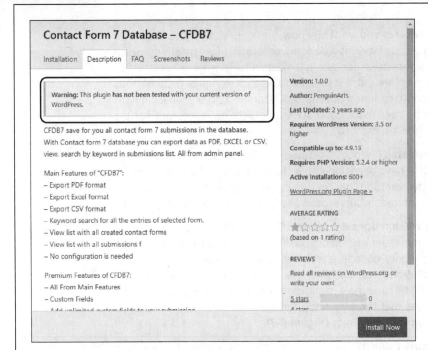

FIGURE 9-7

Sometimes people create plugins, maintain them for a while, and then abandon them. These old plugins keep kicking around the WordPress plugin directory, even though they're of no use to sites that use newer editions of WordPress. When you look at one of these clunkers, WordPress shows a warning box at the top of the description. The plugin may still work, but you don't need to take that risk, because there are always newer plugins out there.

Using the Plugin Links in This Book

In this book, you'll frequently come across links to useful WordPress plugins. These links take you to WordPress's plugin directory, at *http://wordpress.org/plugins*. There, you get extensive information about the plugin, including how it works, the number of times it's been downloaded, its star rating, and its compatibility information (just as you can in the plugin details window shown in *Figure 9-6*).

You can download a plugin from the directory, presumably so you can upload it to your site later, from the Plugins→Add New page. But you don't need to go through this two-step download-and-then-upload process. To install a plugin discussed in this book, log in to your site's administration area, search for the plugin by name, and then click the Install button to transfer it in one step.

Installing a New Plugin

If you decide to go ahead with a plugin you've found, installing it is simple. Just click the Install Now button, which appears next to each plugin in your search results (*Figure 9-5*) and in the details page (*Figure 9-6*). When WordPress prompts you to confirm your installation, click OK.

Typically, the installation takes just a matter of seconds. WordPress gives you minimal feedback as it grabs the plugin's ZIP file, pulls all the plugin files out of it, and transfers them to a new folder in the plugin section of your website. For example, if you install Jetpack on the WordPress site *http://magicteahouse.net*, WordPress will create a folder like *http://magicteahouse.net/wp-content/plugins/jetpack* for the plugin files.

When you install a plugin, it starts out deactivated (which means it can't do anything). You can go to the Plugins→Installed Plugins page and activate it there, as you did with Hello Dolly. Or you can click the Activate button right on the Add Plugins page (*Figure 9-8*).

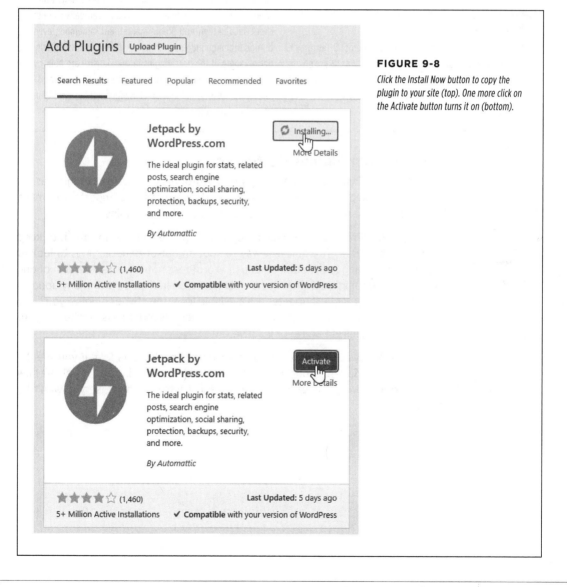

FIGURE 9-8

Click the Install Now button to copy the plugin to your site (top). One more click on the Activate button turns it on (bottom).

How People Sell Premium Plugins

The WordPress plugin directory is the first and best place to look for plugins. It gives you thousands to choose from, and you can be reasonably certain they're safe and stable. And they'll always be free. The most obvious disadvantage is that the WordPress plugin repository is crowded with out-of-date plugins, so you need to get good at sniffing out and avoiding these dead ends.

However, several companies also *sell* plugins. Often they'll release a free version of their plugin in the WordPress plugin directory, but sell a premium version that adds more features or includes support (someone to help you solve plugin setup problems or customize a complex plugin to work exactly the way you want).

In this book, you'll stick to using free plugins from the WordPress directory. However, if you need something more (for example, if you're a professional WordPress site designer who's creating sites for other companies), you'll eventually want to check out the high-end plugin market. But remember, even though premium plugins cost money, they aren't necessarily better. Some of the best plugins in the industry are built by open source developers and companies with WordPress-related businesses, and they don't ask for anything more than an optional donation.

Keeping Your Plugins Up-to-Date

Every time you install a plugin, you extend WordPress's features. And just as it's important to keep WordPress secure and up-to-date, it's also important to make sure you have the latest, most reliable versions of all your plugins.

Although WordPress can keep itself up-to-date (page 37), it's up to you to keep your plugins current. The quickest way to spot an out-of-date plugin is to look at the Plugins heading in the admin menu. If WordPress notices a new version of one or more of your plugins, it shows a number in a circle indicating how many updates await you. For example, if updates are available for three of your plugins, you'll see the number 3 in a circle (*Figure 9-9*, top). This happens regardless of whether your out-of-date plugins are active or inactive.

You should always update your plugins as soon as possible. In fact, if you see the number-in-a-circle, you should probably stop what you're doing and install the new plugins straightaway. Typically, updates take just a few seconds, so you won't be stalled for long.

There are two easy ways to update a plugin. You can do so from the Plugins page by clicking the "update now" link under the plugin (*Figure 9-9*, bottom). Or you can update numerous plugins at once by using the Updates page (click Updates in the admin menu). The Updates page lists all the out-of-date themes and plugins on your site. Check the box next to each plugin you want to update and then click Update Plugins. Blink twice, and WordPress will let you know that everything is up-to-date.

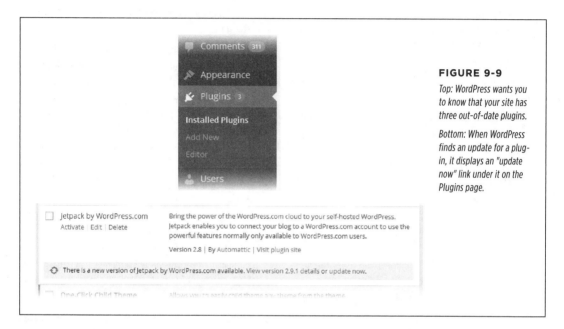

FIGURE 9-9

Top: WordPress wants you to know that your site has three out-of-date plugins.

Bottom: When WordPress finds an update for a plug-in, it displays an "update now" link under it on the Plugins page.

WARNING You must keep *all* your plugins up-to-date, even those that aren't currently active. That's because hackers can still access the inactive plugins on your site and exploit any security flaws they contain.

Removing Old Plugins

You don't need to keep plugins you don't use. If you decide you'd like to trash a particular plugin, removing it is a three-step process:

1. **Choose Plugins→Installed Plugins to call up the plugin list.**

2. **Click Deactivate to switch off a plugin (if it's currently active).**

3. **Click Delete to remove the plugin entirely.**

Why You Don't Want Too Many Plugins

When new WordPress users discover plugins, their first instinct is to load as much free goodness as they can onto their sites. After all, if one plugin is great, two dozen must be mind-blowingly fantastic.

Before you get carried away, it's worth pointing out the many reasons you *don't* want to install every plugin you find:

- **Performance.** As you already learned, plugins use PHP code to carry out their tasks. The more plugins you activate, the more work you're asking WordPress to do. Eventually, this work might add up to enough of an overhead that it begins to slow down your site. The plugins that do the most work (and are thus the most likely to hinder site performance) include those that back up your site, log your traffic, search for broken links, and perform search-engine optimization.

- **Maintenance.** The more plugins you have, the more plugins you need to configure and update (when new versions come out). It's a relatively small job, but pile

on the plugins and you might find yourself with some extra work.

- **Security.** Plugins can have security holes, especially if they're poorly designed or out-of-date. More plugins means more risk.

- **Compatibility.** Sometimes one plugin can mess up another. If your site uses a huge thicket of plugins, it's difficult to track down which one is at fault. You need to resort to disabling all your plugins and then re-enabling them one at a time, until the problem recurs.

- **Obsolescence.** Often plugins are developed by helpful people in the WordPress community who need a given feature and are ready to share their work. But this development model has a downside—it makes it more likely for an author to stop developing a plugin. Eventually, a new version of WordPress may break an old plugin you depend on, and you'll need to scramble to find a substitute.

■ The Jetpack Plugin

You'll meet plenty of plugins in this book. Most plugins are focused on one thing. For example, they might give you a way to collect email subscriptions, or a tool for tracking site statistics. To get all the features you want, you'll probably need several plugins—keeping in mind that sometimes mo' plugins means mo' problems (see the box on page 256 for more about that).

Automattic's Jetpack is a different sort of plugin. It offers a sprawling collection of features in one package. This makes it bigger and bulkier than most other plugins, but it also means that you don't need to find and vet a handful of different plugins from different people—and make sure they all work together. And even if you prefer to pick and choose smaller, targeted plugins that provide just the features you need, you might still want to start out experimenting with Jetpack, because it gives you a taste of so many extensions.

To read a high-level overview of Jetpack's features, check out *http://jetpack.com*. A few highlights include a contact form feature (covered in this chapter), enhanced comments (Chapter 10), Twitter and Facebook sharing (Chapter 12), and statistics about the people who visit your site (Chapter 12).

Signing Up with WordPress.com

You already know how to install the Jetpack plugin—just search the plugin directory for "jetpack" and click the Install Now link.

However, Jetpack has one additional (and slightly irritating) setup requirement. To use it, you need a WordPress.com user ID and password, even though you don't plan to host any sites on WordPress.com.

After you activate Jetpack, the plugin notifies you about this extra requirement with a message asking you to set up Jetpack (*Figure 9-10*).

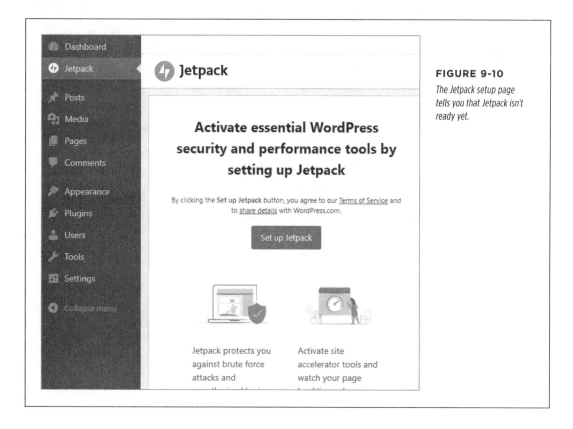

FIGURE 9-10

The Jetpack setup page tells you that Jetpack isn't ready yet.

Here's how to create a WordPress.com account and get Jetpack running:

1. **If you aren't already at the setup page, click Jetpack in the admin menu.**

2. **Click the Set up Jetpack button to set up the connection.**

   The Jetpack plugin takes you to *http://wordpress.com* and invites you to fill in your WordPress.com username and password.

   If you've ever created a site with WordPress.com or signed up for one of its services, you already have these details—you don't need to create a new account, because you can use the same one for as many Jetpack-enabled websites as you want. Skip to step 6 to finish the setup process.

   If you've never used WordPress.com before, now's the time to create an account. Continue to the next step.

3. **Type in an email address, username, and password for your new WordPress. com account-to-be.**

   Your username can be any combination of letters and numbers, so long as another WordPress.com member isn't already using it.

   Your password can be anything, but you should definitely make it different from the password you use to log into your WordPress site.

   Your email address is an essential detail, because WordPress.com sends you an activation email. If you don't get it, you can't activate your account.

4. **Click the "Create your account" button.**

   Now you need to wait for the activation email message.

5. **When the email arrives, click the activation link inside.**

   That brings you back to the Jetpack page, and the link automatically logs you into your WordPress.com account.

6. **Click Approve to authorize Jetpack to use your account (*Figure 9-11*).**

   Now WordPress attempts to upsell you on a paid plan. These plans are a reasonable choice if you want automatic, worry-free backups, hand-holding tech support, or WordPress's own premium themes. But right now you don't need any of that.

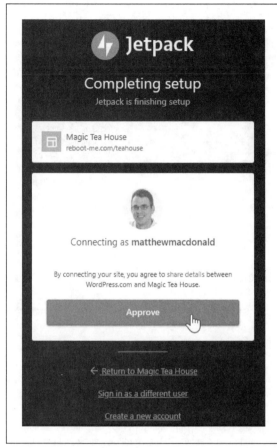

FIGURE 9-11

Before WordPress activates Jetpack, it asks if you want to use your WordPress.com user account (in this example, that's matthewmac- donald's account) to activate Jetpack on this site (Magic Tea House).

7. **Scroll all the way down the page and click the "Start with free" button.**

This completes the setup and brings you to the WordPress.com dashboard (*Figure 9-12*). You can go back to your website and begin using Jetpack.

You'll see many of Jetpack's features in later chapters. In the following sections, you'll try out just a few to get a taste of what the plugin can do.

If you have more than one site with Jetpack, you can switch here ━

Your linked website ━

Upgrade to a paid plan here ━

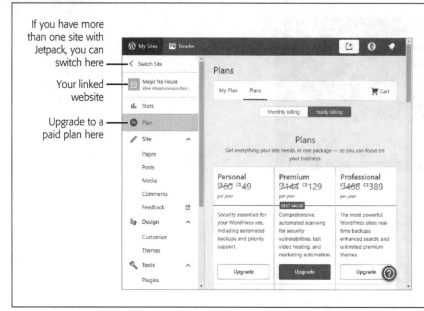

FIGURE 9-12

The WordPress.com dashboard looks a lot like the admin area you're used to (albeit with a different color scheme). But you don't need to use the WordPress.com dashboard, because you're running a full version of WordPress on your own website, not just a free WordPress.com blog.

FREQUENTLY ASKED QUESTION

The WordPress.com Dashboard

Why is WordPress.com giving me another way to manage my site?

The WordPress.com dashboard lets you see all the sites you've registered with Jetpack, and upgrade to one of their paid plans. But here's where life gets a little weird. The WordPress.com dashboard also duplicates many of the features and menus you use to manage your site in the admin area. For example, you can review pages, posts, and comments. You can change themes. You can even browse plugins. Sometimes the WordPress.com dashboard will even send you back to your website's admin interface to finish a job—for example, if you launch the theme customizer from the WordPress.com menu. Confused yet?

There's a reason for this strange behavior. What you see in the WordPress.com dashboard is the admin interface that people use to create free blogs on WordPress.com. But because you have a real website, you don't need to mess around with the WordPress.com dashboard. However, occasionally a Jetpack option might appear in the WordPress.com dashboard but not in the standard admin area you've been using so far. You'll see an example when you use Jetpack to monitor your site (page 261). In this situation, you can see the statistics Jetpack gathers in your site's admin area, but some extra information is available in the WordPress.com dashboard.

Here's our recommendation: don't use the WordPress.com dashboard. Your site's admin area is faster and more familiar. However, if you do come across one of those rare cases where you need to look at something in the WordPress.com dashboard, we'll let you know, and you can take a look.

Turning On Downtime Monitoring

Some of Jetpack's features need to be switched on before they become opera-
tional. You can do that from the Jetpack→Settings page. To try it out, let's switch
on *downtime monitoring*, a service that lets you know when your website has run
into trouble. When monitoring is on, WordPress.com keeps an eye on your site and
notifies you by email if it stops being reachable. Possible causes for a site going
offline include an outage at your web-hosting company, a domain name change, or
a misbehaving plugin that crashes your site.

Here's how to opt in to downtime monitoring:

1. **In the admin area, choose Jetpack→Settings.**

   The Jetpack Settings page appears (*Figure 9-13*).

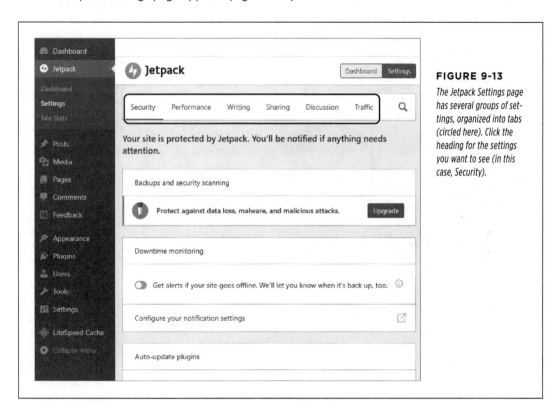

FIGURE 9-13

*The Jetpack Settings page
has several groups of settings, organized into tabs
(circled here). Click the
heading for the settings
you want to see (in this
case, Security).*

2. **Click the Security tab.**

   The Security page has a small group of settings. These include Jetpack's backup
   service (which requires a paid plan), downtime monitoring, automatic plugin
   updates (described in the next section), and brute-force-attack protection. Out
   of all these security settings, only brute-force-attack protection is switched
   on initially. It stops hackers who are trying to crack into your site by guessing
   your password.

3. **Slide the switch next to "Get alerts if your site goes offline."**

   The switch becomes blue, and the text goes from gray to black, showing you that the option is active.

4. **Optionally, click "Configure your notification settings" underneath.**

   This takes you to the WordPress.com website, where you can set a different notification email address. This is the place WordPress.com will send a message if it detects trouble on your site (*Figure 9-14*).

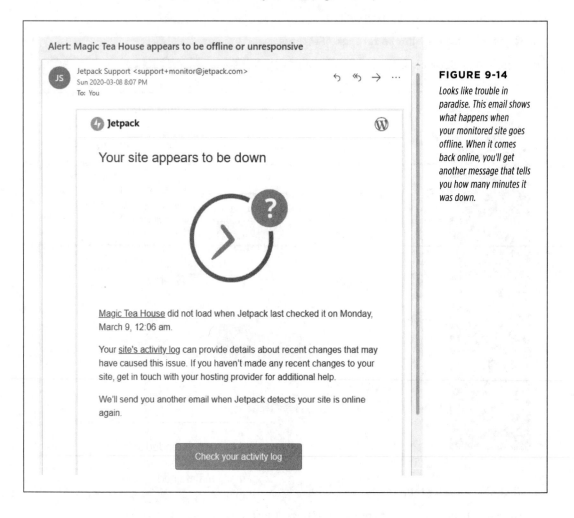

FIGURE 9-14

Looks like trouble in paradise. This email shows what happens when your monitored site goes offline. When it comes back online, you'll get another message that tells you how many minutes it was down.

Turning On Automatic Plugin Updates

As you learned in this chapter, it's important to keep your plugins up-to-date. Ordinarily, the responsibility for plugin updates falls on you, the site maintainer. This is different from core WordPress software, which automatically updates itself to get important updates (page 37).

Jetpack has a feature that can help you out. It's the automatic plugin update, which keeps some (or all) of your plugins up-to-date by installing new releases automatically. Here's how to use it:

1. **Choose Jetpack→Settings.**

2. **Click the Security tab.**

3. **In the "Auto-update plugins" section, click "Choose which plugins to auto-update."**

4. **Choose which plugins Jetpack should update automatically (*Figure 9-15*).**

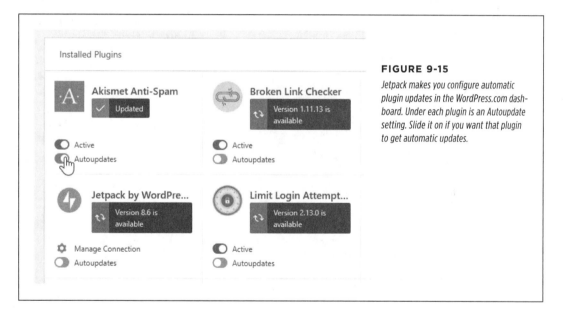

FIGURE 9-15

Jetpack makes you configure automatic plugin updates in the WordPress.com dashboard. Under each plugin is an Autoupdate setting. Slide it on if you want that plugin to get automatic updates.

There's always a tiny chance that a new plugin update could cause unintended side effects. That's why Jetpack makes you configure each plugin separately. It's probably safer to turn on automatic updates for a popular, professionally maintained plugin like Jetpack or Akismet than it is for a small third-party plugin.

> **NOTE** WordPress is planning to add automatic plugin updates as a core feature in the near future. There's a chance that by the time you read this, you'll get the same feature right in the WordPress plugins page, without even needing to install Jetpack at all!

Using Jetpack Blocks

Only some of Jetpack's features turn up on the Settings page. It also makes quiet changes and enhancements elsewhere on your site. One standout example is blocks—the ingredients you use to build the content for a post or page.

You've spent plenty of time using blocks in earlier chapters. You've seen text-based blocks like paragraphs, headings, lists, and quotes, and you've used more exotic blocks for different types of media (like pictures and video) or to embed content from other websites. Jetpack adds to your goodies with its own set of useful blocks.

You don't need to do anything to turn on Jetpack blocks. They automatically show up in the list of blocks when you're working in the post editor (*Figure 9-16*).

FIGURE 9-16

Jetpack puts all its blocks into a separate group. At the time of this writing, Jetpack adds 16 new blocks, but the people at Automattic are always tinkering with the list.

So what can you find in the Jetpack blocks? Some of them are pretty specialized (like the Markdown block that lets you write content in a simple formatting language that maps to HTML). Others don't do much, like the Contact Info and Business Hours blocks. They take your information (which you type into some boxes) and display it in a straightforward, slightly boring table. A few of Jetpack's blocks are tied to WordPress.com and aren't that useful for self-hosted sites like yours (Recurring Payments Button and Star Rating).

Most of the other blocks plug into specific services. Examples include Calendly schedules, Eventbrite tickets, OpenTable reservations, Pinterest boards, and Mailchimp newsletter sign-up forms. They're useful if you already use these services. But if you don't have an account on these sites, there's nothing much you can do with these blocks.

Along with the unusual blocks and the highly specialized ones are a few gems. In the next few sections, we'll introduce you to our favorites.

Creating Slideshows

In Chapter 7, you saw WordPress's underwhelming gallery feature. Using Jetpack, you can create a picture slideshow in the same way—only it looks nicer and has a few more features.

To use the *Slideshow block*, you start by adding the pictures you want to show from the media gallery. Then, use the sidebar to choose whether you want Jetpack to shuffle through your images automatically (*Figure 9-17*).

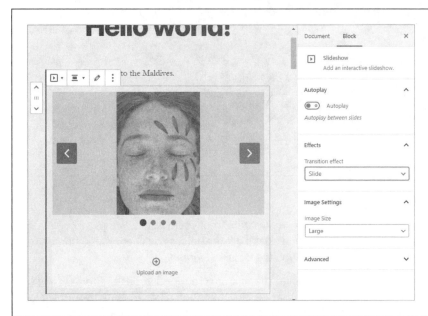

FIGURE 9-17

With a slideshow, you see one picture at a time on your page. You can click through them by using the arrows on the side or the circles at the bottom. (This particular slideshow has four pictures, which is why you see four circles.) Or, you can switch on an automatic fade or slide transition that cycles endlessly from one picture to the next.

Adding Animated GIFs

Animated GIFs—short, usually humorous animations—are the pinnacle of puerile entertainment. Jetpack lets you add them with a minimum of fuss from the free GIPHY site.

To use one, start by adding the GIF block. If you happen to know the web address on GIPHY for the GIF you want, you can paste it in the block. But if you don't, Jetpack lets you perform a quick search right in the post editor. Just type a few words into the search box (like "pizza rat"), hit Enter, and pick the animation that tickles your fancy (*Figure 9-18*). Optionally, you can add your own witty caption.

Write caption...

FIGURE 9-18

*When you search for a GIF,
you get a row of potential
matches, represented by
the thumbnails just under
the search box. Click one,
and you'll see it full size.
And although you can't
tell in this picture, you see
the full animation of the
subway-riding pizza rat,
so you can decide whether
it's the GIF you want.*

Building Forms

A *form* is a tool for collecting information from your visitors. You use a combination of text boxes, checkboxes, and lists to get the details you need. The visitor fills them in, and then Jetpack emails you the details (*Figure 9-19*).

Often people use forms to collect names, email addresses, postal addresses, or support questions. However, Jetpack is clever enough to let you build your *own* form, which you can customize to collect whatever information you want.

To get started, add the Form block. Next, you need to supply your email address and a title (*Figure 9-20*). Once you've done that, click the "Add form" button to get started building your form.

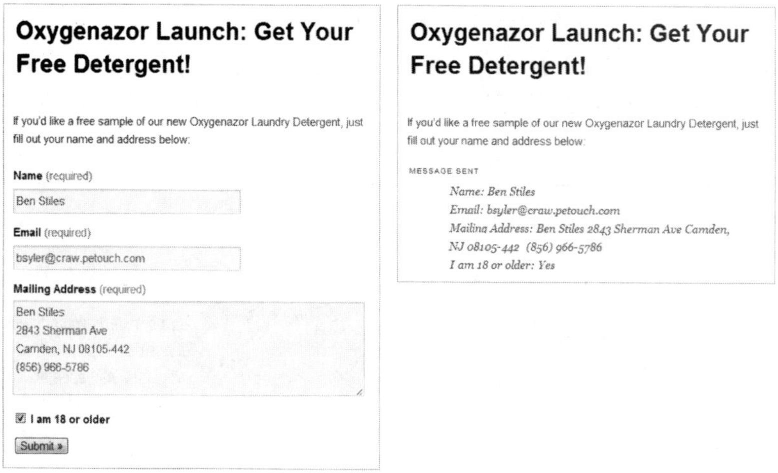

Oxygenazor Launch: Get Your Free Detergent!
If you'd like a free sample of our new Oxygenazor Laundry Detergent, just fill out your name and address below:
Name (required)
Ben Stiles
Email (required)
bsyler@craw.petouch.com
Mailing Address (required)
Ben Stiles
2843 Sherman Ave
Camden, NJ 08105-442
(856) 966-5786
I am 18 or older
Submit »
Oxygenazor Launch: Get Your Free Detergent!
If you'd like a free sample of our new Oxygenazor Laundry Detergent, just fill out your name and address below:
MESSAGE SENT
Name: Ben Stiles
Email: bsyler@craw.petouch.com
Mailing Address: Ben Stiles 2843 Sherman Ave Camden, NJ 08105-442 (856) 966-5786
I am 18 or older: Yes

FIGURE 9-19

You can use a contact form to solicit information from your guests. A reader fills in these blank text boxes (left) and then submits the form (right). WordPress collects the information, and passes it along to you in an email.

Add title
Form
You'll receive an email notification each time someone fills out the form. Where should it go, and what should the subject line be?
Email address
admin@magicteahouse.net
You can enter multiple email addresses separated by commas.
Email subject line
Free Detergent Submission
(If you leave these blank, notifications will go to the author with the post or page title as the subject line.)
Add form
Document
Block
Form
A simple way to get feedback from folks visiting your site.
Email Feedback Settings
Email address
admin@magicteahouse.net
You can enter multiple email addresses separated by commas.
Email subject line
Free Detergent Submission
Confirmation Message
On Submission
Show a summary of submitted fiel
Advanced

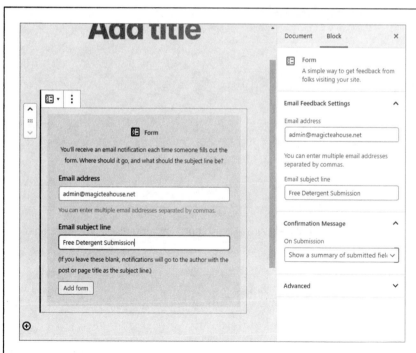

FIGURE 9-20

Before you can start designing your contact form, you need to supply an email address (where Jetpack will send the filled-out forms) and a title (which Jetpack will use as the subject of each email message).

To keep things simple, Jetpack lets you build forms by using the same block editor that you use to write posts and pages. In fact, you can insert all the ingredients you already know how to use, like paragraphs and images. But you can also add *fields*, a special type of block that collects information. For example, a text box is one field you might use in a form. A checkbox is another.

When you start your form, Jetpack automatically adds a few fields to get you started (*Figure 9-21*). But to build the form you really want, you need to add some fields of your own.

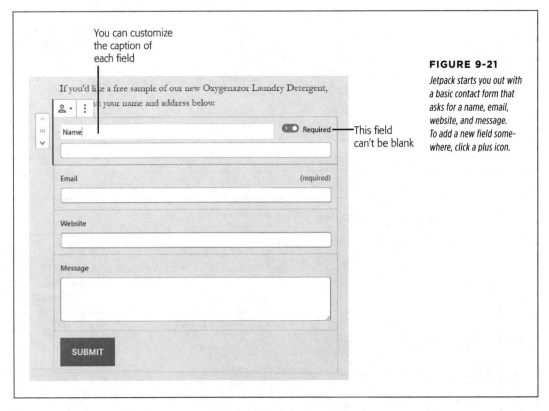

You can customize the caption of each field

This field can't be blank

FIGURE 9-21

Jetpack starts you out with a basic contact form that asks for a name, email, website, and message. To add a new field somewhere, click a plus icon.

Changing your form is easy—it's just like editing anything else in the post editor.

Here's what you can change:

- **Change the caption of a field.** For example, click the word "Message" and type in "What you want to ask us." You can use this technique to edit any of the text you see, including the big Submit button.

- **Change whether a field is required.** If a field is required, the form-filler needs to type something in before submitting the form. In the sample form, Name and Email are required, but nothing else. To change the requirement rule, click the field and slide the Required switch on or off.

- **Remove a field.** Click the field you want, so the tiny toolbar appears over it. Then click the three-dots icon (the More Options command) and choose Remove Block.

To make a truly personalized form, you need to add your own fields. To do that, move your mouse to the end of the form, or between any of the fields that are already there. As you move, the Add Block icon will appear (*Figure 9-22*).

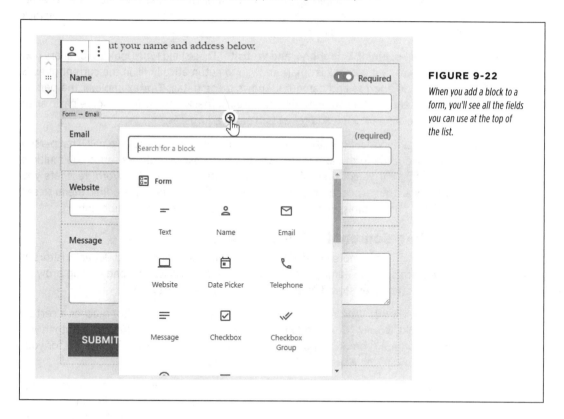

FIGURE 9-22

When you add a block to a form, you'll see all the fields you can use at the top of the list.

Click it and choose the field you want. Jetpack gives you plenty of choices:

- **Text.** This is your standard-issue single-line text box.

- **Message.** This is a bigger text box that can hold whole paragraphs of information. Once you add a message box, you can drag the bottom-right corner to make it as big as you want.

- **Checkbox.** This is a yes-or-no check box—as in, "I would like to receive mailing" or "This message is urgent." If you have a number of related yes-or-no options, you can group them together in a Checkbox Group. For example, use this for a "Choose all that apply" list of preferences.

- **Select.** This is a list of choices. The person filling out the form picks a single value from the list. When you design a form, you can see all the options at once, but when you publish the page, this turns into a drop-down list that shows only what the form-filler picks.

- **Radio.** This lets a person pick from a list of values, like the Select box. The difference is that the form-filler sees all the choices at once on the page. You pick one of the options by clicking it.

- **Name, Email, Telephone, and Website.** These field types look like ordinary text boxes, but Jetpack is smart enough to automatically fill in the current guest's email address and name if it knows those details (for example, if the person recently left a comment on your site). Jetpack also performs some basic error checking to catch and reject bad email addresses.

When someone submits a contact form, WordPress sends you an email. That person can fill out the form more than once (in which case WordPress sends you multiple messages). If you're tired of juggling the notification emails, WordPress lets you review all the responses in a single place. Click Feedback in the admin menu to see a list of form submissions, arranged just like a list of posts or comments.

Setting Content for Repeat Visitors

One of Jetpack's more unusual blocks is an odd ingredient called Repeat Visitors. It keeps track of the number of times each person has visited a page, using browser *cookies* (tiny files stored on the visitor's computer or device).

You set Repeat Visitors to show a special message after a visitor reaches a certain threshold. No, not "Congratulations on visiting the same page 10 times!" Rather, you might offer a promotion, point out another post, or suggest a sign-up link for your email newsletter. *Figure 9-23* shows an example.

You can also flip Repeat Visitors so that it does the reverse—in other words, show a message or some content for the first few visits and then *stop* showing it. To get Repeat Visitors to show and then stop, choose the "Show before threshold" option.

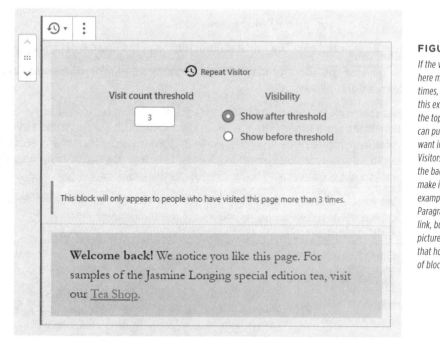

FIGURE 9-23

If the visitor has been here more than three times, Repeat Visitors adds this extra bit of text to the top of the page. You can put any content you want inside the Repeat Visitors block, and change the background color to make it stand out. This example uses an ordinary Paragraph block with a link, but you can add a picture, a list, or a group that holds a combination of blocks.

Using Jetpack Widgets

As you've seen, Jetpack enhances your site with some new blocks. But that isn't the only ingredient it brings to the table. Jetpack also adds about 20 new widgets that you can insert into your theme's widget areas (like sidebars and footers) so they appear all across your site.

Unlike Jetpack's blocks, Jetpack's widgets aren't available unless you specifically switch them on. Here's how you do that:

1. **Choose Jetpack→Settings.**

2. **Click the Writing tab and scroll down to the Widgets section.**

3. **Turn on the "Make extra widgets available for use on your site" setting.**

Once you've got your widgets working, you probably want to see what's there. The easiest way to browse Jetpack's widgets is in the theme customizer (*Figure 9-24*).

Here's how to take a look:

1. **Choose Appearance→Customize.**

2. **Click the Widgets section.**

3. **Pick one of your theme's widget areas.**

4. **Scroll down and click the Add a Widget button at the bottom of the sidebar.**

5. **To see just the widgets that come with Jetpack, type "jetpack" into the search box.**

One Jetpack widget you can start using immediately is the Social Icons widget. Like the Social Icons block (page 158), the Social Icons widget lets you add links to your social profiles—Facebook pages, Twitter timelines, Pinterest boards, and so on. But whereas the Social Icons block can go only inside a post, the Social Icons widget can go into a widget area so it's visible everywhere on your site.

You'll use several Jetpack widgets in Chapter 12. They'll help you show a Facebook Like box, a timeline of tweets from Twitter, and an email sign-up box.

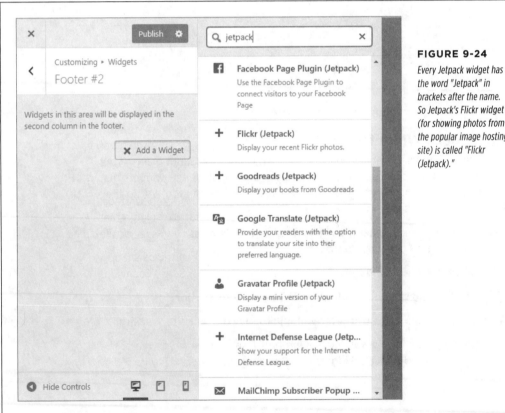

FIGURE 9-24

Every Jetpack widget has the word "Jetpack" in brackets after the name. So Jetpack's Flickr widget (for showing photos from the popular image hosting site) is called "Flickr (Jetpack)."

■ The Last Word

In this chapter, you learned to find and install plugins, and you tried out the all-in-one Jetpack plugin. However, in a directory with thousands of plugins, it's no surprise that many more are worth considering.

You'll learn about some of these plugins in the following chapters (and some you'll need to discover on your own). Here's a list of some of the most popular:

- **Akismet.** This spam-fighting tool comes preinstalled (but not activated) with WordPress. You'll use it in Chapter 10.

- **Yoast SEO.** Search-engine optimization (SEO) is the art of attracting the attention of web search engines like Google, so you can lure new visitors to your site. SEO plugins are among the most popular in the WordPress plugin directory, and Yoast SEO is an all-in-one package by one of the most renowned developers in the WordPress community. You'll take a closer look at it in Chapter 12.

- **WooCommerce.** If you want to sell products on your site, complete with a professional shopping cart and checkout process, this could be the plugin for you. You'll take a peek in Chapter 14.

- **BuddyPress.** If your website is bringing together a tightly knit community—for example, the students of a school, the employees of a business, or the members of a sports team, you can use BuddyPress to add instant social networking features to your site. Users get profiles as well as the ability to message one another, add photos, create content, and talk together in groups. To learn more, visit *http://buddypress.org*.

- **WP-Optimize.** This tool cleans up the junk that's often left cluttering your database. It can remove unapproved junk comments, old post revisions, and other unnecessary data that can expand the size of your database, slowing down your site and bloating your backups. You'll use it in Chapter 14.

- **UpdraftPlus.** Sooner or later, your website needs a good backup strategy. You'll look at the options in Chapter 14, including the UpdraftPlus plugin that can make automatic copies of your website for safekeeping.

- **WP Super Cache.** Speed is essential for any website. To keep things snappy, many WordPress website builders use a caching plugin. WP Super Cache, introduced in Chapter 14, is one of the most popular.

Comments: Letting Your Readers Talk Back

In the chapters you've read up to this point, you learned to create the two most essential ingredients of any WordPress site: posts and pages. You use these ingredients to talk to potential customers or devoted readers. But you haven't considered how your readers can talk back.

The answer is the WordPress commenting system. Used properly, *comments* can change your site from a one-way lecture to a back-and-forth conversation with your readers or customers. Commenting isn't just a way to make fun friends—it's also a serious tool for promoting your work, getting more traffic, turning casual browsers into repeat visitors, and even making money.

In this chapter, you'll learn how to manage comments on your site. You can banish offensive ones, insert yourself into the discussion, and integrate with Facebook and Twitter.

Once you understand the basics of comment management, you'll be ready to confront one of the single biggest hassles that every WordPress site faces: *comment spam*—the messages that dubious marketers and scammers slap across every site they can find. You'll learn strategies for preventing spam without aggravating your readers, and you'll take a side trip to explore the spam-crushing Akismet plugin.

Why Your WordPress Site Needs a Community

Once upon a time, people thought comments belonged only in personal blogs and discussion forums. Serious-minded web publishers ignored them. Small business avoided them—after all, if people really needed to get help or make their opinions known—well, that's what email was for, right?

Today, web commenting is an essential ingredient for sites small and large, fun and serious, casual and moneymaking. Here's what a comments section can do for you:

- **Attract new visitors.** New visitors immediately notice whether a website has a thriving conversation going on or just a single lonely comment. They use that to evaluate a website's popularity. It's crowd mentality, working for you—if new visitors see that other people find a topic interesting, they're more likely to dive in to check out your content for themselves.

- **Build buzz.** If you've taken to the web to promote something—whether it's a new restaurant, a book, a community service, or whatever—you can do only so

much to persuade people. But if you get your fans talking to other people, the effect is exponential. Comments help you spread the word, getting your readers to talk up your products or services. And once they're talking on your site, it's just a short hop away for other people to post about you on *their* sites.

- **Build loyalty.** A good discussion helps make a site *sticky*—in other words, it encourages people to return. People may come to your site for the content, but they stay for the comments.

- **Encourage readers to help other readers.** Often readers will want to respond to your content with their own comments or questions. If you ask them to do that by email and your site is popular, you readers will easily overwhelm you. But with comments, your audience can discuss among themselves, with you tossing in the occasional follow-up comment for all to see. The end result is that your site still has that personal touch, even when it's big and massively popular.

■ Allowing or Forbidding Comments

If you haven't changed WordPress's factory settings, all your posts have comments turned on. You've probably already noticed that when you view an individual post (but not a page), there's a large Leave a Reply section just below your content (*Figure 10-1*).

Allowing comments on posts isn't an all-or-nothing decision. As you'll soon see, you can pick and choose which content allows comments and which content doesn't. This is good, because it doesn't always make sense to allow comments on everything you publish. For example, you might want to disable comments if you write on a contentious subject that's likely to attract an avalanche of inflammatory or insulting feedback. News sites sometimes disable comments to avoid legal liability (for comments that might be libelous or might reveal information that's under a publication ban).

When it comes to pages, WordPress assumes you don't want a discussion. After all, you probably won't get a great conversation going on an About Us page, and comments on this sort of content can make a page look less professional. However, you don't need to follow WordPress's standard assumptions. If you want comments on a page, you can selectively enable them.

FIGURE 10-1

The exact appearance of the comment section depends on your theme. In Twenty Twenty, it appears when you view a full post, just under the links that let you skip to the previous or next post.

> **NOTE** Comments apply equally to posts and pages. For convenience, most of the discussion in this chapter refers to posts, but everything you'll learn applies equally to pages.

Changing Comment Settings for a Post

So how do you decide which posts and pages get comments and which don't? The answer is in the comment settings, which you can adjust whenever you create (or edit) a post or page. However, you might not have noticed these settings, because WordPress tucks them away in the sidebar.

Let's say you're editing a post. Here's what you need to do to change whether it allows comments:

1. **If the sidebar isn't visible, click the Settings (the gear) icon in the top-right corner of the post editor.**

2. **Click the Document tab in the sidebar, to see the settings that apply to your entire post (or page).**

3. **Expand the Discussion section (*Figure 10-2*).**

4. **If you want comments, make sure the Allow Comments checkbox is checked.**

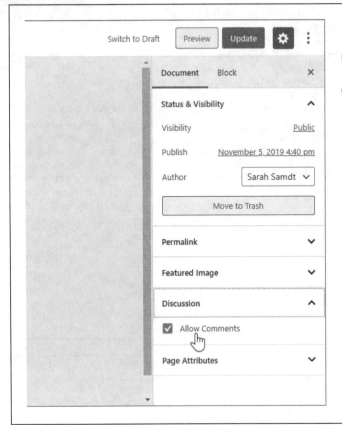

FIGURE 10-2

You can opt out of comments for a single post or page by turning off the Allow Comments checkbox.

This works perfectly well, but it's not the most efficient approach if you have a pile of posts you want to change. For example, let's say you have a dozen old posts that allow comments, and you want to remove the comment feature from all of them. To accomplish this task, you need to use the WordPress bulk editing feature:

1. **Choose Posts→All Posts to see all your posts.**

2. **Turn on the checkbox next to each post you want to change.**

3. **Click the Bulk Actions box and choose Edit.**

4. **Click the Apply button to show the edit panel with all the settings you can change.**

5. **Look for the Comments box. Change it to "Do not allow."**

6. **Click Update to make the change.**

You can use the same trick to turn commenting back on and to change the comment settings on your pages.

Changing the Default Comment Settings Site-Wide

To create a site that's mostly or entirely comment-free, you probably don't want to fiddle with the Discussion settings for every post. Instead, you should create a universal setting that applies to all new posts and pages:

1. **Choose Settings→Discussion in the administration menu.**

2. **Turn off the checkmark next to "Allow people to submit comments on new articles."**

3. **Scroll down to the bottom of the page and click Save Changes.**

Now all new posts and pages will be comment-free. You can add the comment feature back to specific posts by turning on Allow Comments in the Discussion box, as shown back in *Figure 10-2*.

Many more options in the Settings→Discussion page change the way comments work. You'll learn to use them in the rest of this chapter.

■ The Life Cycle of a Comment

The easiest way to understand how WordPress comments work is to follow one from its creation to the moment it appears on your site and starts a conversation.

Depending on how you configure your site, comments travel one of two routes:

- **The slow lane.** In this scenario, anyone can leave a comment, but you need to approve it before it appears on the post. You can grant an exemption for repeat commenters, but most people will find that the conversation slows down significantly, no matter how quickly you review new comments.

- **The fast lane.** Here, each comment appears on your site as soon as someone leaves it. However, unless you want your website drowned in thousands of spam messages, you need to use some sort of spam-fighting tool with this option—usually, it's an automated program that detects and quarantines suspicious-looking messages.

For most sites, the second choice is the best approach, because it allows discussions to unfold quickly, spontaneously, and with the least possible extra work on your part. But this solution introduces more risk, because even the best spam-catcher will miss some junk, or allow messages that aren't spam but are just plain offensive. For that reason, WordPress starts your site out on the safer slow lane instead.

In this chapter, you'll consider both routes. First, you'll learn the slow-lane approach. Then, when you're ready to step up your game with more powerful spam-fighting tools, you'll consider the fast-lane approach.

Leaving a Comment

Leaving a comment is easy, which is the point—the more convenient it is to join the conversation, the more likely your visitors are to weigh in.

Assuming you haven't tweaked any of WordPress's comment settings, visitors need to supply two pieces of information before they can make their thoughts known: their name and their email address. They can optionally include a website address too (*Figure 10-3*).

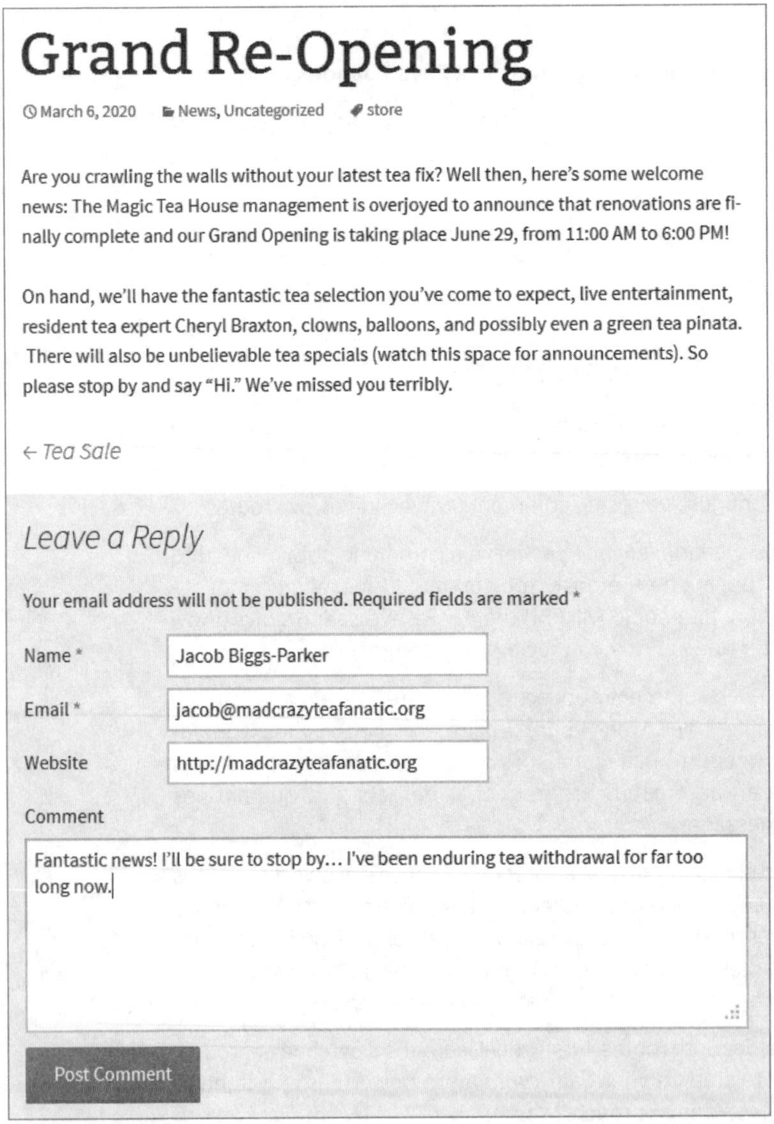

Grand Re-Opening

🕘 March 6, 2020   🏷 News, Uncategorized   🏷 store

Are you crawling the walls without your latest tea fix? Well then, here's some welcome news: The Magic Tea House management is overjoyed to announce that renovations are finally complete and our Grand Opening is taking place June 29, from 11:00 AM to 6:00 PM!

On hand, we'll have the fantastic tea selection you've come to expect, live entertainment, resident tea expert Cheryl Braxton, clowns, balloons, and possibly even a green tea pinata. There will also be unbelievable tea specials (watch this space for announcements). So please stop by and say "Hi." We've missed you terribly.

← *Tea Sale*

Leave a Reply

Your email address will not be published. Required fields are marked *

Name *   Jacob Biggs-Parker

Email *   jacob@madcrazyteafanatic.org

Website   http://madcrazyteafanatic.org

Comment

Fantastic news! I'll be sure to stop by… I've been enduring tea withdrawal for far too long now.

Post Comment

FIGURE 10-3

Ordinarily, a commenter needs to include a name and email address (although WordPress doesn't verify either). Optionally, commenters can include a website address or leave this box blank.

Here's the catch: if you're logged in to your website as the administrator, you won't see the commenting layout shown in *Figure 10-3*. Instead, you'll see just the box for comment text, because WordPress already knows who you are. This won't help you understand what life is like for ordinary readers, however, so before you go any further, you need to make yourself look like an ordinary internet visitor.

You could just log out (click "Log out" above the comment box), but then you'll need to log back in to moderate the comment. Another option is to go to the page from another computer or using a different browser.

> **TIP** An even more convenient option is to open the page in an "incognito" browser window. (In Chrome, open the menu and choose "New incognito window." In the Edge browser, choose "New InPrivate window." Every browser words this option a little differently.) Now your site will treat you like a stranger, and you'll see the same commenting boxes your visitors see.

Here's what WordPress does with the information it gets from commenters:

- **Name.** It displays the commenter's name prominently above the comment, thereby identifying them to other readers. The commenter can pick whatever name they want.

- **Email address.** WordPress doesn't display this publicly, so commenters don't need to worry about spam. In fact, WordPress won't stop visitors from inventing imaginary email addresses (although it will prevent them from typing in gibberish that obviously doesn't make sense). WordPress won't even send would-be commenters one of those pesky "Confirm this is your address" email messages. However, email addresses are important if you want to display a tiny picture of each commenter next to each comment (see page 296 for details).

- **Comment text.** This is the meat of the comment.

- **Website.** If your commenter includes this detail, WordPress turns the commenter's name, which appears above posts, into a link. Other readers can click it to travel to the commenter's site.

To see how comments work, try typing in one of your own. First, make sure you aren't logged in as the administrator. (If you are, you'll bypass the moderation process described below, because WordPress figures you'll always allow your own comments.) Assuming you're logged out and you see the text boxes shown in *Figure 10-3*, type in a comment and then click Post Comment.

Now, WordPress plays a slight trick on you. When you submit a comment, WordPress immediately adds it below your post (*Figure 10-4*), making it look as though your comment has been published. But in reality, when you use the slow-lane commenting route, no one can see the comment until the site owner (that's you) reviews it and formally approves it. This process is called *moderation*.

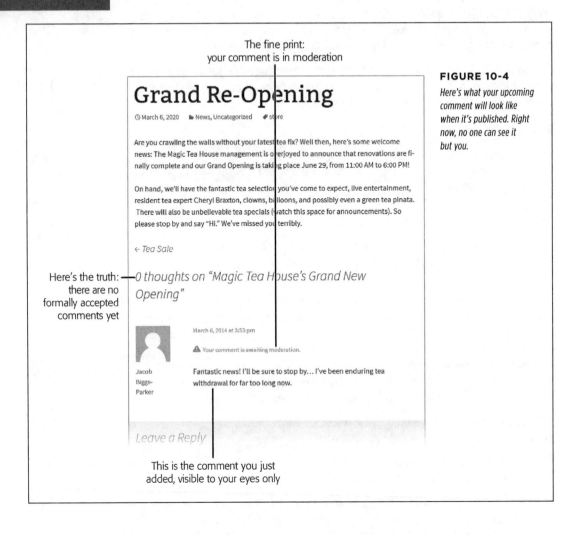

The fine print:
your comment is in moderation

Grand Re-Opening

Here's the truth:
there are no
formally accepted
comments yet

This is the comment you just
added, visible to your eyes only

FIGURE 10-4

*Here's what your upcoming
comment will look like
when it's published. Right
now, no one can see it
but you.*

Comments That Use HTML

Most people who comment on a post or page will type in one or more paragraphs of ordinary text. However, craftier commenters may include a few HTML tags to format their comments.

For example, you can use the *<b>* and *<i>* elements to bold and italicize text. Type this:

    I'm <i>really</i> annoyed.

and your comment will look like this:

I'm *really* annoyed.

You can also add headlines, line breaks, bulleted and numbered lists, and even tables. You could use the *<a>* element to create a link, but that's not necessary—if you type in text that starts with *www.* or *http://*, WordPress automatically converts it to a clickable link.

Now that you know you can use HTML in a comment, the next question is, *should* you? Most site owners don't mind the odd bit of bold or italic formatting, but they may trash messages that include shamelessly self-promotional links or ones that attempt to steal focus from the conversation with wild formatting—it's like an attention-starved kid throwing a grocery-store tantrum.

You can edit comments that use HTML inappropriately, but that takes time and effort. As a safeguard, some site owners don't allow HTML elements at all. You can ban HTML by creating a custom theme, an advanced task detailed in Chapter 13. Once you do, you need to edit its *functions.php* file (page 400) and add these instructions anywhere after the first line (which holds the *<?php* marker that starts the code block):

```
add_filter( 'comment_text',
    'wp_filter_nohtml_kses' );
add_filter( 'comment_text_rss',
    'wp_filter_nohtml_kses' );
add_filter( 'comment_excerpt',
    'wp_filter_nohtml_kses' );
```

Now WordPress strips out any HTML tags from comments and disables the linking capability of web addresses.

Moderating Comments Through Email

When a comment awaits moderation, the discussion on your site stalls. WordPress takes two steps to notify you of waiting comments:

- It sends you an email message, with information about the new comment (and the links you need to manage it).

- It adds an eye-catching number-in-a-circle icon to the Comments button in the admin menu, where you can manage all your comments.

These two actions underlie the two ways you moderate WordPress comments: either by email or through the admin area. First, you'll consider the email approach.

Email moderation is, for practical purposes, an option only for a small site that receives a relatively small number of comments. For each new comment, you get an email message that you can use to approve it or discard it (*Figure 10-5*). The notification comes mere minutes after the comment is left.

Email moderation is a great idea, but it's increasingly impractical for the websites of today. The problem is comment spam—advertisements for Viagra and Cialis, porn, shady discount deals, Nigerian princes who need help moving their fortunes, and so on. If you use email moderation, you'll receive an ever-increasing load of notifications as a host of suspicious characters try to insert their junk onto your pages. Not only is it difficult to manage the sheer number of messages you get, but it's often difficult to quickly verify that a message is legitimate, because spammers try to make their comments sound real. Often the only way to confirm that a comment is bogus is to visit the commenter's site, where you usually find ads unrelated to anything in the comment.

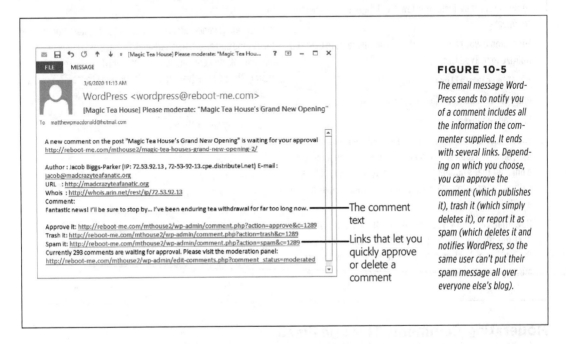

FIGURE 10-5

The email message Word-Press sends to notify you of a comment includes all the information the commenter supplied. It ends with several links. Depending on which you choose, you can approve the comment (which publishes it), trash it (which simply deletes it), or report it as spam (which deletes it and notifies WordPress, so the same user can't put their spam message all over everyone else's blog).

For these reasons, few people use email moderation to manage comments. You can try it, and it may work wonderfully at first, but you'll probably need to abandon it as more and more spammers discover your site, or you'll need to supplement it with one of the antispam plugins you'll learn about on page 309. That way, your plugin can take care of the massive amounts of obvious spam, while you concentrate on moderating the comments that make it past the spam filter.

NOTE Don't fall into the trap of thinking that you're safe because your audience is small. Most spammers don't target WordPress sites by popularity. Instead, they try to spread their junk everywhere they can. And their site-discovering techniques are surprisingly sophisticated. Even if you haven't told anyone about your site and you've configured it so it's hidden from search engines, you'll *still* get spam comments, usually within days of the site's creation. But here's the happy news: any plugin that blocks automated spam should reduce comment moderation to a manageable task.

WordPress comes with email moderation turned on. If you decide you don't want to be notified because you're receiving too many spam messages, you can easily switch it off. Choose Settings→Discussion, find the "Email me whenever" section, and clear the checkmarks next to "Anyone posts a comment" and "A comment is held for moderation."

FREQUENTLY ASKED QUESTION

Where Are My Emails?

I have the comment notification settings switched on, but I'm not getting any emails.

Ironically, email programs often misinterpret the notifications that WordPress sends as junk mail. The problem is that the messages contain quite a few links, which is a red flag suggesting spam. To find your missing messages, check your junk mail folder.

To avoid having your comment notifications identified as junk mail, tell your email program to always trust the address that sends them. The sending address is *wordpress* followed by your website domain, as in *wordpress@magicteahouse.net*.

Moderating Comments in the Admin Area

The other way to moderate comments is through the admin area. The disadvantage here is that you need to open a browser, visit your site, and log in. The advantage is that you'll see all your site's comments in one place, and you can accept or discard them en masse.

If you have comments awaiting moderation, you'll see a circle-with-a-number icon in the menu. This circle looks like the one that notifies you of WordPress and plugin updates (page 38), except that it appears over the Comments menu and indicates the number of unreviewed comments you have (*Figure 10-6*).

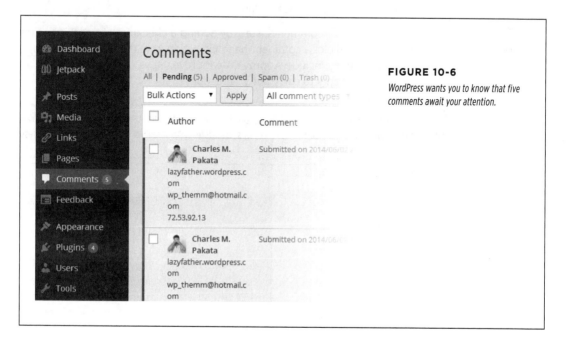

FIGURE 10-6

WordPress wants you to know that five comments await your attention.

To review comments, click Comments in the menu. Initially, you see a list of all the comments left on all the posts and pages of your site, ordered from newest to oldest. Click the Pending link above the comment list to focus on just the comments you need to review (*Figure 10-7*).

Here's what to do after you examine a comment:

- **If it's spam**, click the Spam link. Do *not* click Trash. Yes, both links remove the comment from your list, but only Spam reports the spammer to WordPress, which can help intercept the spam before it hits other sites.

- **If it's a valid comment**, click Approve to publish it. If the same person returns and posts another comment using the same email address, WordPress lets it through automatically, no moderation required. (This works because WordPress automatically turns on the "Comment author must have a previously approved comment" setting.)

- **If it's a valid comment that you don't want to allow**, click Trash. For example, if someone read your post and replied in an abusive manner, you don't need to publish the comment—it's up to you.

TIP Remember, if you accidentally put a comment you want in the Spam or Trash bin, you can still get it out. Click the Spam or Trash link above the comments list to see a list of removed comments, which you can then restore.

You don't need to deal with comments one at a time. You can use a handy bulk action to deal with multiple comments at once. This is particularly useful if you need to clear out a batch of suspicious-looking junk.

To deal with a group of comments, start by adding a checkmark to each one you want to process. Then pick a comment-handling action from the Bulk Actions dropdown list. Your options include Approve, Unapprove, Move to Trash, and Mark as Spam. Finally, click Apply to carry out your action.

TIP If you want to review comments on the go, consider trying out WordPress's mobile apps, which you can use on a smartphone or tablet. To learn more, visit *https://wordpress.org/mobile*, or just search for the WordPress app in your device's app store.

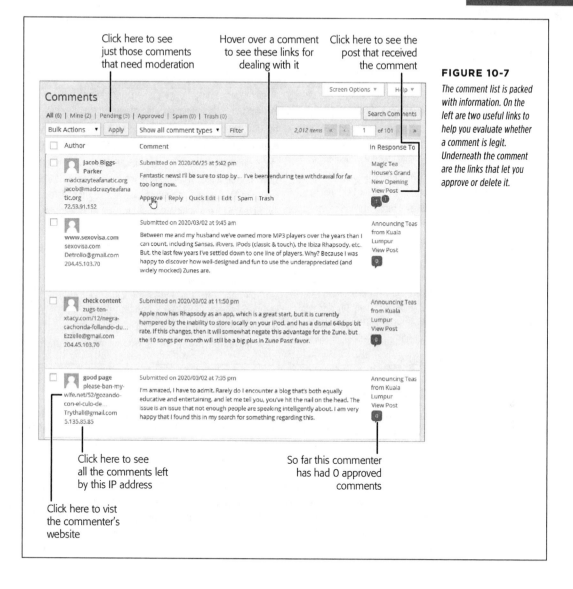

Click here to see
just those comments
that need moderation

Hover over a comment
to see these links for
dealing with it

Click here to see the
post that received
the comment

FIGURE 10-7

*The comment list is packed
with information. On the
left are two useful links to
help you evaluate whether
a comment is legit.
Underneath the comment
are the links that let you
approve or delete it.*

Click here to see
all the comments left
by this IP address

So far this commenter
has had 0 approved
comments

Click here to vist
the commenter's
website

Evaluating Comments

When you review comments, your goal is to separate the well-meaning ones from the offensive ones (which you may not want to allow) and to delete spam (which you definitely don't want). Be careful, because spammers are often crafty enough to add a seemingly appropriate comment that actually links to a spam site. They may identify keywords in your posts and cobble them together in their comments. They may report imaginary errors in your blog, claiming links don't work or pictures don't load. Often they'll throw in some flattery in a desperate attempt to get approved.

For example, in *Figure 10-7*, the last three comments are real spam comments, received on the actual Magic Tea House sample site. The second and third comments were posted together, and they appear to strike up a fictitious conversation. But the clues abound that something isn't right. The comments discuss a product that hasn't existed in years (Microsoft's Zune player) and has nothing to do with the post topic (teas from Kuala Lumpur). The fourth comment is a more typical example of spam: vague but effusive praise for the site that always manages to avoid stating anything specific.

The acid test for spam is to view the commenter's website. To do that, click the corresponding link (to the left of the comment in the comment list). But if you're in doubt and you suspect spam, it's safest not to click. Sometimes just looking at the URL is enough. In *Figure 10-7*, a careful examination exposes at least two of the spam comments as come-ons for X-rated websites.

Once you identify one spam message, you may be able to detect others sent from the same spammer by using the message's IP address (a numeric code that uniquely identifies web-connected computers). For example, in *Figure 10-7* two spam messages come from the same IP address (204.45.103.70). WordPress even gives you a shortcut—click the IP address, and it shows you only the comments that originated from that address. You can then flag them all as spam in a single bulk action.

Sanitizing Comments

By now, you're well acquainted with your role as supreme comment commander. Only you can decide which comments live to see the light of day, and which ones are banished to the trash or spam folder.

WordPress gives you one more power over comments that may surprise you. You can crack open any comment and edit it, exactly as though it were your own content. That means you can delete text, insert new bits, change the formatting, and even add HTML tags. You can do this by clicking the Edit link under the comment, which switches to a new page named Edit Comment, or you can edit it more efficiently by clicking the Quick Edit link, which opens a comment-editing text box right inside the list of comments.

You might use this ability to remove something objectionable from a comment before you allow it, such as profanity or off-site links. However, few site administrators have the time to personally review their readers' comments. Instead, they get WordPress to do the dirty work.

One way to do that is to use the Comment Moderation box. Choose Settings→Discussion and fill the box with words you don't want to allow (one per line). If a comment uses a restricted word, WordPress adds it to the list of comments that need your review, even if you approved an earlier comment from the same person, and even if you disabled moderation (page 281). However, mind the fact that WordPress checks not only whole words, but *within* words as well, so if you disallow *ass*, WordPress won't allow jack*ass* or *Ass*yria. If you want to be even stricter, you can use the Comment Blacklist box instead of the Comment Moderation box. You again provide a list of offensive words, but this time WordPress sends offending comments straight to your spam folder.

With the right plugin, you can use a gentler approach, one that *replaces* objectionable words but still allows the comment. For example, the WP Content Filter plugin (*http://tinyurl.com/wpcontentfilter*) changes words you don't want (like jackass) with an appropriately blanked-out substitution (like j******, j*****s or *******). Of course, crafty commenters will get around these limitations by adding spaces and dashes (jack a s s), replacing letters with similar-looking numbers or special characters (jacka55), or just using creative misspellings (jackahss). So if you have a real problem with inappropriate comments and can't tolerate them even temporarily (in other words, before you have the chance to find and remove them), then you need to keep using strict moderation on your site so you get the chance to review every comment before it's published.

The Ongoing Conversation

You've now seen how a single, lonely comment finds its way onto a WordPress post or page. On a healthy site, this small step is just the start of a long conversation. As readers stop by, more and more will add their own thoughts. And before long, some people will stop replying to your content and start replying to each other.

WordPress keeps track of all this in its *comment stream*, which is similar to the stream of posts that occupies your site's home page. WordPress sandwiches the comment stream between your content (the text of your post or page), which sits at the top, and the Leave a Reply box, which sits at the bottom. Unlike the post stream, the comment stream starts with older comments, followed by newer ones. This arrangement makes it easy to follow an unfolding conversation, where new comments refer to earlier ones.

TIP If you have lots of comments and want to emphasize the newest ones, you can flip the order, so that the newest comments appear first. Choose Settings→Discussion, find the setting that says "Comments should be displayed with the older comments at the top of each page," and pick "newer" instead of "older."

Threaded Comments

The most interesting part of the comment stream is the way it *threads* comments— it orders the comments that visitors post in reply to other comments. When new visitors read your post and join the conversation, they have two options: they can reply directly to your post by scrolling to the Leave a Reply section at the bottom of the page, or they can reply to one of the existing comments by clicking the Reply button (or link) next to the comment.

When a guest comments on another comment, WordPress puts the reply underneath the original note, indented slightly to show the relationship (*Figure 10-8*).

> **TIP** WordPress has a handy shortcut that lets you, the site owner, join a conversation straight from the admin area. When reviewing a comment on the Comments page, click the Reply link, fill in some text, and then click the Reply button (or "Approve and Reply" if you're responding to a comment you haven't approved yet).

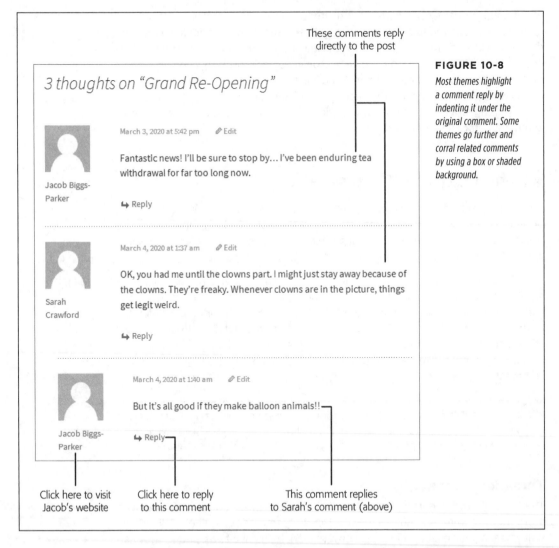

These comments reply directly to the post

3 thoughts on "Grand Re-Opening"

Jacob Biggs-Parker

March 3, 2020 at 5:42 pm   ✎ Edit

Fantastic news! I'll be sure to stop by... I've been enduring tea withdrawal for far too long now.

↳ Reply

Sarah Crawford

March 4, 2020 at 1:37 am   ✎ Edit

OK, you had me until the clowns part. I might just stay away because of the clowns. They're freaky. Whenever clowns are in the picture, things get legit weird.

↳ Reply

Jacob Biggs-Parker

March 4, 2020 at 1:40 am   ✎ Edit

But it's all good if they make balloon animals!!

↳ Reply

Click here to visit Jacob's website

Click here to reply to this comment

This comment replies to Sarah's comment (above)

FIGURE 10-8

Most themes highlight a comment reply by indenting it under the original comment. Some themes go further and corral related comments by using a box or shaded background.

If several people reply to the same comment, WordPress arranges the replies underneath the comment and indents them, either from oldest to newest (the standard) or newest to oldest (if you changed the discussion settings as described in the Tip on page 290).

Comment replies can go several layers deep. For example, if Sarah replies to your post, Jacob can reply to Sarah's comment, Sergio can reply to Jacob's comment, and then Sarah can reply to Sergio's reply, creating four layers of stacked comments (*Figure 10-9*).

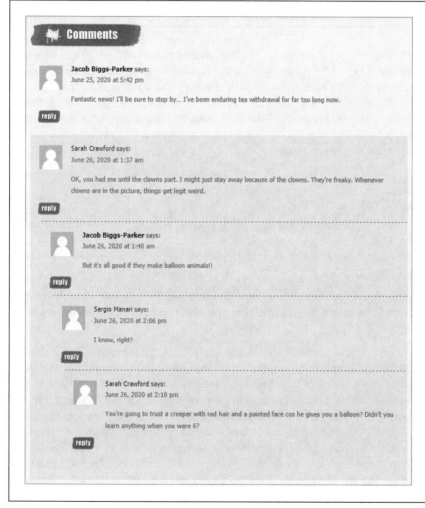

FIGURE 10-9

If you expect to get piles of comments, the Word-Press year themes might not be the best choice for you. They tend to spread comments out with plenty of whitespace in between, which makes for more visitor scrolling. Many other themes pack comments tightly together, like the Greyzed theme shown here.

WordPress allows this replying-to-replies madness to continue only so far; once you get five levels of comments, it no longer displays the Reply button. This prevents the conversation from becoming dizzyingly self-referential, and it stops the ever-increasing indenting from messing with your site's layout. However, you can reduce or increase this cap (the maximum is 10 levels) by choosing Settings→Discussion, finding the setting "Enable threaded (nested) comments 5 levels deep," and then picking a different number. Or turn off the checkmark for this setting to switch off threaded comments altogether, which keeps your conversations super-simple, but looks more than a bit old-fashioned.

Author Comments

Don't forget to add your voice to the discussion. Authors who never take the time to directly engage their readers lose their readers' interest—quickly.

Of course, it's also possible to have too much of a good thing, and authors who reply to every comment will seem desperate (at best) or intrusive (at worst). They'll suffocate a conversation like a micromanaging boss. The best guideline is to step in periodically, answering obvious questions and giving credit to good feedback (while ignoring or deleting the obvious junk). Do that, and your readers will see that your comments section

is well cared for. They'll know that you read your feedback, and they'll be more likely to join in.

WordPress makes site owners' comments stick out from those of the riffraff so your readers can easily spot your contributions. The way it does so depends on the theme, but most change the background color behind your comment. If you're willing to edit your theme, you can make your replies even more obvious. The trick is to tweak the formatting that the *bypostauthor* style applies. Page 412 explains how.

Paged Comments

WordPress provides a comment-organizing feature called *paging* that divides masses of comments into separate pages. The advantage is that you split awkwardly long discussions into more manageable (and readable) chunks. The disadvantage is that readers need to click more links to follow a long discussion.

To use pages, choose Settings→Discussion and then turn on the checkbox next to "Break comments into pages." You can type in the number of comments you want included on each page (the factory setting is 50).

You can also choose the page that readers begin on—the standard setting is "last," which means that new readers will start on the last page of comments first, seeing the most recent chunk of the conversation before they see older exchanges. But the overall effect is a bit weird, because the very latest comment appears at the *bottom* of the first page. What you probably want is the latest comment to appear at the *top* of that page. To get this effect—paged comments, with the most recent comment at the top of the list on the first page—change "last" to "first" (so the setting shows "and the first page displayed by default") and change "older" to "newer" (so the setting shows "Comments should be displayed with the newer comments at the top of each page").

Advertising a Post's Comments

As you've seen, comments appear right underneath the post they refer to. They don't appear at all in the reverse-chronological list of posts that acts as the home page for most WordPress sites. You can think of it this way: each post is like a separate room at a party, with its own conversation. The same guests can wander between rooms and join different conversations, but the conversation from one room doesn't intrude on the conversation in the next.

However, WordPress does attempt to alert readers to the *presence* of comments in the post list, if not their content. If a post has at least one comment, WordPress shows the comment *count* next to that post. If a post doesn't have any comments, WordPress displays a link that says "No comments" or "Leave a comment" or something slightly different, depending on your theme (*Figure 10-10*).

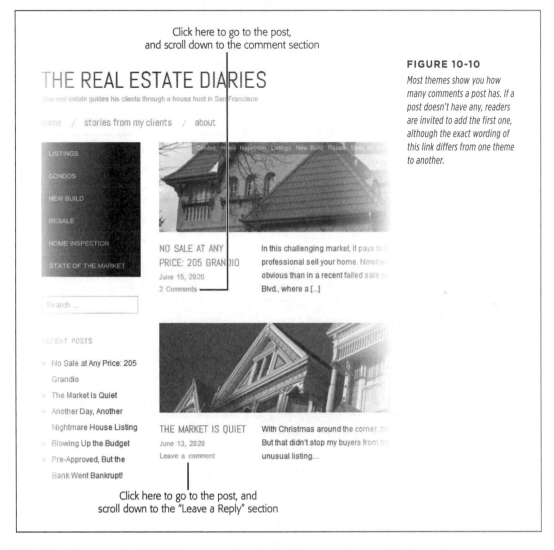

Click here to go to the post,
and scroll down to the comment section

THE REAL ESTATE DIARIES

One real estate guides his clients through a house hunt in San Francisco

home / stories from my clients / about

LISTINGS
CONDOS
NEW BUILD
RESALE
HOME INSPECTION
STATE OF THE MARKET

Search ...

RECENT POSTS

- No Sale at Any Price: 205 Grandio
- The Market Is Quiet
- Another Day, Another Nightmare House Listing
- Blowing Up the Budget
- Pre-Approved, But the Bank Went Bankrupt!

NO SALE AT ANY PRICE: 205 GRANDIO
June 15, 2020
2 Comments

In this challenging market, it pays to professional sell your home. Nowher obvious than in a recent failed sale Blvd., where a [...]

THE MARKET IS QUIET
June 13, 2020
Leave a comment

With Christmas around the corner, th But that didn't stop my buyers from tr unusual listing...

Click here to go to the post, and
scroll down to the "Leave a Reply" section

FIGURE 10-10

Most themes show you how many comments a post has. If a post doesn't have any, readers are invited to add the first one, although the exact wording of this link differs from one theme to another.

Here's another way to highlight comments on your home page: use the Recent Comments widget, which highlights the most recent comments made on *any* post or page (*Figure 10-11*). When you add this widget, you can choose the number of recent comments it lists. The standard setting is 5.

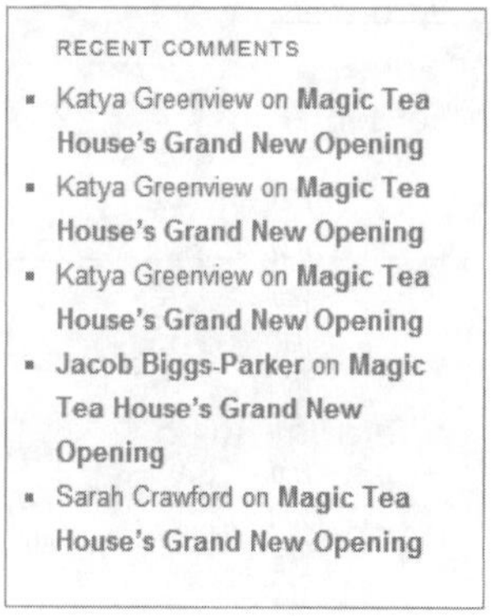

RECENT COMMENTS

- Katya Greenview on **Magic Tea House's Grand New Opening**
- Katya Greenview on **Magic Tea House's Grand New Opening**
- Katya Greenview on **Magic Tea House's Grand New Opening**
- Jacob Biggs-Parker on **Magic Tea House's Grand New Opening**
- Sarah Crawford on **Magic Tea House's Grand New Opening**

FIGURE 10-11

The Recent Comments widget tells you who's commenting on what post. However, it doesn't show you any of the comment content, which is a shame. Readers can click a comment link to see both the comment and the corresponding post.

Linkbacks

There's one type of comment you haven't seen yet: the *linkback*, a short, automatically generated comment that lets you know when somebody is talking about your post. *Figure 10-12* shows what a linkback looks like—but be warned, it's not particularly pretty.

> **NOTE** Linkbacks *are* comments. They appear in the comment list and need your approval before WordPress publishes them, just as any other comment does.

The interesting thing about linkback comments is that WordPress creates them *automatically*. Here's how the linkback in *Figure 10-12* came to be:

1. First, you published the "Community Outreach Fridays" post on the Canton School site.

2. Then, the Time for Diane site created the "Fun at Glenacres Retirement" post. Although it isn't shown in *Figure 10-12*, that post included a link to your "Community Outreach Fridays" post.

3. When the Time for Diane site published the "Fun at Glenacres Retirement" post, the site sent a notification to the Canton School site, saying "Hey, I linked to you" in computer language. (The person who wrote the "Fun at Glenacres Retirement" post doesn't need to take any action, and probably doesn't even know that a notification is being sent.)

4. On the Canton School site, WordPress springs into action, adding the linkback comment shown in *Figure 10-12*.

NOTE Linkbacks aren't a WordPress-only feature. Many web publishing platforms support them, and virtually all blogs can send linkback notifications and add linkback comments.

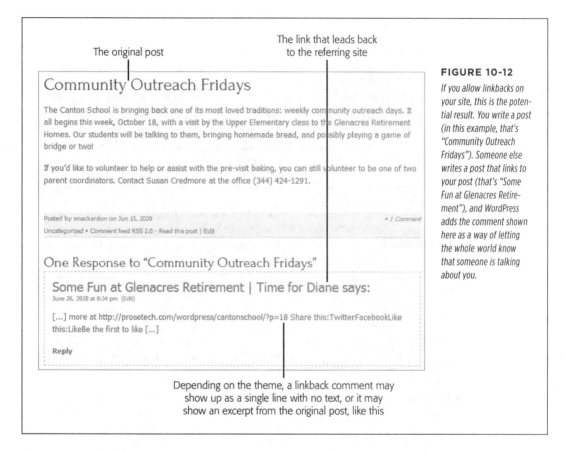

The original post

The link that leads back
to the referring site

Community Outreach Fridays

The Canton School is bringing back one of its most loved traditions: weekly community outreach days. It all begins this week, October 18, with a visit by the Upper Elementary class to the Glenacres Retirement Homes. Our students will be talking to them, bringing homemade bread, and possibly playing a game of bridge or two!

If you'd like to volunteer to help or assist with the pre-visit baking, you can still volunteer to be one of two parent coordinators. Contact Susan Credmore at the office (344) 424-1291.

Posted by smackerdon on Jun 15, 2020   • 1 Comment
Uncategorized • Comment feed RSS 2.0 - Read this post | Edit

One Response to "Community Outreach Fridays"

Some Fun at Glenacres Retirement | Time for Diane says:
June 26, 2020 at 8:34 pm (Edit)

[...] more at http://prosetech.com/wordpress/cantonschool/?p=18 Share this:TwitterFacebookLike this:LikeBe the first to like [...]

Reply

Depending on the theme, a linkback comment may
show up as a single line with no text, or it may
show an excerpt from the original post, like this

FIGURE 10-12

If you allow linkbacks on your site, this is the potential result. You write a post (in this example, that's "Community Outreach Fridays"). Someone else writes a post that links to your post (that's "Some Fun at Glenacres Retirement"), and WordPress adds the comment shown here as a way of letting the whole world know that someone is talking about you.

The purpose of linkbacks is twofold. First, they show your readers that people are seeing and discussing your content, which makes it seem more popular and more relevant. Second, it provides your readers with a link to the post that mentioned your post. That means readers on your site (say, Canton School) can click a linkback comment to head to the referring post on the other site (Time for Diane). In an ideal world, this is a great way to network with like-minded sites.

In the distant past, a certain faction of bloggers cared dearly about linkbacks and saw them as an important community-building tool. Nowadays, popular opinion has shifted. People don't use them as much, and many sites don't allow them at all. Here are some reasons you might not want to allow linkbacks on your site:

- **Clutter.** Extra comments, no matter how brief, can end up crowding out real conversation. Some themes (like Bueno) separate linkbacks from the main comment stream, but most mix them together. If you have a popular topic that gets plenty of mention on other sites, your linkbacks can split up the more interesting human feedback and push it out of sight.

- **Why risk spam?** More comments equals more spam, and shady advertisers can send linkbacks to your site just as often as they send other types of comment spam.

- **Links to reward your commenters.** If someone writes a good comment, they can include a link in their comment text ("I was frustrated with the stains my kids left on everything, so I wrote a post with my favorite stain tips in it. Check it out at *http://helpfatheroftwelve.com*."). And if the commenter included their website address in the Leave a Reply form (page 280), WordPress automatically turns the username at the top of the comment into a clickable link. With all this intra-post linking going on, why reward someone who hasn't even bothered to comment on your site with a linkback?

> **NOTE** In short, most people find that linkbacks aren't worth the trouble. To disable them, choose Settings→Discussion and remove the checkmark next to the setting "Allow link notifications from other blogs (pingbacks and trackbacks)." Technically, WordPress supports two linkback mechanisms: pingbacks and trackbacks. The technical details about how pingbacks and trackbacks send their messages aren't terribly interesting. The important thing is that if you allow linkbacks (and, unless you change the factory settings, your site does), you may start getting comments like the one in *Figure 10-12*.

Optionally, you can clear the checkmark next to the setting "Attempt to notify any blogs linked to from the article." When this setting is on and you write a post that links to another post on someone else's site, WordPress automatically sends a notification to that site, and its administrator can choose whether to display the linkback.

> **NOTE** Oddly enough, if you have the "Attempt to notify any blogs linked to from the article" setting switched on, WordPress notifies even *your own* site if you create a post that includes a link to one of your other posts. It creates a linkback comment in the initial post that points to the referring post, just as though the posts were on two different sites. (Of course, you're free to delete this comment if it bothers you.)

■ Making Comments More Personal with Gravatars

On a really good website, you won't feel like you're debating current affairs with anonymous_guy_65. Instead, you'll have the sense that you're talking to an actual person who exists in the real world, beyond the pixels on your screen.

Often, all you need to do to personalize comments is include a few small details in the right places. One key enhancement is including a user-supplied profile picture with that person's comment. WordPress gives you two ways to do that—you can get pictures from its handy Gravatar service, or you can take them from a person's

Facebook or Twitter account. The Facebook and Twitter support needs the Jetpack plugin (and we'll get to that a bit later in this chapter). But first, let's take a closer look at WordPress's built-in support for Gravatar.

The Gravatar Service

To give comments a personal touch, you can display a tiny picture next to each person's thoughts. This picture, called an *avatar,* could be an actual photograph of the person or something quirkier, like a mythical creature or cartoon character the person has chosen to represent them. The idea is that the avatar helps your guests see, at a glance, which comments belong to the same person, and it just might give them a taste of the author's personality (*Figure 10-13*). Avatars also add a visual complement to web discussions, making a page of comments seem just a bit more like a real conversation.

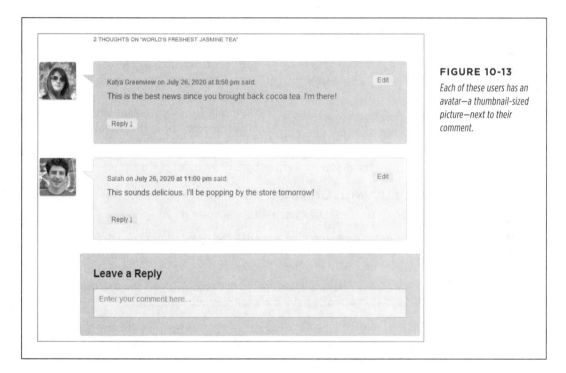

FIGURE 10-13

Each of these users has an avatar—a thumbnail-sized picture—next to their comment.

WordPress uses an avatar service called Gravatar, which is short for "globally recognized avatar." The idea is that ordinary people can use Gravatar to set up an avatar and include some basic personal information. They can then use that image and profile info on sites throughout the web. Originally, Gravatar was a small service cooked up by a single person, but these days Automattic runs the service, making it freely available to virtually any blogging platform or website-building framework. A Gravatar-supplied avatar goes by the name *gravatar*.

You don't need to take any special steps to enable avatars; WordPress uses them automatically. As you already know, every would-be commenter has to enter an email address. When they do, WordPress contacts the Gravatar service and asks if it has a picture affiliated with that address. If it does, WordPress displays the picture next to the comment. If it doesn't, WordPress shows a featureless gray silhouette instead.

Why Gravatars Make Good Sense

The obvious limitation to gravatars is that you won't see personalized images unless your readers sign up with the Gravatar service. And unless your visitors are web nerds, they probably haven't signed up yet—in fact, they probably haven't even heard of Gravatar.

However, this dilemma isn't as bad as it seems, for the following reasons:

- **Gravatars are optional.** Some people use them; others don't. There's no downside to allowing gravatars on your site. And if someone notices that another commenter gets a personalized picture, that person just might ask about how to get the same feature.

- **Gravatars can be autogenerated.** As page 300 explains, you can replace the boring gray silhouettes for non-

Gravatar users with an autogenerated gravatar. The neat thing about autogenerated gravatars is that they're unique and consistent, which means they can help people identify comments left by the same person.

- **Gravatar can coexist with Facebook and Twitter pictures.** As you'll learn later in this chapter, you can get comment pictures from Facebook and Twitter accounts. In this case, Gravatar is just one more picture-gathering option that works in harmony with the others.

- **You can remind your readers to get a gravatar.** For example, you can edit the *comments.php* file in your theme (page 419) to add a reminder, like a link that says "Sign up for a Gravatar and get a personalized picture next to your comment." Just don't expect that many people will follow your recommendation.

Signing Up with Gravatar

If *you* aren't a Gravatar user yet, here's how you sign up:

1. **Go to *http://gravatar.com* and click the Create Your Own Gravatar button.**

   The sign-up page appears.

2. **If you already have a WordPress.com account, you can use that with Gravatar. Click "Already have a WordPress.com account."**

   For example, you might have already signed up to get the Jetpack plugin described on page 256. If you have a WordPress.com account, you'll need to fill in your email address and password to link everything up. Then skip to step 4.

3. **If you don't have a WordPress.com account, you can sign up now. Fill in your email address, pick a username and a password, and then click the "Sign up" button.**

   When you get the confirmation email, click the activation link inside to complete the sign-up.

4. **If you haven't already done so, sign in to Gravatar with your email address and password.**

You arrive at the Manage Gravatars page, which informs you that you don't yet have any images associated with your account.

5. **Click the "Add a new image" link.**

   Gravatar gives you a number of ways to find an image. You can upload it from your computer's hard drive (click "Upload new"), which is good if you have a nice photo on hand or one of those bespoke Bitmoji cartoons that everyone loves so much. Alternatively, you can snag it from a website address (From URL), or snap a new one from your computer's webcam, if you have one (From webcam).

6. **Click the appropriate button for your image (for example, "Upload new") and follow the instructions to find and crop your picture.**

   Gravatars are square. You can use an image as big as 512 × 512 pixels, and Gravatar will shrink it down to a thumbnail-size tile and display it next to each comment you leave.

7. **Choose a rating for your gravatar (see *Figure 10-14*).**

   Ordinarily, WordPress sites show only gravatars that have a G rating. If you want to tolerate more friskiness on your site, go to Settings→Discussion. Scroll to the Avatars section and ratchet up the Maximum Rating setting to PG, R, or X.

Choose a rating for your Gravatar

Site owners are given the option to choose how mature a gravatar may be before allowing it to appear on their site. Set a rating of more than G if your gravatar may not be suited for younger or sensitive audiences.

FIGURE 10-14

Some sites may not display gravatars that are mildly naughty (PG), violent or sexually explicit (R), or over-the-top disturbing (X). It's up to you to pick the rating that best represents your image, but if you use an ordinary headshot, G is the right choice.

○ **rated G** This gravatar is suitable for display on all websites with any audience type.

○ **rated PG** This gravatar may contain rude gestures, provocatively dressed individuals, the lesser swear words, or mild violence.

○ **rated R** This gravatar may contain such things as harsh profanity, intense violence, nudity, or hard drug use.

○ **rated X** This gravatar may contain hardcore sexual imagery or extremely disturbing violence.

Set Rating

TIP   Are you concerned about inappropriate gravatars? You can disable gravatars altogether from the Settings→Discussion page. In the Avatars section, turn on the "Don't show Avatars" radio button.

8. **Now Gravatar associates your avatar with your email address.**

   All new comments you leave will include your new picture, and comments you already left will get it too (assuming you haven't changed your email address since you posted the comment). If, in the future, you decide you want a different picture, log back into Gravatar and upload a new one.

Changing the "Mystery Man" Gravatar

Ordinarily, if a commenter doesn't have a gravatar, WordPress displays the infamous gray silhouette that it calls Mystery Man. You can replace Mystery Man with one of several other pictures from the Settings→Discussion page. Scroll down to the Avatars section and change the Default Avatar option.

The alternate possibilities include no image at all (select Blank from the Default Avatar list) or a stock Gravatar logo (select Gravatar Logo). More interestingly, you can give mystery commenters a tailor-made, unique gravatar (for your site only). WordPress creates it by taking your guest's email address, using it to generate semi-random computer gibberish, and then translating that into a specific type of picture. You can choose from four autogenerated gravatar types: Identicon (geometric patterns), Wavatar (cartoon-like faces), MonsterID (whimsical monster drawings), and Retro (video-game-style pixelated icons). *Figure 10-15* shows two examples.

FIGURE 10-15

Algorithmically generated gravatars add fun to your site, even if your readers don't have real profile pictures. Here are two examples: Wavatar (left) and Retro (right). Notice that Sarah Crawford's gravatar remains consistent for both her comments.

Gravatar Hovercards

The tiny comment pictures that Gravatar provides add a personal touch to your comments section, but the service can provide more than just pictures. It can also smuggle in a bit of personal information about each commenter. This information shows up as a *hovercard*—a small box that pops up when someone points to an avatar (*Figure 10-16*).

FIGURE 10-16

A hovercard is like a virtual business card. It displays your personal information, no matter what Gravatar-enabled site you visit.

But there's a catch: hovercards appear only if your site is using Jetpack (the useful free plugin you learned to install on page 256). If you've installed and activated Jetpack, you can enable hovercards like this:

1. **Choose Jetpack→Settings.**

2. **Click the Discussion tab.**

3. **Switch on the option "Enable pop-up business cards over commenters' Gravatars."**

Hovercards are a small but nice feature. They help readers learn a little bit about your commenters. You might assume that the hovercard details are part of your visitor's WordPress profile, but they're not. (In fact, hovercards work even if guests don't have a WordPress account.) Instead, hovercards get their information from the profile that Gravatar users can optionally set up.

This design is both good and bad. The advantage is that it makes hovercard information portable—it travels with the avatar, no matter what Gravatar-enabled site you visit (even if the site *doesn't* run WordPress). The disadvantage is that if your readers don't bother to fill out the Gravatar profile information, hovercards won't appear at all (*Figure 10-17*).

To make sure *your* hovercard looks good, you need to fill in the profile information too. Visit the Gravatar site (*http://gravatar.com*), click the My Account button, and then choose Edit My Profile. There's plenty of information you can fill in, but the details that appear on the hovercard are your full name (Display Name), where you live (Location), and a short blurb that describes yourself (About Me), which the hovercard truncates after the first couple of sentences. When you finish, click Save Profile to store your information with your Gravatar, allowing it to appear on hovercard-supporting sites everywhere.

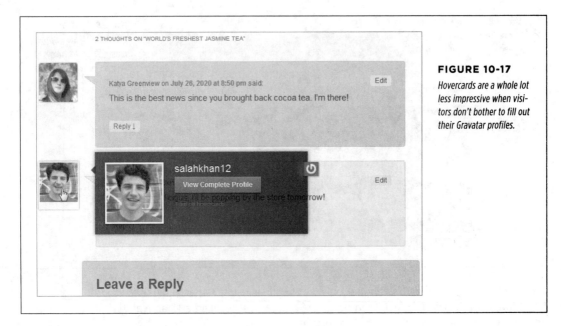

FIGURE 10-17

Hovercards are a whole lot less impressive when visitors don't bother to fill out their Gravatar profiles.

■ Using Facebook and Twitter Comments

Gravatar works well because it's an optional enhancement. If your visitors use Gravatar, they get a better commenting experience. If not, your site still works.

However, if *no one* who visits your site has a gravatar, you're stuck with either the useless Mystery Man image or one of the quirky autogenerated gravatars. This isn't much of an improvement. Fortunately, another option lets people take their carefully cultivated social media identity from Facebook or Twitter, and use that when they post comments on your site. The only drawback is that it isn't part of the basic WordPress package.

To get support for social comments, you can use one of several popular plugins. The current most popular plugin is Super Socializer (*https://tinyurl.com/super-s*), which supports 10 social platforms and has plenty of foofy frills. But if you're looking for something a little less fancy that just works, you can try out the built-in social comment support in Jetpack, which you'll see in the next section.

Social Logins with Jetpack

When you use Jetpack's social sign-in, users choose the social network they want to use and log in. There are four options: Facebook, Twitter, Google, or WordPress. com. When a signed-in person posts a comment, WordPress grabs the guest's profile picture and displays it next to the comment (*Figure 10-18*).

FIGURE 10-18

In this example, Charles Pakata is a WordPress. com user who has signed up with the Gravatar service. But Lisa and Rakesh are Facebook users. As long as they log in with Facebook, WordPress uses their Facebook profile pictures, without forcing them to sign up with Gravatar or take any extra steps.

To enable social comments with Jetpack, first make sure it's installed and activated, and then follow these steps:

1. **Choose Jetpack→Settings.**

2. **Click the Discussion tab.**

3. **Switch on the option "Let visitors use a WordPress.com, Twitter, Facebook, or Google account to comment."**

The new social-powered Leave a Reply section completely replaces the old one (*Figure 10-19*). It looks similar, but there are differences in the way the text boxes are sized and the way the Post Comment button is styled. You'll also see the social icons that let visitors sign in, and a couple of additional notification options that let people keep track of new comments. (You'll learn how to use or hide these options in Chapter 12.)

TIP You might find that once you enable Jetpack comments, your comment section gets a new background that doesn't blend in with the rest of your page. To fix this, choose Settings→Discussion, scroll down to the Jetpack Comments section, and try different options under Color Scheme. You can pick Light, Dark, or Transparent; finding the best fit is a trial-and-error process.

Use these boxes to enter
a comment the normal way

Use these icons to log in
to a social network

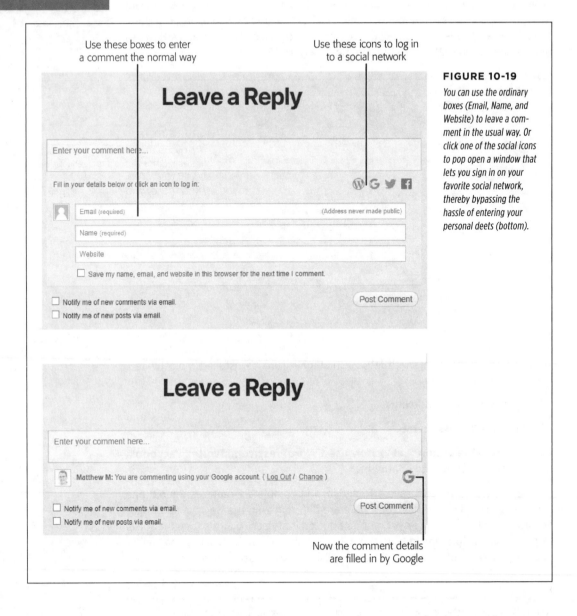

FIGURE 10-19

You can use the ordinary boxes (Email, Name, and Website) to leave a comment in the usual way. Or click one of the social icons to pop open a window that lets you sign in on your favorite social network, thereby bypassing the hassle of entering your personal deets (bottom).

Now the comment details
are filled in by Google

Requiring Social Logins

Some people turn on social comments and enable the "Users must be registered and logged in to comment" setting (which you can find at Settings→Discussion). This creates a site that *requires* commenters to provide a social identity. When a site owner takes this step, they're usually thinking something like this:

> I've been flexible, and now I want something in return. I've given my readers several good options for establishing their identity (Facebook, Twitter, and Google). By making them use one, I can lock out spammers and force people to bring their virtual identities to my site.

Think carefully before you take this step. First, it only partly protects your site against spam, because many spambots have fake Facebook identities. Second, it guarantees that you'll scare away at least some potential commenters, including those who don't have a social media account, those who can't be bothered to log in, and those who don't want to reveal their social identities to you.

▮ Stamping Out Comment Spam

So far, you've focused on the comments that are *supposed* to be on your site—the ones your visitors leave in response to your posts. Up to now, this discussion has skirted a disquieting fact: on the average WordPress site, spam comments outweigh legit comments by a factor of 10 to 1. And spammers don't discriminate—they don't attempt to chase the most popular blogs or the ones that cover their favorite topics. Instead, they spew their dreck everywhere.

Understanding Spam

You're no doubt familiar with the idea of *email spam*—trashy chain letters and hoaxes that try to get you to download malware or send your banking information to a Nigerian gentleman with a cash flow problem. *Blog spam* is a different creature altogether. While email spam tries to lure you in, blog spam tries to slip right past you. That's because blog spammers aren't after you—they're targeting your *readers.* The goal is for them to sneak their advertisements onto your site, where they can attract the attention of people who already trust your blog. Every bit of blog spam is trying to lure a reader to travel to the spammer's site, either by clicking the commenter's name or a link in the comment text.

In the past, spammers were crude and their messages easy to identify. Today, they're trickier than ever. They attempt to disguise themselves as actual readers to fool you into allowing their comment (with its link to their site). Or they pretend to sell real products (which they never deliver).

Some WordPressers tell horror stories of receiving hundreds or thousands of spam messages a day. The problem is severe enough that, if you're not careful, you can wind up spending more time dealing with spam than managing the rest of your site. Fortunately, you can use the tools and strategies discussed in the next sidebar to fight back.

Caught in the Wild

Spammers take great care to make their messages look as natural as possible. The spammer's payload is a link, which is submitted with the comment and hidden behind the commenter's name.

Here are some of the spam messages that we caught on this book's example sites. Would any have fooled you?

"Glad to know about something like this."

"Perhaps this is one of the most interesting blogs that i have ever seen. interesting article, funny comment. keep it up!"

"i was exactly talking about this with a friend yesterday, and now i found about it in your blog. this is awesome."

"Could you tell me when you're going to update your posts?"

"hello"

"I've also been thinking the identical thing myself lately. Grateful to see another person on the same wavelength! Nice article."

"We're a bunch of volunteers and opening a brand new scheme in our community. Your site offered us with valuable info to paintings on. You have done an impressive job and our whole community can be grateful to you."

Spam-Fighting Strategies

You can defend against spam in several ways:

- **Forbidding all comments.** This is obviously a drastic, ironclad approach. To disable comments, you turn off the "Allow people to post comments on new articles" checkbox on the Settings→Discussion page. But be warned that if you do, you'll sacrifice the lively conversation your visitors expect.

  Verdict: An extreme solution. The fix is worse than the problem.

- **Using moderation.** This is the default WordPress approach, and it's the one you learned about in this chapter. The problem is that you just can't keep moderating a site that's growing in size and popularity—it becomes too labor-intensive. It also has a distinct drawback: it forces new commenters to wait before their comment appears on your site, by which point they may have lost interest in the conversation.

  Verdict: Not practical in the long term, unless you combine it with a spam-catching tool (like Akismet, which you'll meet in a moment).

- **Forcing commenters to log in.** To use this approach, you need to add an account for each visitor to your WordPress sites. This approach definitely isn't suitable for the average public blog. However, it may work if you have a small, captive audience—for example, if you're building a site for family members only, or for a team of coworkers.

  Verdict: For special cases only. You'll learn about multiuser blogs in Chapter 11.

- **Making commenters log in, but allowing third-party logins.** A third-party login verifies your guests through an authentication service—for example, one provided by Facebook or Twitter. This requirement may work, because many people already have Facebook or Twitter accounts that they don't mind using (whereas they definitely won't bother creating a new account just to leave a single comment). Still, forcing logins may drive away as many as half of your would-be commenters. And it's still not truly spam-proof, because clever spambots can create Facebook accounts, just as real people can.

  Verdict: A good idea, but not a complete spam-fighting solution.

- **Using Akismet or another spam-catching plugin.** Many WordPress administrators swear that their lives would not be livable without the automatic spam-detecting feature of Akismet. It isn't perfect—some site owners complain that legitimate comments get trashed, and they need to spend serious time fishing them out of the spam bucket—but it usually gives the best spam protection with the minimum amount of disruption to the commenting process.

  Verdict: The best compromise for most people. It's also essential if you turn off moderation.

The pros and cons of managing comments by moderation versus spam-fighting are a lot to digest, even for seasoned webheads. But the evidence is clear: most WordPress pros eventually start using a spam-catching tool. They may use it in addition to moderation, or—more likely—instead of it.

NOTE If you don't have a spam filter, *you* are the spam filter. And given that an ordinary WordPress site can attract dozens of spam messages a day, you don't want to play that role.

If you're ready to ditch comment moderation in favor of a livelier, more responsive, and less controlled discussion, choose Settings→Discussion and turn off the checkboxes next to these settings: "An administrator must always approve the comment" and "Comment author must have a previously approved comment." Then click Save Changes at the bottom of the page.

Now continue to the next section to make sure you have a proper spam-blocker in place.

WordPress's Other Spam-Catching Options

WordPress has a few built-in spam-fighting options on the Settings→Discussion page. In the past, they were a practical line of defense that could intercept and stop a lot of junk comments. Unfortunately, spamming evolved in the intervening years, and now these settings are only occasionally useful. They include the following:

- **"Hold a comment in the queue if it contains 2 or more links."** Use this setting to catch posts that have a huge number of links. The problem is that spammers are on to this restriction, so they've toned down their links to make their spam look more like real comments.

- **The Comment Moderation and Comment Blacklist boxes.** Try these boxes, described earlier (page 288), as a way to keep out offensive text. They also double as a way to catch spam. However, don't rush to put in obvious spammy keywords, because you'll just end up doing a clumsier version of what Akismet already does. Instead, consider using these boxes if you have a spam problem that's specific to your site—for example, a certain keyword that keeps coming up when spammers target your posts.

- **"Automatically close comments on articles older than 14 days."** Unless you set it, this option isn't switched on. However, it's a potentially useful way to stop spammers from targeting old posts, where the conversation has long since died down. And you don't need to stick to the suggested 14 days. You can type in any number, even making the lockout period start a year after you publish a post. With fewer posts accepting comments, there's less potential spam for you to review.

Understanding Akismet

Akismet is one of many spam-fighting plugins developers created for WordPress. However, it has a special distinction: Automattic, the same folks who built WordPress, makes it.

Akismet works by intercepting each new comment. It sends the details of that comment (including its text and the commenter's website, email, and IP addresses) to one of Akismet's web servers. There, the server analyzes it, using some crafty code and a secret spam-fighting database, to attempt to determine whether it's legitimate. Any one of a number of details can betray a spam message, including links to known spam sites, a known spammer IP address, phrases commonly found in spam messages ("free Viagra" for instance), and so on. Akismet quickly makes its decision and reports back to your website. Your site then either publishes the comment or puts it in the spam folder, depending on Akismet's judgment.

WordPress experts report that Akismet's success rate hovers at around 97%. Usually, when Akismet errs, it does so by flagging a safe comment as spam (rather than allowing real spam through). However, Akismet's success depends on the site and the timing. When spammers adjust their tactics, it may take Akismet a little time to catch up, during which its accuracy will drop.

Akismet is free, mostly. Personal sites pay nothing (unless you volunteer a small donation). However, small businesses and moneymaking blogs are expected to contribute $5 per month. Large publishers that want to spam-proof multiple sites are asked for $50 a month.

NOTE Akismet uses an honor system, and plenty of sites earn a bit of money but don't pay the Akismet fee. If you want a totally free business-friendly solution, you need to find a different plugin. Several good alternatives are described in the following box.

FREQUENTLY ASKED QUESTION

Akismet Alternatives

I need a spam-catching tool, but I don't want Akismet. Are there other options?

There's no shortage of spam-fighting plugins. Unlike Akismet, many are free for almost everyone. (Some plugin developers collect donations, charge for only the highest-traffic sites, or make extra money charging support fees to big companies. Others do it simply for the prestige.)

Two caveats apply. First, if you plan to use Jetpack's social commenting feature (page 303), which lets visitors comment by using their Facebook and Twitter identities, your options are limited. Currently, Akismet is the only spam fighter that works with these identities.

Second, it's impossible to know which antispam tool is the best for your site—you need to try them out yourself. Antispam developers and spammers are locked in an ever-escalating arms race. The spam blocker that works perfectly this week might falter the next week when clever spammers work around its detection rules.

Good Akismet alternatives include these three:

- Titan Anti-spam (*http://tinyurl.com/wp-anti-spam*)
- Antispam Bee (*http://tinyurl.com/spam-bee*)
- CleanTalk Spam Protection (*https://tinyurl.com/cleantalk*)

The Akismet plugin is so valuable that Automattic bundles a copy with every WordPress site. However, it isn't activated, which means it's just an idle file sitting on your web server. To make Akismet spring to life, you need to activate it and get an *Akismet key*—a secret code (like *0292d4c598d7*) that authorizes you to use the spam-fighting plugin.

Here's where your path takes a turn. The process of getting your Akismet key depends on how your site is set up. If you've already installed and activated the Jetpack plugin, you already have a WordPress.com account, and that WordPress.com account includes an Akismet key. This means the Akismet setup process will be quick and painless. But if you aren't using Jetpack, you need to create a new WordPress.com account and get the Akismet key as part of the Akismet setup process. It's not difficult, but it will take several extra steps.

The next two sections have you covered. First figure out whether you're using Jetpack and then pick the right one.

Setting Up Akismet (with Jetpack)

Already using Jetpack? Here's the express setup process you need to complete to get Akismet running:

1. **Choose Plugins in the admin menu.**

2. **In the list of plugins, click the Activate link under Akismet.**

3. **In the Akismet setup page, click Connect with Jetpack (*Figure 10-20*).**

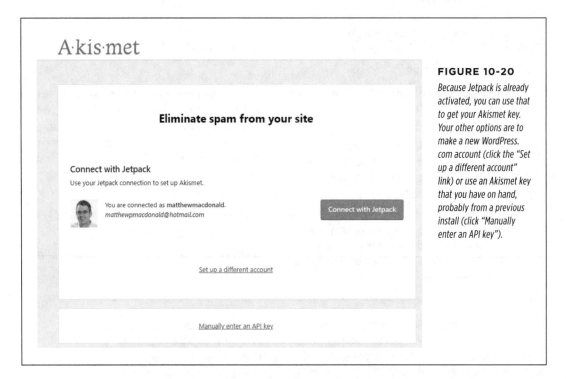

FIGURE 10-20

Because Jetpack is already activated, you can use that to get your Akismet key. Your other options are to make a new WordPress. com account (click the "Set up a different account" link) or use an Akismet key that you have on hand, probably from a previous install (click "Manually enter an API key").

Akismet is now running and keeping a close eye on all your site's comments. It shows a final confirmation page with your API key (*Figure 10-21*). You'll also see a few extra settings that you can tweak:

- **"Show the number of approved comments"** tells Akismet to add an extra piece of information to the comments list. This is a count with the number of comments you previously approved from each would-be commenter. Presumably, if you've approved plenty of messages from the same person, you can trust their newest contributions.

- **"Silently discard the worst and most pervasive spam"** tells Akismet to trash messages that it believes are obvious spam, rather than store them in the spam folder. If they're in the spam folder, you have the chance to take a final review before you click Empty Spam to banish them forever. But the sheer weight of bad spam can clog up your database and bury other, potentially misidentified spam comments. For those reasons, it's generally a good idea to leave this setting on.

- **"Display a privacy notice under your comment forms"** tells Akismet to add a disclaimer about how it uses comment data. This disclaimer may be required depending on your country (for example, it's probably necessary to comply with the European Union's data protection laws). The disclaimer is a single line of text under the comment form that says "This site uses Akismet to reduce spam" and offers a link with more information that potential commenters can read.

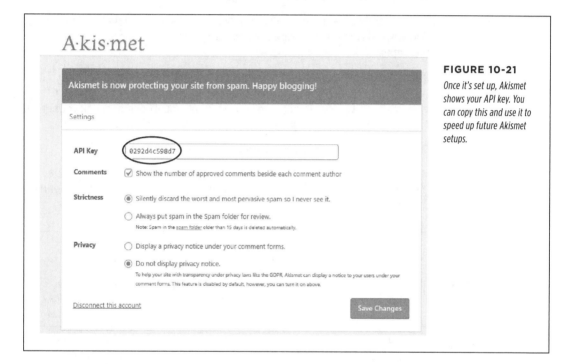

FIGURE 10-21

Once it's set up, Akismet shows your API key. You can copy this and use it to speed up future Akismet setups.

You can revise Akismet's settings later (or upgrade to a paid plan). If you have the Jetpack plugin running, you'll find them at Jetpack→Akismet Anti-Spam. If you don't have Jetpack, you'll see the same settings in a slightly different spot: Settings→Akismet Anti-Spam.

Setting Up Akismet (from Scratch)

If you don't have the Jetpack plugin activated on your site, you need to go through a few more steps to satisfy Akismet. Here's the process you follow:

1. **Choose Plugins in the admin menu.**

    You'll see a list of all your plugins. Akismet should be near the top.

2. **Click the Activate link under Akismet.**

    Now you get to the Akismet setup page.

3. **Click the "Set up your Akismet account" button.**

   This opens a new browser tab and sends you to the Akismet website to finish signing up.

4. **On the Akismet site, you need to click a similarly named "Set up your Akismet account" button.**

   Now Akismet asks you to create a WordPress.com account.

5. **Fill in the details on the sign-up page.**

   If you already have a WordPress.com account, you can use that login information with Akismet. Click "Already have a WordPress.com account" and fill in the details.

   If you don't have a WordPress.com account, now's the time to sign up. Fill in your email address, pick a username and a password, and then click the "Create your account" button. To activate your account, you'll need to check your email for a confirmation message, and click the activation link inside.

6. **Once you have your account, sign in on the Akismet site.**

   Before Akismet will give you a key, it checks to see if you're willing to pay for the privilege. Akismet shows three sign-up options (*Figure 10-22*), depending on the type of site you have.

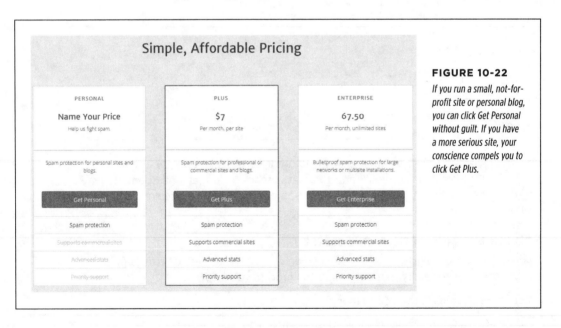

FIGURE 10-22

If you run a small, not-for-profit site or personal blog, you can click Get Personal without guilt. If you have a more serious site, your conscience compels you to click Get Plus.

7. Click the appropriate sign-up button.

If you picked the personal plan, Akismet still asks for a donation (*Figure 10-23*). You choose an amount using a slider below the question "What is Akismet worth to you?" (In fairness to freeloaders everywhere, it's difficult to answer this question *before* you actually start using Akismet.)

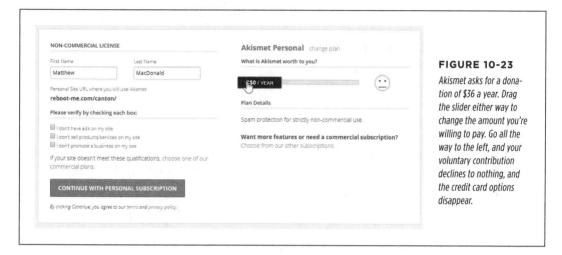

FIGURE 10-23

Akismet asks for a donation of $36 a year. Drag the slider either way to change the amount you're willing to pay. Go all the way to the left, and your voluntary contribution declines to nothing, and the credit card options disappear.

8. Fill in your name and click Continue.

If you elected to pay for Akismet, you need to enter your credit card or PayPal information as well.

If not, you need to check several checkboxes confirming that you don't have ads on your site and you aren't selling products.

After a brief pause, Akismet shows your API key (*Figure 10-24*).

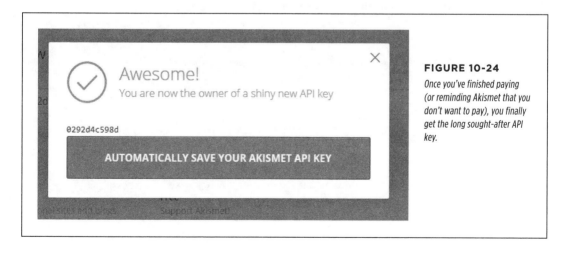

FIGURE 10-24

Once you've finished paying (or reminding Akismet that you don't want to pay), you finally get the long sought-after API key.

9. **Click the Automatically Save Your Akismet API Key button.**

   Akismet navigates to the admin area of your site and opens the Akismet setup page, which your API key filled in. You're ready to go!

   If for some reason you don't want to log in to your site right now, you can copy the key code and use it later. When you return to your site, head to the Akismet setup page (Settings→Akismet Anti-Spam), click the "Manually enter an API key" link, and paste or type in your key code.

10. **Configure the Akismet settings.**

    Once Akismet has your API key, it lets you tweak a few additional settings (*Figure 10-21*). You can get the lowdown on these settings in the bulleted list on page 310.

Using Akismet

Akismet integrates so seamlessly into WordPress's comment system that you might not even realize it's there. It takes over the comments list, automatically moving suspicious comments to the spam folder and publishing everything else.

To give Akismet a very simple test, sign out of your site and then try adding a few comments. If you enter ordinary text, the comment should sail through without a hiccup. But type in something like "Viagra! Cialis!!" and Akismet will quietly dispose of your comment.

Just because you disabled moderation and started using Akismet doesn't mean your comment-reviewing days are over. Once your site is up and running with Akismet, you should start making regular trips to the Comments section page. Only now, instead of reviewing pending comments that haven't been published, you should click the Spam link and check for any valid comments that were accidentally removed. If you find one, point to it and click the Not Spam link. If you find several, you can restore them all with a bulk action—first, turn on the checkboxes next to the comments, pick Not Spam from the Bulk Actions list, and then click Apply. You'll soon get a feeling for how often you need to check for stray messages.

What to Do When Your Blog Is Buried in Pending Comments

Spammy comments are a danger to any blog. If visitors find your site choked with spam, they're far less likely to keep reading or make a return visit.

But even if spam comments aren't approved, they can still pose a problem for your site. First, they clog your Comments page, making it harder for you to find the real comments. And because WordPress stores them in its database, they can bloat it with meaningless content, wasting space on your web host and making it more difficult and time-consuming to back up your site.

The solution seems obvious—just delete all the spam—but it's not always so easy. If your site has the misfortune to fall victim to an automated spam-spewing tool, you can find yourself with thousands or even *hundreds of thousands* of spam comments in short order. (It's happened to us.) So what's a WordPress administrator to do?

If you use a spam-catching plugin like Akismet, spam comments end up in the spam folder. The good news is that you can clean out all your spam with just a few clicks. In the menu, click Comments, and then click the Spam link at the top of the list. Finally, click the Empty Spam button. (Even better, get your spam catcher to automatically clean out old spam, as explained on page 310.)

If you're not using a spam-catching plugin, you have a bigger problem on your hands. That's because the spam comments will be pending comments, and WordPress doesn't provide a way to delete a huge number of pending comments at once. Even bulk actions can act on no more than a single page of comments at a time. At that rate, deleting thousands of spam comments is a several-day affair.

To tackle this problem, you can use a plugin that removes all pending comments. One example is WP-Optimize (*http://tinyurl.com/wp-opti*), which is described in more detail in Chapter 14.

■ The Last Word

In this chapter, you looked at a single feature—WordPress comments. Though it may sound simple, there's a lot to learn about comments. You can control where comments show up, what they look like, and whether they show a gravatar or a profile picture from Facebook or Twitter. But the most important decision you'll make is how to manage bad comments and spam.

As you saw in this chapter, WordPress gives you two basic paths. You can moderate comments, which means you choose which comments live and which ones die. Or you can use Akismet, the spam-fighting plugin that can trash bad comments automatically, before they darken your inbox. You can even combine both, so you use the trusty Akismet plugin to get rid of the obvious garbage, and then review the rest yourself through moderation. In the end, it's a balance of priorities between speed (how quickly a comment is approved) and safety (how likely it is that a spam comment will slip through undetected). Ultimately, you may need to live with Akismet and comment moderation for a while before you really decide which approach suits your workflow and your site.

Collaborating with Multiple Authors

W hen you first create a WordPress site, it's a solo affair. You choose your site's theme, write every post and page, and put every widget in place. Your readers can add comments, but you're in charge of starting every conversation.

You might like this arrangement—and if so, that's fine. But WordPress also makes it possible for you to have friends, colleagues, family members, and even complete strangers contribute to your site. You can, for example, create a site where several people post content. Or you could be more selective, letting some people write content and other people review and edit it. You can also implement an approval system to check the work of contributors before it goes live. You can even create an entirely private site that only the people you approve can view.

In this chapter, you'll learn how to enable all these features by registering new people—not new *visitors*, but new WordPress *users* who have special privileges on your site. You'll also consider WordPress's more ambitious multisite feature that lets other people create their own sites on *your* web hosting account. For example, big companies can use the multisite feature to give each employee a personal blog on the same company website.

■ Adding People to Your Site

A new WordPress website starts with only one member: you. You assume the role of administrator, which means you can do anything from write a post to vaporize the entire site. Eventually, you may decide to make room for company. Usually, you make that decision because you want to work with coauthors, who will write posts for your site.

Before you add someone to your site, though, you need to decide what privileges that person will have. WordPress recognizes five roles, listed here in order of most to least powerful:

- **Administrator.** Administrators can do absolutely everything. For example, if you add a friend as an administrator to your site, they can remove you, delete all your posts, and switch your site to a Hawaiian beach theme. Usually, every site should have just one administrator, which prevents power struggles.

- **Editor.** Editors have full control over all posts and pages. They can create their own posts, and they can edit or delete any post, even ones they didn't create. Editors can also manage post categories and tags, upload files, and moderate comments. They can't change site settings, tweak the site's layout and theme, or manage users.

- **Author.** Authors have control over their own posts only. They can create new ones and upload pictures, and they can edit or delete their posts anytime. Everyone else's content is off limits.

- **Contributor.** Contributors are a more limited form of author. They can create draft posts but can't publish them. Instead, contributors submit their work for review, and an editor or administrator approves and publishes it. Sadly, contributors can't upload pictures, even for their own posts.

- **Subscriber.** These people can read posts and add comments, just like ordinary anonymous people. There's no feature for notifications or newsletters (for that, you need an email list, as described on page 367). So why would you bother to create subscribers? Most of the time, you won't. But the sidebar on page 319 explains some special cases where subscribers make sense.

> **NOTE** In WordPress lingo, all of these types of people—administrators, editors, authors, contributors, and so on—are users. Technically, a *user* is any person who has an account on your site. This account identifies the person and determines what they're allowed to do. Everyone else is an ordinary, anonymous visitor.

The Role of a Subscriber

Why would I add subscribers? Can't everyone read my posts and make comments?

Ordinarily, there are no limits to who can read posts and write comments on a WordPress site. Subscribers don't get any special privileges over regular, unregistered guests.

However, the situation changes in these special cases:

- If you restrict comments with the "Users must be registered and logged in to comment" setting (page 305), and you don't allow Facebook or Twitter logins (page 303), only subscribers can leave comments. This is a pretty severe restriction, and few sites use it.

- If you create a private site (page 336), every reader needs a subscriber account. Without one, they won't be able to see anything on your site. This is the most dramatic reading restriction you can use.

- If you use a social plugin like BuddyPress (*http://buddypress.org*), people with subscriber accounts get extra features, like the ability to share content with friends and chat in discussion groups. In this situation, you want to hand out as many subscriber accounts as possible, so your site becomes more like a community. But this works only if you have a large, dedicated user base. For example, it might make sense if you're creating a site for all the employees in a business or all the students in a school.

Adding New People to a Site

Using the admin area, you can register new users, one at a time. You supply a few key details (like a username, password, and email address), and let your users take it from there.

Here's what to do:

1. **Choose Users→Add New.**

   The Add New User page opens (*Figure 11-1*).

2. **Choose a good username for the person you're inviting.**

   The best approach is to use a consistent system you can apply to everyone you add. For example, you might choose to combine a person's first and last name, separated by an underscore (like sam_picheski).

3. **Type in the person's email address.**

   WordPress uses that address to send the person important notifications, including password resets.

4. **Optionally, specify the person's first name, last name, and website.**

   These are three descriptive details that become part of the person's profile.

   You don't need to fill in the password, because WordPress will randomly generate a temporary password (it's the usual random gibberish, like xlyHw@X!G^AOELF3$o@KA2&Q). If you want to see the password—or even change it—click the Show Password button.

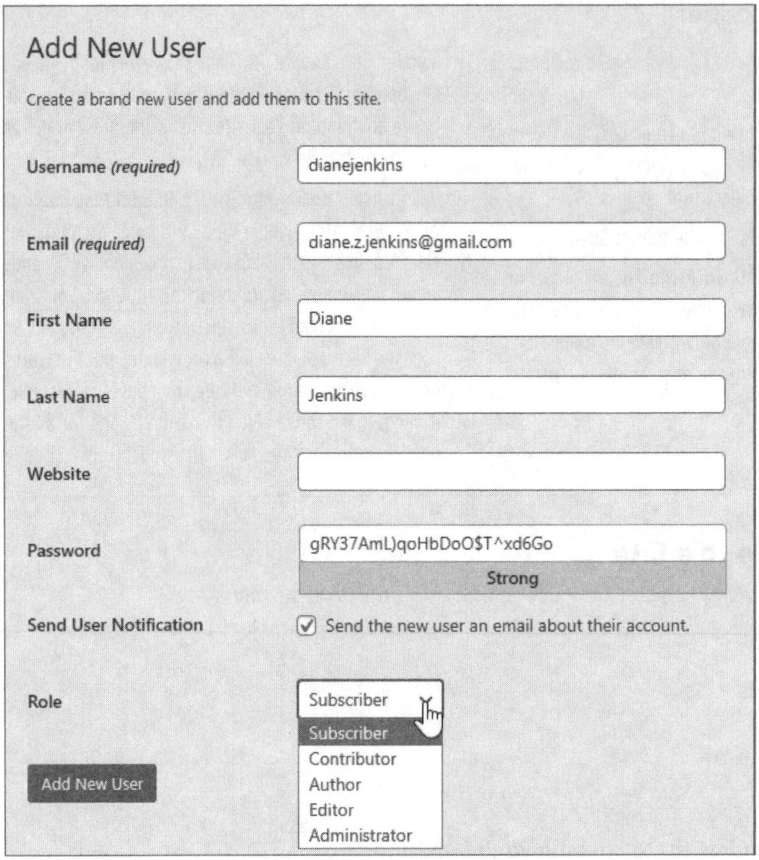

Add New User

Create a brand new user and add them to this site.

Username *(required)*	dianejenkins
Email *(required)*	diane.z.jenkins@gmail.com
First Name	Diane
Last Name	Jenkins
Website	
Password	gRY37AmL)qoHbDoO$T^xd6Go
	Strong
Send User Notification	☑ Send the new user an email about their account.
Role	Subscriber

Subscriber
Contributor
Author
Editor
Administrator

Add New User

FIGURE 11-1

There's nothing mysterious about the Add New User page. Here, the site's administrator has typed in the sign-up information for a person with the username dianejenkins and is specifying her site privileges using the Role drop-down list.

NOTE The emails WordPress sends often end up in an email account's junk folder. You may need to tell new users to check their junk mail to find the messages with their WordPress user details.

5. **Make sure the "Send the new user an email about their account" option is checked.**

   This way, WordPress will send the new user an email with a link they can use to choose a new password (*Figure 11-2*). If you don't choose this option, users can still reset their passwords, but they'll need to click the "Lost your password?" link on the login page.

No matter what you choose for the "Send the new user..." setting, WordPress always sends you, the administrator, an email with a record of the new user's name and email address.

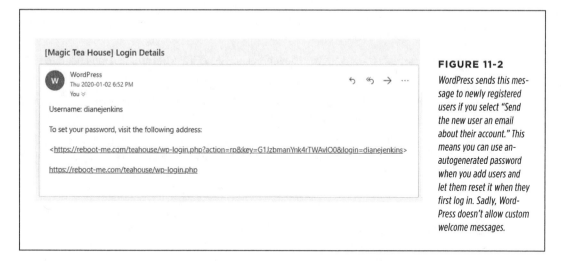

FIGURE 11-2

WordPress sends this message to newly registered users if you select "Send the new user an email about their account." This means you can use an autogenerated password when you add users and let them reset it when they first log in. Sadly, WordPress doesn't allow custom welcome messages.

6. **Pick a role from the drop-down list.**

   You can use any of the roles described earlier.

7. **Click Add New User.**

   WordPress creates the account, sends a notification email (if you chose the "Send the new user..." option), and takes you to the Users→All Users section of the admin area.

   On the Users page, you'll see a list of everyone you've ever added to your site, along with a quick count of how many posts they've written (*Figure 11-3*).

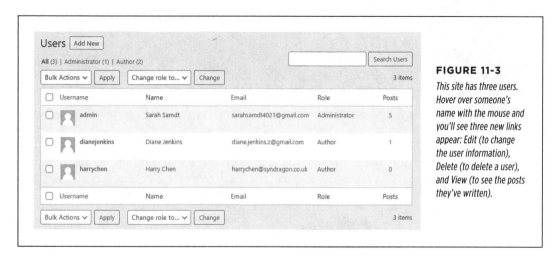

FIGURE 11-3

This site has three users. Hover over someone's name with the mouse and you'll see three new links appear: Edit (to change the user information), Delete (to delete a user), and View (to see the posts they've written).

How Users Log In

To log in to your site, users need to go to the login page. That means that if your site is at *http://cantonschool.org*, they need to visit *http://cantonschool.org/login*. Alternatively, people can go straight to the admin area by requesting the *wp-admin* page (as in *http://cantonschool.org/wp-admin*). In that case, WordPress asks them to log in before they can continue.

If you have a lot of users who haven't used WordPress before, you may need to help them find the login page. Here are two good options:

- You can create your own welcome email message that contains a link to the login page, and send it to everyone.

- You can add a link that goes to your login page. You can put that link in the main site menu or add it to a custom home page. Make sure you give your link descriptive text that clearly explains why the person needs to log in, like "Log in to write your own posts" or "Log in as a contributor."

Every type of WordPress user can log in and visit the admin area. However, Word-Press tailors the admin area to the person's role, so that people see only the menu commands they can use. For example, if you log in as a contributor, your admin area isn't very "admin." Instead, you see the stripped-down interface shown in *Figure 11-4*.

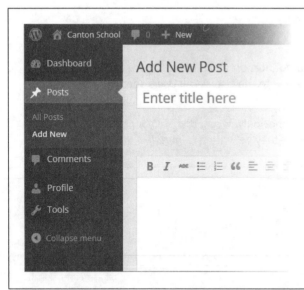

FIGURE 11-4

Different people see different versions of the WordPress admin area. In this example, a contributor has logged in to the Canton School site. The menu has commands for creating posts, reading comments (but not moderating them), and changing the contributor's profile. It doesn't offer commands for changing the site's theme, its widgets, or its settings.

Subscribers get just one option in the menu—the Profile command that lets them change their preferences and personal information. The only type of user who can see the full, unrestricted version of the admin area is an administrator—and, ideally, that's just you.

Working with Authors

The most common reason to add new users to your WordPress site is to get more content from more people. After all, new and interesting content is the lifeblood of any site, and by recruiting others to help you write it, you increase the odds that your site will grow and flourish.

As you already learned, every type of WordPress user except subscribers can write posts. Whether you're an administrator, editor, author, or contributor, the first step is the same: to write a post, you have to log in to the admin area.

The Post Approval Process for Contributors

Administrators, editors, and authors can add posts and pages to your site in the usual way, by choosing Posts→Add New from the menu. When they finish writing, they simply click Publish to make their content go live.

Contributors have more limited powers. They can create posts but not publish them, which gives you a broad safety net—there's no chance that bad content can get on your site, because you get to review it first. When contributors create posts, they have two options: they can save the post as a draft so they can return to it and edit it later, or they can submit it for review (*Figure 11-5*).

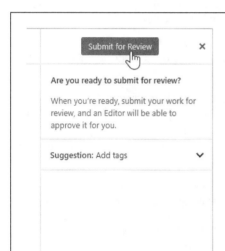

FIGURE 11-5

When a contributor logs in to your site, WordPress changes the familiar Publish button to a Submit for Review button.

When a contributor submits a post this way, WordPress assigns it a special status, called *pending*. A pending post won't appear on your site until an editor or administrator approves it. Here's how you do that:

1. **In the admin area, choose Posts→All Posts.**

   You'll see a list of all the posts you've created.

2. **Click the Pending link (*Figure 11-6*).**

   This limits the list to show pending posts only.

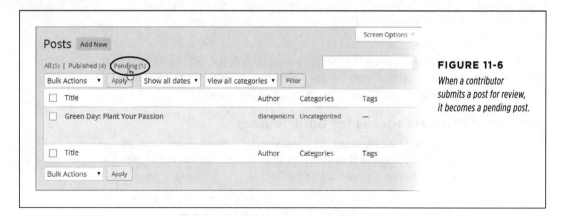

FIGURE 11-6

*When a contributor
submits a post for review,
it becomes a pending post.*

3. **If you see a pending post you want to review, click its title.**

   WordPress opens the post in the Edit Post window. Make any changes you want, from minor corrections to adding completely new content.

4. **If the post is ready for prime time, click Publish.**

   This is the same way you publish your own posts. In this case, however, the newly published post will have the original author's name, even if you edited the post (which is what you want).

   If you make changes to a post but you're not quite ready to publish it, click Save as Pending to store the edited version. You might do this to add questions or comments to the author's work, for example. You can wrap your comments in square brackets [like this]. Then the post author can make changes and resubmit the post.

NOTE Don't be confused by the way WordPress uses the term "author." Even though WordPress has a specific type of user called author, WordPress experts often use the same term to refer to anyone who writes a post on a WordPress site. Thus, administrators, authors, editors, and contributors can all act as post authors.

Better Workflow for Reviewing Posts

WordPress gives you the basic procedures you need to get multiple contributors working together on the same site, but the process has a few rough edges.

The problem is the workflow—the way a task passes from one person to another. Right now, it's up to editors or administrators to go looking for pending posts. WordPress makes no effort to notify you that content is waiting for review. Similarly, if an administrator edits a post but decides it needs more work, there's no easy way to let the contributor know that you need a rewrite. And when you do publish a pending post, WordPress once again fails to notify the original author.

WordPress's creators are aware of these gaps, and they may fix them in future versions of the program. But because the contributor feature is a bit of a specialized tool, those fixes are low on the list of WordPress priorities.

If you want to implement a better system, consider the PublishPress plugin (*https://tinyurl.com/publishp*). It adds the structure you need to manage a multistage review process, including the following:

- **Custom status notices.** Instead of designating a post as Pending or Published, you can give it a status that reflects its stage in your organization's workflow. For

example, if you run a news site, you might want posts to go from "Pitched" to "Assigned" to "Pending Review" to "Published."

- **Editorial comments.** PublishPress lets people attach brief notes to a post as it whizzes back and forth between them (as in "I love your post, but can you expand on paragraph 3?")

- **Email notifications.** PublishPress can send notifications at key points in the review process—when authors submit new posts for review, when an editor publishes a post, when an editor places a comment on a post asking for changes, and so on. On a bustling site, these emails keep the post review process running quickly and efficiently.

- **Calendar.** If you want to make long-term content plans to ensure that there's always something new on your site, the PublishPress calendar can help you plan authors' contributions.

In the past, Automattic made a similar plugin called Edit Flow, which is still in use on many WordPress sites. However, Automattic recently abandoned the project, which makes PublishPress the better choice for the future.

Post Locking

As you already know, authors and contributors are limited to editing their own work. But editors and administrators have more sweeping powers. They can dip into any page or post and make changes.

This setup creates the possibility of conflicts. An editor could start editing a post that a contributor is still writing. Or two editors could attempt to revise the same work at the same time. WordPress doesn't have any post collaboration features—instead, only one person at a time can revise a post and save changes. To prevent one person's edits from wiping out another person's work, WordPress uses a simple post-locking system.

Here's an example of how it works. If Diane starts editing a post, WordPress takes note. Every 15 seconds, it sends a message from Diane's browser to the server, which essentially says, "I'm still working!" (This message is delivered by an internal system called the Heartbeat API.) If Diane stops editing—say, she saves her work and navigates to another page, or she just closes the browser window without a second thought—the messages stop and WordPress realizes (within 15 seconds or so) that Diane has stopped working.

If you want to see which posts are currently being edited, choose Posts→All Posts (*Figure 11-7*).

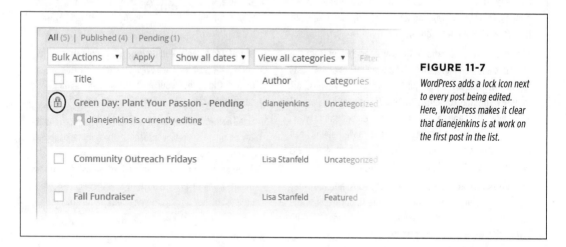

FIGURE 11-7

WordPress adds a lock icon next to every post being edited. Here, WordPress makes it clear that dianejenkins is at work on the first post in the list.

Life gets more interesting when someone else joins the editing party. For example, consider what happens if Lisa attempts to edit the post Diane is working on. Word-Press notices the conflict the moment Lisa picks up the post, and shows a warning message (*Figure 11-8*, top). Now it's up to Lisa to decide whether she wants to wait for Diane to finish (by clicking Exit the Editor) or push her out of the way and take over (by clicking Take Over; see *Figure 11-8*, bottom).

NOTE When it comes to taking over a post, everyone is created equal. For example, if Lisa wants to take over the post that Diane is editing, it doesn't matter if Diane is an administrator or a lowly contributor. Nor does it matter who created the post. As long as Lisa is allowed to edit the post, she'll be able to wrest control from anyone who's currently working on it.

Diane has no chance to stop Lisa's post takeover. Once it happens, her editing session ends. All she can do is click All Posts to return to the post list. However, once Diane is back at the post list, nothing stops her from editing the same post again. If she does, she'll see the same warning message that Lisa saw, informing her that someone else is working on the post. If she forges ahead, Lisa will be kicked out of her editing session.

Of course, the point of the post-locking system isn't for editors and authors to become locked in a series of dueling edits. Instead, it's for times when someone still has a post open but probably isn't editing it anymore. For example, if Diane starts revising a post but walks away from her computer for lunch, Lisa can still get some work done by assuming control of Diane's work in progress.

FIGURE 11-8

Top: Lisa tried to edit a post that Diane is editing. Now she has a choice.

Bottom: If Lisa clicks Take Over, WordPress boots Diane out of the editing window with this message.

Revision Tracking

WordPress has another tool that can help you manage successive edits. It's called *revision tracking*, and it saves old versions of every post. You can step through a post's edit history to see who changed a post, when the edit took place, and exactly what that person changed.

> **NOTE** Revision tracking is particularly useful on sites that have multiple authors. (And it's a lifesaver when an overeager author overwrites another person's work.) However, revision tracking works just as well on single-author sites, where it lets you review your own edits.

You don't need to turn on revision tracking—it's always at work. Every time someone saves a post as a draft, publishes it, or updates it, WordPress takes a snapshot of the post's content and adds it to the revision history. WordPress may also take a snapshot of a post-in-progress as you edit it, but it won't keep more than one copy of your post in this state. After all, you wouldn't want to clutter your revision list with thousands of in-progress autosaves.

Here's how to see the edit history of a post:

1. **Start editing the post.**

2. **If the sidebar isn't visible, click the Settings gear icon (it's in the top-right corner of the post editor) to show it.**

3. **Click Document.**

4. **Click the Revisions section (*Figure 11-9*).**

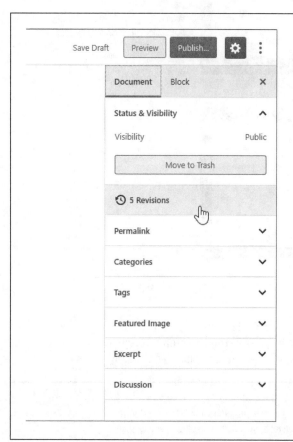

FIGURE 11-9

The Revisions section tells you how many versions WordPress has stored for the current post. This post has five separate edits in its version history. To take a closer look at each snapshot, click the Revisions section.

When you click the Revisions section, WordPress takes you to the Compare Revisions page (*Figure 11-10*). Initially, it compares the differences between the current version of a post (shown in green on the right) and the previously saved version (shown in red on the left).

Once at the Compare Revisions page, you can dig deeper into the post's edit history. First, grab hold of the circle that sits in the small slider near the top of the page. To see an older snapshot, drag the circle to the left. For example, if you drag the circle one notch left, WordPress compares the previous version of the post (which it shows on the right) to the version it saved just before that (which it shows on the left). Keep dragging and you'll go further and further back in time (*Figure 11-11*).

Alternatively, you can use the buttons on either side of the slider button. Click Previous to go one step backward in time, and Next to go one step forward.

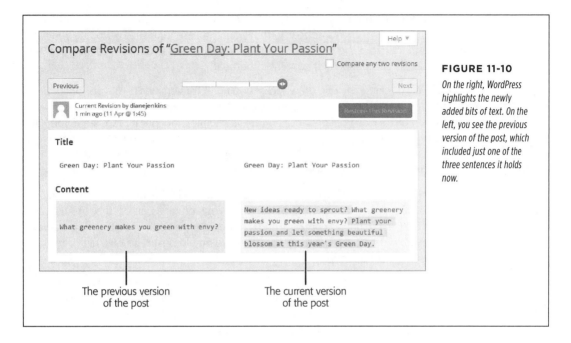

FIGURE 11-10

On the right, WordPress highlights the newly added bits of text. On the left, you see the previous version of the post, which included just one of the three sentences it holds now.

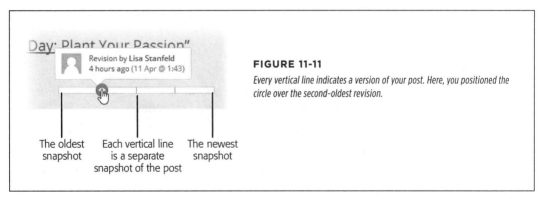

FIGURE 11-11

Every vertical line indicates a version of your post. Here, you positioned the circle over the second-oldest revision.

To revert to an older version of a post, drag the slider back until the version you want appears on the right, in green. Then, click Restore This Revision, which appears just above the snapshot's content.

When you drag the circle, WordPress compares two successive versions of a post (the one you pick, and the one just before that). But WordPress also allows curious administrators to perform an in-depth comparison that puts any two versions of a post under the microscope. For example, you could compare the oldest version of a post with the newest.

To do that, turn on the "Compare any two revisions" checkbox in the top-right corner of the page. Two circles appear in the slider, representing the two posts you want to compare. Drag the two circles to the two snapshots you want to examine, and WordPress compares their content underneath.

Revision tracking has one drawback: keeping extra versions of all your posts takes space. If you're an obsessive sort who revises the same post over and over, your revision history can balloon into hundreds of snapshots per post. Even if you have room for all these revisions (which you almost certainly do), it's never a good idea to waste space in your WordPress database. An unnecessarily big database makes backups slower and can even slow down the overall performance of your pages.

Unfortunately, there's no way to delete old revisions in the admin area. Instead, you need a plugin to clear out the bloat. Numerous plugins can do the job, but WP-Optimize (*http://tinyurl.com/wp-opti*) is a popular, versatile choice. It clears several types of old data out of your database, including old post revisions, old drafts, and unapproved comments.

Alternatively, you can tell WordPress to save fewer snapshots. For example, you could set WordPress to store a maximum of five revisions for each post. Unfortunately, this setting isn't included anywhere in the admin area, so you have to edit the *wp-config.php* configuration file to make the change.

If you decide to edit the *wp-config.php* file, you need to add a line like this to the end of the file in your website folder:

```
define('WP_POST_REVISIONS', 5);.
```

This tells WordPress to store a maximum of five revisions per post. However, you can replace the 5 with whatever number you want. Or you can tell WordPress to never store any revisions by substituting this line of code:

```
define('WP_POST_REVISIONS', false );
```

Now you'll never have to worry about database bloat from old revisions, but you'll also lose the safety net that lets you recover content if you accidentally erase or mangle it.

FREQUENTLY ASKED QUESTION

Coauthoring Posts

What if several people edit the same post? Can they all be credited as authors?

Revision tracking is neat, but your readers don't see any of that information. All they see is the name of the post author—the person who initially created the post. Even when someone else edits a post, it remains the property of the initial author. An editor or administrator can edit a post and attribute it to someone else, but you can't credit two people as authors of the same post—unless you're willing to fiddle with your theme.

To create true coauthored posts, you need to take two steps. First, you need to add the Co-Authors Plus plugin (*http://tinyurl.com/co-authors-plus*), which lets you designate multiple authors for any post or page. Second—and this is the hard part—you have to get your posts and pages to actually display the names of the authors who worked on them. To make that possible, you need to edit your theme, as the Co-Authors Plus plugin explains (see *https://tinyurl.com/tumatg6*).

Browsing an Author's Posts

Once you've added a few authors to your site and figured out a way for them to work without stepping on each other's toes, it's worth thinking about what your multiauthored site looks like to visitors. How can they browse the work of specific authors, or find out more about the writers they like best?

The first issue—reading the work of a specific author—is the easiest to resolve. Just as you can view all the posts in a particular category or all the posts that have the same tag, you can also browse all the posts by a specific author. The easiest way to do that is to click the author's name, which appears just before or after the post content, depending on the theme (*Figure 11-12*).

FIGURE 11-12

By clicking an author's name (in this case, lisachang), you can dig up all the posts that the author wrote.

NOTE The name WordPress uses to sign posts is the author's display name, which might be the person's WordPress username, a nickname, or a combination of the first and last name. Users can configure their display names by editing their profile settings (Users→My Profile). Or administrators can configure any user's display name by going to the Users page (Users→All Users), pointing to a user, and clicking Edit.

WordPress uses special web addresses to make it easy to browse posts by author. For example, if you click lisachang, WordPress adds */author/lisachang* to the end of the site address, like this:

http://therealestatediaries.com/**author/lisachang**

You might notice that this is very similar to the way that you browse category web addresses (page 96).

Now that you know how to get the posts for a specific author, you can make it easier for visitors to get them as well. For example, you could create a menu that has a link for each author, and then display that menu in the main navigation area of your site (*Figure 11-13*). Or you could put it in a sidebar or footer, with the help of the Navigation Menu widget (page 228).

FIGURE 11-13

This menu includes a link to each author's posts, making it easy for readers to browse content by author.

Adding Author Information

Ordinarily, WordPress keeps author information to a minimum. Even though it stores a few key details in each user's profile—including a basic "Biographical info" box—none of these details show up in a post. All your readers see is the author's name.

In some cases, you might prefer to showcase your author. For example, you might want to add a more detailed byline or include a brief bio that highlights the author's achievements. The low-tech solution is to add this information to the bottom of the post (consider setting it in italics to make it stand apart from the rest of the content). But another option is to tackle the challenge with a plugin.

The WordPress plugin repository is overflowing with author info widgets and bio boxes. One decent starting point is the WP Post Author plugin (*https://tinyurl.com /post-author*), which automatically adds an author box to the bottom of every post. It also adds several new text boxes to every user's profile, where authors can enter links to sites where they have public pages (like Facebook, Twitter, and LinkedIn). The author's bio will then include these links (*Figure 11-14*).

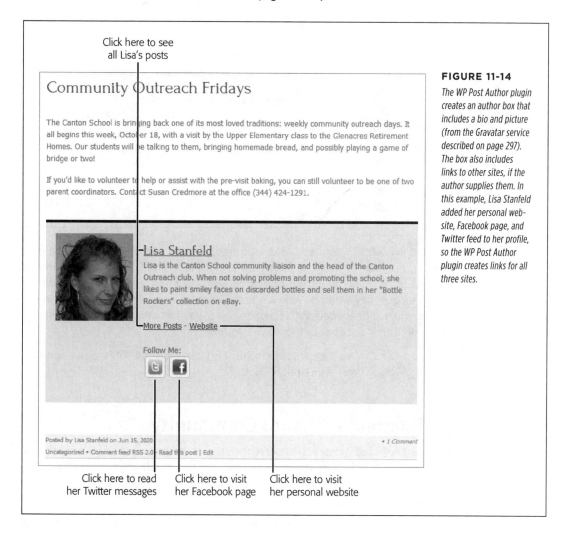

Click here to see
all Lisa's posts

FIGURE 11-14

The WP Post Author plugin creates an author box that includes a bio and picture (from the Gravatar service described on page 297). The box also includes links to other sites, if the author supplies them. In this example, Lisa Stanfeld added her personal website, Facebook page, and Twitter feed to her profile, so the WP Post Author plugin creates links for all three sites.

Click here to read
her Twitter messages

Click here to visit
her Facebook page

Click here to visit
her personal website

The WP Post Author plugin also includes a widget. Use it to take the author details out of a post and tuck them into a sidebar (if your theme provides one) or a footer.

Removing Authors (and Other Users)

As your site evolves, the group of people you work with may change. You already know how to add users, but at some point you may decide to remove one. To do that, you first need to view your site's user list by choosing Users→All Users.

The user list includes everyone who can publish on your site. You remove authors by clicking the Delete link. Deleting someone's account is a fairly drastic step, because it completely wipes the traces of that person off your site. WordPress will ask you what to do with any posts that belong to the newly deleted author (*Figure 11-15*). You can either delete the posts or assign them to another person.

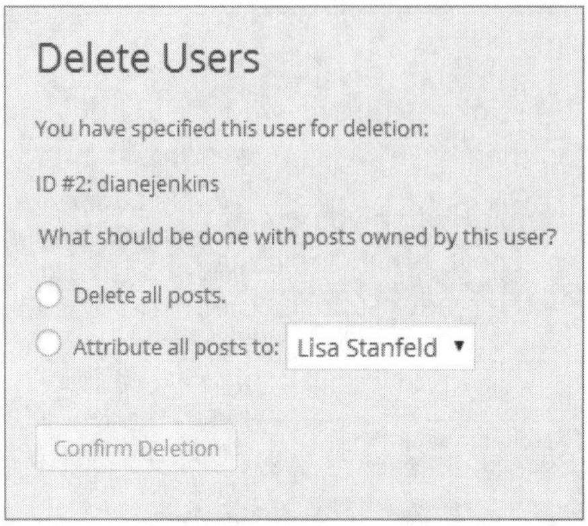

Delete Users

You have specified this user for deletion:

ID #2: dianejenkins

What should be done with posts owned by this user?

○ Delete all posts.

○ Attribute all posts to: Lisa Stanfeld ▾

Confirm Deletion

FIGURE 11-15

When you delete an author, WordPress asks you what to do with their posts. (If you have a last-minute change of heart, click somewhere else, and WordPress abandons the delete operation.)

If you don't want to take such a drastic step, you can demote a user, so the user remains on your site but has fewer privileges. For example, you could change a contributor to an ordinary subscriber. That way, their existing posts will remain on your site, but they won't be able to create any new ones. And if, sometime in the future, you decide to reenlist this person's help, you can simply change their status from subscriber back to contributor.

To change a role, find the person in the user list and click Edit. Then pick a new role from the Role list and click Update User.

■ Building a Private Community

So far, you've used WordPress's user registration features to open up your site to new contributors. Ironically, those same features are also an effective way to close the door to strangers. For example, you can prevent unregistered guests from reading certain posts, or even stop them from seeing any content at all.

Before you build a private site, however, make sure you have enough interested members. Transforming an ordinary site into a members-only hideout is a sure way to scare off 99.9% of your visitors. However, a private site makes sense if you already have a locked-in group of members. Your site might be the online home for a group of related people in the real world—for example, a team of researchers planning a new product, or a local self-help group for cancer survivors. But if you hope that people will stumble across your site and ask to sign up, you're in for some long and lonely nights.

> **NOTE** The goals of a private site are very different from the goals of the average public site. A public site aims to attract new faces, gain buzz, and grow ever more popular. A private site is less ambitious—it allows certain types of discussions or collaboration in a quiet space.

Hiding and Locking Posts

You don't need to make your site entirely private; WordPress gives you two ways to protect individual posts, so the wrong people can't read them.

The first technique is *password protection*. The idea is simple—when you create a post, you pick a password that potential readers need to know. When someone tries to read the post, WordPress refuses to display it until the reader supplies the right password (*Figure 11-16*).

FIGURE 11-16

WordPress adds the text "Protected" to the title of every password-protected post. To read the post, you need to type in the correct password.

The nice part about password protection is that it's straightforward: You either know the password or you don't, and the password is all you need—protected posts don't require that a reader be registered with the site. Of course, administrators and editors can edit any post, so password protection doesn't affect them.

> **NOTE** Use password-protected posts sparingly. If your site includes a mix of public and password-protected posts, frustrated readers are likely to give up on you altogether. Password-protected posts make sense if your site isn't really on the web, but hosted on an organization's internal network (a.k.a. an intranet).

WordPress's second post-protection technique is *private posts*, which are hidden from everyone except logged-in administrators and editors. When other people visit the site, WordPress scrubs every trace of your private posts. They won't appear in the post list, show up in searches, or appear when you browse by category, tag, date, or author.

To see your options for hiding posts, click Document in the sidebar of the post editor. Then click the Status & Visibility heading. Initially, you'll see the word Public appear next to the Visibility setting (*Figure 11-17*). Click this, and choose Private or Password Protected.

 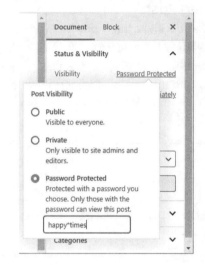

FIGURE 11-17

Left: You can choose from three Visibility options for posts: Public (the standard), Private (the title, post, and comments are hidden from everyone except editors and administrators), or Password Protected (the post's title is visible to everyone, but only people with the right password can read the content).

Right: Here, a new password is being set on the current post.

Private Sites

The problem with private posts is that they're too private. You need to be an editor or administrator to view them, and that may be more power than you want to give other people.

WordPress doesn't have a built-in way to restrict posts to subscribers or specific people, although plenty of plugins offer this sort of functionality. A simple example is Page Restrict (*http://tinyurl.com/page-restrict*), which lets you prevent people from accessing specific pages—or your entire site—until they log in. Page Restrict also lets you pick a suitable message that explains the issue to anonymous visitors, such as "This content is private. To view it, you must log in." A more advanced (and not-free) security plugin is Paid Member Subscriptions (*https://tinyurl.com/paid-member*), which aims to help you protect and sell member-only content. But if that's the road you want to pursue, make sure you also consider the wildly popular WooCommerce plugin introduced in Chapter 14, which supports membership along with plenty of other ecommerce features.

No matter how you go about hiding content on your site, it's important to realize that WordPress privacy settings should never be used for truly sensitive data, like corporate secrets, financial reports, or legal documents. That's because WordPress's privacy settings generally work by hiding content rather than making it inaccessible. Even if you can't browse to a locked post, there may be other ways to see its content. For example, a crafty hacker might find a way to dig up the pictures the post uses, which are stored as ordinary files on the web server. Or they could hack into your website database. (If they do, they'll see not just your password-protected posts, but also the password you set to protect them.)

NOTE Think of WordPress's privacy features as a way to keep ordinary people from reading the wrong content. Don't assume it's an impenetrable barrier against hardcore hackers.

Letting People Register Themselves

You're in complete control of your WordPress site and its users. If you're feeling a bit daring, you can open the floodgates to your site and let your readers register themselves.

This strategy might seem a bit dangerous—and if you don't think it through, it is. Giving random web visitors extra powers on your site is an extreme step for even the most trusting person. However, in several scenarios, self-registration makes a lot of sense. Here are the most common:

- You're creating a private blog and want to prohibit anonymous contributors, but you don't want to make your restrictions onerous—you simply want to deter spammers and other riffraff. Often the process of signing up is enough to keep out these troublemakers. And if you let readers sign themselves up, you save yourself the task of doing so, and save visitors the need to wait for your approval.

- You're creating a site that welcomes community contributors. You're ready to let anyone sign up as a contributor, but you want to approve their content before it gets published (page 323). Be aware, however, that this is no small task—reviewing other people's content and sniffing out spam makes comment moderation seem like a day at the spa.

- You've restricted comments to people who have registered and logged in to the site (page 305), but you're willing to let people comment if they go through the trouble of creating an account. Sometimes, site owners take this step to lock out spammers, and typically it works well, although it also drives away legitimate commenters who can't be bothered signing up. In most cases, it's better to allow Facebook and Twitter authentication (page 303), and to use Akismet to fight spam (page 308).

- Your WordPress site isn't really on the web; it's on the internal network of a business or organization. Thus, you can assume that the people who reach your site are relatively trustworthy. (Of course, you still shouldn't grant them any privileges more powerful than a contributor account without your personal review.)

Flipping on the self-registration feature takes just a few seconds. In the admin menu, choose Settings→General. Add a checkmark next to "Anyone can register," choose a role in the New User Default Role box below, and then click Save Changes.

> **WARNING** You should set the role for new users to subscriber or contributor—subscriber to welcome new readers to a private blog, or contributor to let potential authors sign themselves up. Never allow new people to sign themselves up as authors or editors, unless you want spammers to paste their ads all over your site.

When you turn on self-registration, WordPress adds an extra link to the login page (*Figure 11-18*).

FIGURE 11-18

This blog lets people register themselves. They simply click the Register link (left), enter an email address and password (right), and then wait for an activation link to arrive by email.

If you allow self-registration on a public website, you'll eventually have spammers creating accounts. Usually, the offender is an automated computer program called a *spambot*. It searches the web for WordPress sites and attempts to sign up on every one it finds, in the hope that the site will grant the spambot author or editor permissions. If a site is unwise enough to do so, the spambot immediately gets to work spewing spam into new posts. As long as you limit new users to the role of contributor or (powerless) subscriber, the spambot won't be able to do anything.

Creating a Network of Sites

So far, you've learned how to transform your site from a lonely one-man-band to a collaborative workspace full of authors, editors, and contributors. This transformation keeps you in control of your site but allows new recruits so you can expand your content, extend your reach, and attract new visitors.

Now you'll take a step in a different direction. Instead of looking at adding multiple people to a crowded site, you'll see how to create multiple WordPress sites that coexist on the same web server. Think of it as a way to empower your users to do even more. Now, each author gets a separate site, complete with its own web address, admin area, theme, and reverse-chronological list of posts. Your web server hosts all these sites alongside your own, much like children living in their parent's home.

For example, say you create a WordPress multisite network at *http://EvilCompany OfDoom.com*. An employee named Gareth Keenan might create a site at *http:// EvilCompanyOfDoom.com/garethkeenan*. Similarly, another employee might add a site at *http://EvilCompanyOfDoom.com/dawntinsley*. Of course, you don't need to create sites based on individual people—you can just as easily create sites that represent departments, teams, clubs, or any other group of people who need to blog together.

> **NOTE** The multisite feature works well if you have a community of people who need to work independently, keep their content separate from everyone else's, and have complete control over the way their content is organized and presented. For example, the Canton School site might use the multisite feature to give each teacher their own site. Teachers could then use their sites to post assignments and answer student questions. The multisite feature isn't very useful if you want people to team up on the same project, share ideas, or blog together—in all these cases, a single site with multiple users makes more sense.

The multisite feature is particularly convenient when it comes to administration. When you build a network of sites, you become its network administrator—a special sort of administrator with sweeping powers over all the sites in the network. Using these powers, you can choose what themes your users can install and what plugins they use. And when a new version of WordPress comes out, you can update all the sites in a single step.

Before going any further, be aware of one thing: building a network of sites is significantly more complex than adding people to an existing one. Expect to spend more time feeling your way around and learning how to configure sites and users. Furthermore, be careful with the plugins you use, because some won't work in a multisite network.

By the end of this chapter, you'll know how to set up a network of sites, add sites, and perform the basic configuration that holds it all together. However, significant aspects of the multisite feature remain beyond the scope of this book, like using it with subdomains (see the following Note).

> **NOTE** There are two ways to create addresses for the sites in a network. You can give each site its own subfolder (as in *http://OrilliaBaseballTeams.com/madcats*), or you can give each site its own subdomain (as in *http://madcats.OrilliaBaseballTeams.com*). The latter is the way WordPress.com works with its free hosting service. It's slightly more complicated, because it requires additional settings on your web host. In this chapter, you'll stick to the subfolder approach.

Creating a New Multisite Network from Scratch

The easiest way to create a multisite network is to create a new WordPress site from scratch, using an autoinstaller that supports the multisite feature. For example, if you use Softaculous, the installation process is almost exactly the same as the one you used in Chapter 2. The difference is that somewhere in your autoinstaller, you need to find a setting named something like Enable Multisite and switch it on (*Figure 11-19*).

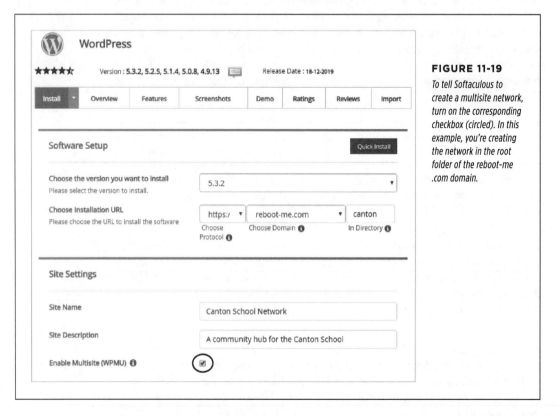

FIGURE 11-19

To tell Softaculous to create a multisite network, turn on the corresponding checkbox (circled). In this example, you're creating the network in the root folder of the reboot-me .com domain.

Once you install your site, you can go straight to the admin area and look around. Skip ahead to the section "Your Multisite Network: A First Look" on page 341.

If you don't have an autoinstaller that supports the multisite feature, you'll need to install a normal WordPress site first and then go through the somewhat awkward conversion process outlined in the next section.

Converting an Existing Site to a Multisite Network

Converting an existing WordPress site into a multisite network is trickier than creating a new network from scratch. If you use subfolders (rather than subdomains) in your network, the conversion process will break any links within posts. For that reason, it's best to convert a newly created WordPress site, rather than one you've been using (and that other people have been reading) for some time.

But if you know how to use an FTP program and you're undaunted by the challenge, it is possible to transition from an ordinary site to a multisite network. WordPress has the full and rather technical step-by-step instructions at *http://tinyurl.com/2835suo*. The process involves modifying two files in your site—*wp-config.php* and *.htaccess*—and changing a few related settings in the admin area. But because you can't directly edit the files on your site, you need to download them to your computer (that's where the FTP program comes in), make your changes in a text editor, and then upload the new, modified files. If you've never fiddled with a WordPress installation before, it's a bit tedious.

> **WARNING** Make sure you really want a multisite network before you forge ahead, because there's no easy way to change a multisite network back to a single site after you make the jump.

Your Multisite Network: A First Look

When you create a multisite network, WordPress starts you out with a single home site in the root of the installation folder. For example, if you install a multisite network at *http://prosetech.com*, the first site is at *http://prosetech.com*. This is exactly the same as when you create a standalone site. When you create additional sites, however, WordPress places them in subfolders. So if you add a site named teamseven, WordPress creates it at *http://prosetech.com/teamseven*. (You might think that it makes more sense to write "TeamSeven" rather than "teamseven," but to WordPress it's all the same. No matter what capitalization you use, WordPress shows the site name in lowercase letters when you manage it in the admin area.)

> **NOTE** If you're using subfolders (not subdomains) to arrange your multisite network, you'll find one quirk in WordPress's naming system. When you view a post or page on your home site, WordPress adds */blog* to the address. For example, WordPress puts a post that would ordinarily be found at *http://prosetech.com/2021/06/peanut-butter-prices-spike* at *http://prosetech.com/blog/2021/06/peanut-butter-prices-spike*. This slightly awkward system makes sure that WordPress can't confuse your home site blog with another site in the network, because it doesn't allow any other site to use the name blog.

When you finish creating your multisite network, you find yourself in the admin area for your home site. But if you attempt to augment your site's features, you'll find a new restriction. Even though you can activate an existing plugin or theme, WordPress won't let you install new ones. On a fresh WordPress install, you'll probably get the latest year theme and two basic plugins (the essential Akismet spam-catcher, and the pointless Hello Dolly example).

If you haven't already guessed, your home site has these new and slightly unwelcome limitations because it's now part of your multisite network. These rules can be frustrating, but they have sound logic behind them. First, the theme limitations guarantee that your sites share a consistent look. Second, the theme and plugin restrictions act as safeguards that prevent inexperienced users from uploading spam-filled extensions, which could compromise your entire network.

That said, you'll probably want to tweak these restrictions to make them better suit your site. For example, you may want the sites on your network to use a different standard theme, or you may want to allow site creators to choose from a small group of approved themes. You might also have trusted plugins that you want to run on everyone's site. You'll learn how to make these changes shortly. But first, you need to understand how to add new sites to your network.

Adding a Site to Your Network

To add a site, you need to enter network administration mode. This is a step that only you, the network administrator, can take. Other administrators on your network will be able to manage their own sites, but they won't be able to change the network settings—or even look at them.

NOTE In WordPress parlance, a *network administrator* (also known as a *super admin*) is the person who manages a multisite network and has full power over all the sites inside. By contrast, site administrators oversee a single site—the site you create for them.

To start managing the network, point to the My Sites menu, which is in the navigation bar, just above the admin menu. Then click Network Admin (*Figure 11-20*).

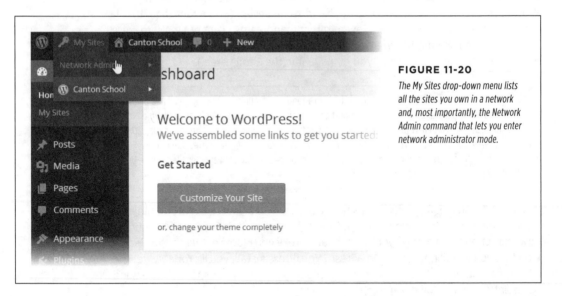

FIGURE 11-20

The My Sites drop-down menu lists all the sites you own in a network and, most importantly, the Network Admin command that lets you enter network administrator mode.

In network administration mode, the admin area changes. Because you're no longer managing a specific site, the Posts, Pages, Comments, Links, and Media menus all disappear. In their place is a smaller set of commands for managing sites, users, themes, plugins, network settings, and updates.

TIP You can go straight to the network administration page by adding */wp-admin/network* to the end of your home site address, as in *http://prosetech.com/wp-admin/network*.

Once in network administration mode, you can create a new site:

1. **Choose Sites→Add New from the admin menu.**

   The Add New Site page opens (*Figure 11-21*).

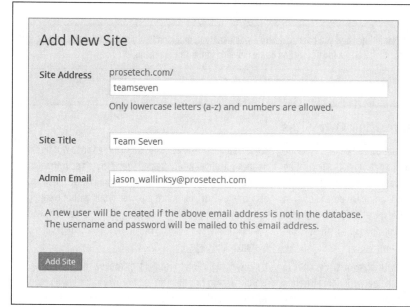

Add New Site

Site Address   prosetech.com/

teamseven

Only lowercase letters (a-z) and numbers are allowed.

Site Title   Team Seven

Admin Email   jason_wallinksy@prosetech.com

A new user will be created if the above email address is not in the database.
The username and password will be mailed to this email address.

Add Site

FIGURE 11-21

WordPress knows that a net-work administrator may need to create dozens of sites. To keep your life simple, it asks for just three pieces of information: the site address, the site title, and the email address of the person who will become the site's administrator. Here, WordPress will create the new site at http://prosetech.com /teamseven.

2. **Type the site's folder name in the Site Address box.**

   WordPress adds the folder name to the address of your multisite network. For example, if you use the folder name *drjanespears* and your multisite network is at *http://StMarciMarguerettaDoctors.org*, the new site has the address *http:// StMarciMarguerettaDoctors.org/drjanespears*.

3. **Give the site a title.**

   The site administrator can change this later.

4. **Supply the email address of the person who will own the site.**

   That person will become the site's administrator.

   Adding people to a multisite network is different from adding people to a stand-alone site in one important respect: you don't need to pick the password for new users. WordPress knows you're busy, and it generates a random password and emails it to the new administrator.

5. **Click Add Site.**

   WordPress creates the site and adds two links to the top of the Add New Site page: Visit Dashboard (which takes you to the new site's dashboard page in the admin area) and Edit Site (which lets you change the site's settings). You can also log in to the admin area in the usual way—just add */wp-admin* to the end of the site address.

Ideally, you won't need to visit the new site's admin area, because the newly christened administrator will take it from there.

> **TIP** If, sometime later, you need to delete a site, modify it, or assign it to a new administrator, start at the list of sites in the Sites→Add Sites section of the network administration menu.

GEM IN THE ROUGH

Letting People Create Their Own Sites

Ordinarily, it's up to you to create every site in a multisite network. WordPress helps you out by automatically creating an account for the new administrator, so you can create a site in one step instead of two. But if you have dozens or even hundreds of users who want sites, manually creating each one is tedious. WordPress gives you another option—you can choose to let people create their own sites.

To allow your users to create their own sites, choose Settings→Network Settings in the network administration dashboard. Then, next to "Allow new registrations," choose "Logged in users may register new sites." Make sure "Send the network admin an email notification every time someone registers a site or user account" is also turned on so WordPress notifies you about newly created sites. (There's no restriction on the number of sites, so if someone can create 1, they can also create 12. If you notice a power-drunk author creating too many sites, you need to step in, delete some, and send the miscreant a stern email.) Finally, click Save Changes at the bottom of the page.

New users might not realize that they're allowed to create sites. WordPress won't tell them unless they ask for the sign-up page, by requesting *wp-signup.php* in the root site (as in *http:// prosetech.com/wp-signup.php*). *Figure 11-22* shows the page.

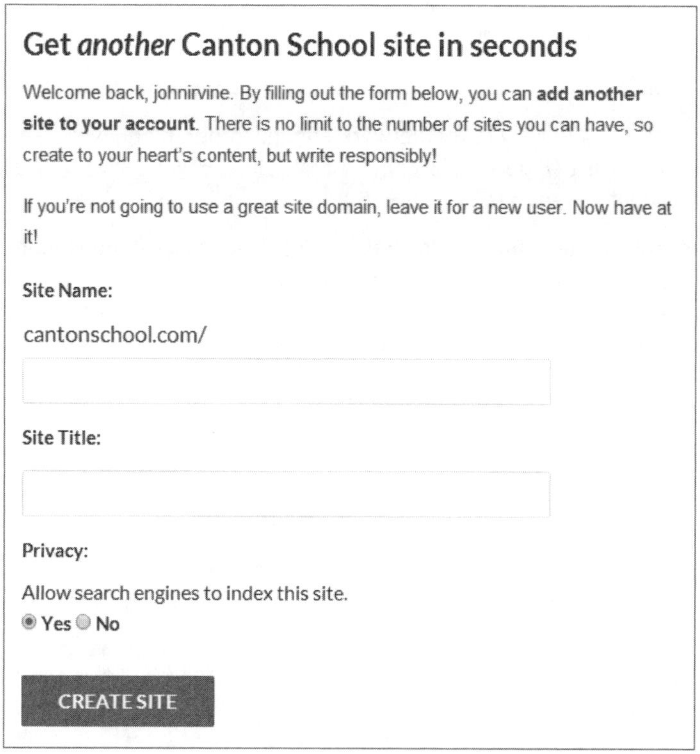

Get *another* Canton School site in seconds

Welcome back, johnirvine. By filling out the form below, you can **add another site to your account**. There is no limit to the number of sites you can have, so create to your heart's content, but write responsibly!

If you're not going to use a great site domain, leave it for a new user. Now have at it!

Site Name:

cantonschool.com/

Site Title:

Privacy:

Allow search engines to index this site.
◉ Yes ○ No

CREATE SITE

FIGURE 11-22

To create a new site, a logged-in user needs to supply the site folder name and the site's title on the sign-up page, and then click the big Create Site button. The sign-up text shown here is WordPress boilerplate. Once you learn how to edit a theme (Chapter 13), you'll be able to customize this text.

Understanding How Users Work in a Multisite Network

You can create as many sites as you want in a multisite network. In each site, you can add as many users as you need.

Sometimes the same person needs to work on more than one site. For example, one person might need to contribute to different blogs maintained by different people. Or an administrator who manages one site in a network might also want to contribute to another.

To understand how to deal with this, you need to realize that a multisite network maintains a master list of all the users who belong to any site in the network. Each of those people has subscriber privileges on every site. (As you learned earlier, subscribers are the lowest class of WordPress user—they can't do anything more than read posts and write comments.)

In addition, you can give people special privileges for specific sites. For example, you might make someone an administrator on one blog and an author on another.

In this case, there's still just one record for that user, but now it's registered with two different sets of capabilities on two different sites.

> **NOTE** Happily, WordPress makes people log in only once. When visitors move from one site to another in the same network, WordPress remembers who they are and determines what privileges they should have on each site.

If you choose Users→Add New in the network administration menu, you can add people to the master list (*Figure 11-23*, top). But WordPress won't give new users any special privileges for any site.

Life is different for ordinary site administrators. Consider what happens if an administrator named Suzy logs in to manage her site. When she chooses Users→Add New, she's not given the option to create an account for someone else. Instead, she can invite an existing user to take on a more powerful role on her site (*Figure 11-23*, bottom).

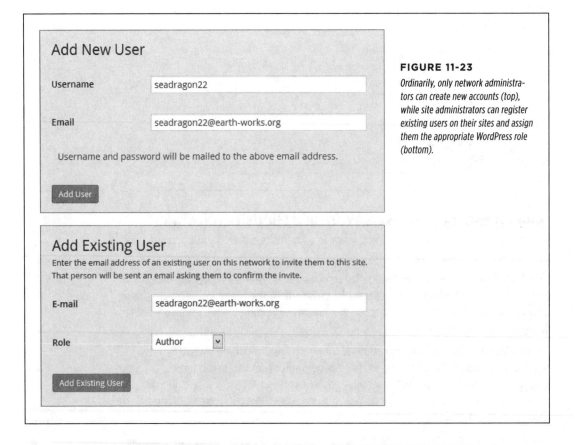

FIGURE 11-23

Ordinarily, only network administrators can create new accounts (top), while site administrators can register existing users on their sites and assign them the appropriate WordPress role (bottom).

One potential problem with the user registration system is that it can create a lot of extra work. For example, if a site administrator needs to add someone new, they need to ask you, the network administrator, to create the account first. To circumvent this restriction, go to Settings→Network Settings, choose "Allow site administrators to add new users to their site," and then click Save Changes. Now site administrators can add new people to the master list.

Another problem occurs if one person contributes to several sites. In that case, someone needs to visit each admin area and invite the user separately to each site. If you're not the sort of person who likes to spend all weekend tweaking WordPress settings, you may want to enlist the help of a plugin like User Role Editor (*https:// tinyurl.com/user-role*). It lets you set a default role for each site in a multisite network. Then, when you create a new user, that user is automatically registered on each site with the default role you chose.

Rolling Out Updates

One advantage of a multisite network is that it streamlines certain management tasks. For example, you can update WordPress on all the sites in your network in a single operation from the network administration page.

To get started, choose Updates→Available Updates from the network administration menu. You'll see, at a glance, what themes, plugins, and WordPress system updates are available. If you're not up-to-date, start by installing your updates on this page.

When you update themes or plugins, the changes take effect on all the sites in your network immediately. That's because a multisite network stores only a single copy of each theme and each plugin.

When you install a new version of WordPress, you need to take one more step. Choose Updates→Update Network, and then click the Update Network button to upgrade all your sites at once.

Adding Themes and Plugins

In an ordinary WordPress website, the site administrator controls the themes and plugins the site uses. But in a multisite network, this approach would be too risky, because a single malicious plugin could steal sensitive data from any site in the network, or wipe out the database of your entire network.

Instead, multisite networks use a more disciplined system. You, the network administrator, can pick the themes and plugins you want to allow. Site administrators can then choose from the options you set.

A typical multisite installation begins with a few standard year themes, but only the latest is network enabled (the latest is Twenty Twenty). That means Twenty Twenty is the only theme the sites in your network can use. In fact, site administrators can't see the other themes at all.

To add a new theme and make it accessible to the sites in your network, follow these steps:

1. **Choose Themes→Add New, and search for the themes you want.**

   If you need a refresher, page 107 has the full story on theme searches.

   To activate a standard year theme already on your network (but not enabled), such as Twenty Nineteen, jump straight to step 3.

2. **When you find a suitable theme, click Install Now.**

   This downloads the theme to your multisite network but doesn't actually make it available to any sites.

3. **Click Network Enable.**

   The Network Enable link takes the place of the Activate link you see when you install a theme on an ordinary, standalone WordPress website. You can click Network Enable immediately after you install a theme, or you can view all your themes (*Figure 11-24*) and then click Network Enable next to the ones you want.

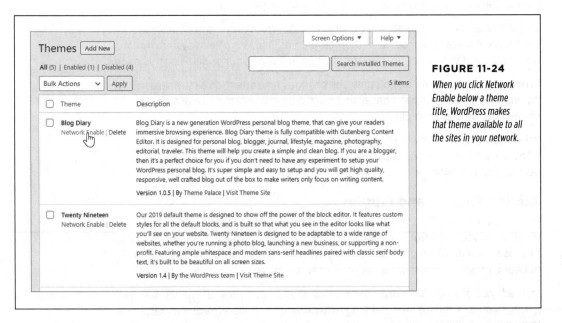

FIGURE 11-24

When you click Network Enable below a theme title, WordPress makes that theme available to all the sites in your network.

You can also enable a theme for some sites but not others, although it's awkward. First, make sure your theme isn't network-activated. Then choose Sites→All Sites and click the Edit link under the site where you want to apply the theme. When the Edit Site page appears, click the Themes tab. In the list of disabled themes, click Enable next to the ones you want to add.

The process for installing plugins across networked sites is similar but subtly different. First, choose Plugins→Add New and then search for the plugin you want. When you find it, click Install Now. Now you have a choice:

- **Make the plugin optional (don't do anything).** Once you install a plugin, WordPress makes it available to every site administrator in your network. Each administrator can log in, visit the Plugins page in the admin area, and choose what plugins to activate.

- **Activate the plugin for every site.** To do this, click the Network Activate link. When you network-activate a plugin, that plugin automatically runs on every site in your network. However, site administrators won't see the plugin in the Plugins page, and they won't be able to deactivate it. That's your job (click Network Deactivate to switch off a network-activated plugin).

> **NOTE** Not all plugins work properly when network activated, so it's worth contacting the plugin maker to ask, or testing a new plugin the first time you network activate it.

GEM IN THE ROUGH

Setting a Storage Limit

Themes and plugins aren't the only restrictions that come into play on a multisite network. You can also set storage limits to restrict how many pictures, documents, and other files people can upload to their sites. These settings prevent space hoggers from swallowing gigabytes of hosting room, leaving your web server starved for space.

Ordinarily, your network has no site restrictions. To put one into effect, choose Settings→Network Settings on the network administration area. Scroll down to the "Site upload space" heading and switch on "Limit total size of files uploaded." That caps the amount of space for posts, pages, pictures, and uploaded files on a site to 100 MB. However, you can type in whatever maximum you want. You can also change the "Max upload file size" to set the maximum size of an individual file (usually 1.5 MB). However, there's a catch—there's often a similar setting at the web server level that also needs to be ratcheted up. If you increase the WordPress limit but still aren't allowed to upload really big files, contact your web host to find out what setting you need to change.

Site administrators can use the dashboard to keep an eye on the size of their sites. Choose Dashboard→Home, and look at the At a Glance box. At the bottom, you find the key details: the maximum size allotment, the current size of the site, and the percentage of space used so far.

■ The Last Word

In this chapter, you learned how to invite other people to work with you on a WordPress site. First, you considered the usual approach—adding users and giving them the ability to write posts (but not tamper with other parts of your site). This works well if you have contributors who work independently, and you just need a way to let them publish their work. But if you need something more regimented, like a formal edit and review process, consider stepping up to a plugin like PublishPress, as described in the box on page 325.

In the second half of this chapter, you tackled the WordPress multisite feature. With a multisite network, your collaborators aren't just writing posts on your site—they're able to create their own linked sites and to manage their own group of contributors. Multisites are one of WordPress's more advanced but less commonly used features. But considering it underpins big hosting projects like WordPress.com, it's safe to say that the feature will live on for years to come.

Attracting a Crowd

N ow that you know how to build a fantastic WordPress site, you need to show it off to the world. That means you need to spend some serious time *promoting* your site.

Web promotion can be grueling work, and many WordPressers would rather avoid the subject altogether. Not only does it take a significant amount of effort, but the benefits aren't always clear, and you often need to pursue a promotional strategy without knowing how well it'll work. The best approach is to make web promotion as easy and natural as possible. That means weaving it into your daily routine and integrating it into the way your website works. It also means using honest promotional strategies rather than search engine ploys and other trickery. Follow the guidelines here, and you'll still have plenty of time to pursue your real job—publishing fabulous content.

In this chapter, you'll learn a common-sense approach to web promotion. You begin with the best type of advertising a site can have: word-of-mouth recommendations. That doesn't mean waiting for your site to crop up in casual conversation. Instead, it involves learning how to help your readers rate, "Like," and tweet your content through social media services such as Facebook and Twitter.

Next, you'll help existing readers bond with your site. You'll notify them when you publish new posts and alert them when their comments receive a reply. Done right, these steps build long-term relationships with your fans and increase the number of repeat visitors.

After that, you'll consider a few basics of SEO (search engine optimization). You'll learn how to make your site more Google-friendly, and you'll take a look at web statistics, so you can assess how well your promotional strategies are working.

■ Encouraging Your Readers to Share

There's a gaping chasm of difference between commercial promotion and personal recommendations. If you can get your readers to share your posts and recommend your site to friends, you'll accomplish far more than the average ad campaign.

Usually, sharing means enlisting the help of Facebook and Twitter, two social media sites that are all about exchanging information, from gossipy chitchat to breaking news. With the right WordPress settings and widgets, you can make it easy for your visitors to recommend your site to their friends and followers.

A Word About Jetpack

Most of the features in this chapter use the Jetpack plugin, which you first met in Chapter 9. Jetpack isn't the only option for features like social media buttons, email notifications, and site statistics. However, it's the only plugin that bundles all these useful features into one place.

As you become more experienced with WordPress, you may decide to add fewer features to your site, and you may switch to other, more focused plugins. (We'll mention some of them on the way.) Or you may decide to stick with Jetpack for the long haul. Either way, the concepts you'll learn here are the same.

So if you haven't done it already, now's a good time to get Jetpack up and running. Page 256 explains how.

How Sharing Buttons Work

Sharing is often an impulsive act. You stumble across a site, it catches your easily distracted mind for a few seconds, and you pass the word out to a few choice friends. You're more likely to share a site if the process is quick and easy—for example, if the site provides a handy link that does the bulk of the job for you. If a guest has to compose an email or switch over to Facebook or Twitter, they might just defer the task for another time—and then forget about it altogether.

The best way to make sharing easy, quick, and convenient is to add buttons that reduce the task to a couple of mouse clicks (*Figure 12-1*). That way, your readers can share your site before they move on to their next distraction.

Facebook sharing needs no introduction—if you've been on the world's most popular social network today, you've probably already scrolled past a few dozen shared posts from friends and family. When a guest clicks the Facebook button, WordPress pops opens a Facebook window that shows what the shared post will look like in Facebook land (*Figure 12-2*). The guest can then add a personal comment, fine-tune who gets to see it, and click Post to Facebook to make it happen.

FIGURE 12-1

This site has the three most popular sharing buttons. They appear after the post and just before the comments section. Readers can share a link to this post by email, Twitter, or Facebook.

Add a comment
to make it personal

FIGURE 12-2

Two mouse clicks and an optional comment are all it takes to share a post with your Facebook friends. (Assuming you're currently logged in to Facebook. If you aren't, the pop-up window will get you to sign in first.)

Share it with your
fellow Facebookers

Twitter sharing is a great way to get the word out into the ever-chattering Twitter-verse. Serious Twitter fans are always looking for small tidbits of interesting material, and your Tweet button will be a hard temptation for them to resist. Twitter sharing works in the same way as Facebook sharing—when you click the Tweet button, a small Twitter window opens where you can customize the tweet before sending it along to your legions of followers.

Email sharing is the old-fashioned-but-classic approach (really just one step above printing the page and posting it in the mail). It's great for guests who may not use social media. Best of all, it works even if your visitor doesn't have an email program handy, because WordPress sends the message on your guest's behalf.

To share a post by email, a visitor starts by clicking the Email link. A box drops down where the visitor can type in the recipient's email address. Click Send Email, and WordPress delivers a short message that looks something like this:

> Jason Minegra (*jackerspan4evs@gmail.com*) thinks you may be interested in the following post:
>
> The Felon I Didn't Represent
>
> *http://thoughtsofalawyer.net/the-felon-i-didnt-represent*

Adding Sharing Buttons to All Your Posts

A fresh install for WordPress doesn't include any social media smarts. To add sharing buttons to posts or pages, you need to use a plugin. In this chapter, you'll stick to the familiar Jetpack plugin. But if you prefer to work with a plugin that offers super customizable sharing buttons and nothing else, check out the Social Media Share Buttons plugin (*http://tinyurl.com/sm-buttons*).

NOTE Don't get misled by WordPress's Social Icons block. As you learned earlier (page 158), the Social Icons block is just a handy way to link to *your* social media accounts—for example, if you want to give your readers a chance to follow you on Twitter or connect on LinkedIn. The Social Icons block doesn't help people share your posts, although the makers of WordPress might add that feature in the future.

To get started using social media buttons with Jetpack, follow these steps:

1. **Choose Jetpack→Settings from the admin menu.**

2. **Click the Sharing tab at the top of the settings page.**

3. **In the "Sharing buttons" section, switch on "Add sharing buttons to your posts and pages" (*Figure 12-3*).**

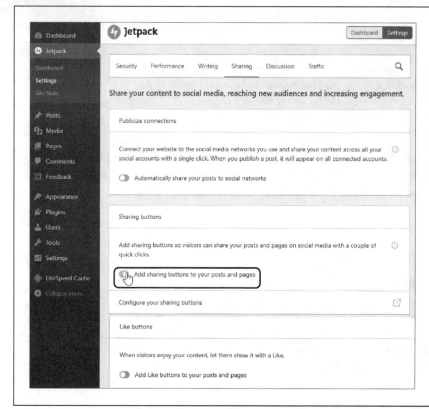

FIGURE 12-3

Jetpack gives you one setting to switch on social media buttons. If you want to customize how they work, you need to click "Configure your sharing buttons," which takes you to the WordPress.com admin area.

When you turn on sharing, all your posts and pages get Facebook and Twitter buttons. Just hit the Refresh button in your browser to reload your site. Sometimes there's a short delay before Jetpack shows sharing buttons for the first time, so don't panic if you need to wait a minute and click Refresh again before they appear.

You may not see the Email sharing button—it all depends on your antispam settings. That's because the creators of Jetpack discovered that malicious users could hijack the email sharing system to send spam messages. The technique was to bury a junk mail message in an email share. That way, their message would be delivered by the WordPress.com mail servers, and potentially avoid the scrutiny of automated spam filters.

Jetpack's solution to spam was to hide the Email sharing button, unless you have Akismet active on your site. That's because Akismet does double duty, not only catching bad spam comments (as you learned in Chapter 10), but also intercepting spammy email shares. If your site allows comments, you're probably already using Akismet. But if not, flip back to Chapter 10 for the lowdown, so you can get set up with email sharing.

Adding More Sharing Buttons

The social media world is bigger than Facebook and Twitter, and Jetpack actually has options for several more social sites. But to add them, you need to do a bit more configuration.

Awkwardly, you can't configure your social media buttons directly inside the admin area for your website. Instead, you need to go to the administration pages on WordPress.com. Here's what to do:

1. **In the admin area on your site, choose Jetpack→Settings.**

2. **Click the Sharing tab at the top of the page.**

3. **Click Jetpack's "Configure your sharing buttons" link (*Figure 12-3*).**

   This takes you straight to the right part of the WordPress.com admin area.

NOTE The social media options are buried in the great mass of WordPress.com options. If you're ever searching for them on your own, here's how to find the right place. Pick your website, and choose Tools→Marketing in the menu, which takes you to the Marketing and Integrations page. Then click the Sharing Buttons tab at the top of the page.

---

4. **Click "Edit sharing buttons."**

   A panel pops open with a list of all the social media services that Jetpack supports (*Figure 12-4*).

5. **Click to select the social media buttons you want.**

   You can tell which buttons are currently in use by their color—unselected buttons are shaded light gray.

TIP Facebook, Twitter, and email aren't your only sharing options, but they're three of the best. Another good choice is the Print button, which gives people an easy way to print out your post and take it to their friends on foot. But the best advice for sharing buttons is to use just a few of the most useful ones (ideally, cap it at four or five). Too many buttons can overwhelm your readers and make you look needy.

---

6. **Optionally, click Reorder to change the order of your buttons.**

   Now you'll see just the sharing buttons you picked. To get them in the right order, click a button and drag it somewhere else.

7. **When you're finished adding and rearranging, click Close.**

   But don't go anywhere yet! You still need to make your changes permanent.

8. **Optionally, add a More button.**

   The More button allows you to tuck some sharing buttons out of site. Jetpack adds a More button, and if you click it, a panel appears with extra, less common sharing options (*Figure 12-5*). Truthfully, the More button is probably just a clumsy way to complicate your page. But if you feel compelled to stuff your page with a few more sharing buttons, click "Add 'More' button" and click away.

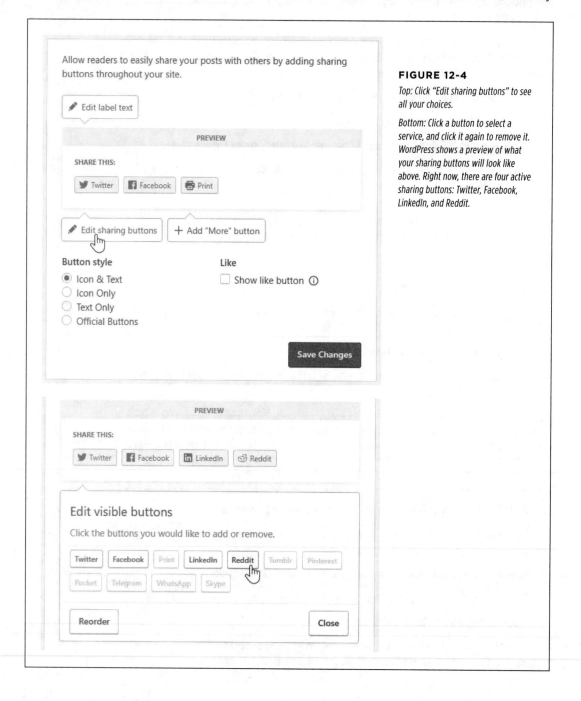

FIGURE 12-4

Top: Click "Edit sharing buttons" to see all your choices.

Bottom: Click a button to select a service, and click it again to remove it. WordPress shows a preview of what your sharing buttons will look like above. Right now, there are four active sharing buttons: Twitter, Facebook, LinkedIn, and Reddit.

FIGURE 12-5

This site crams in nine sharing buttons. However, the less commonly used ones start off hidden. When someone clicks More, the additional options appear in the drop-down panel shown here.

9. **Optionally, pick a different style for your buttons.**

The Button style setting has four choices:

Icon & text. This is the standard button style you've been looking at all along.

Icon only. This gives you tiny picture buttons with no text. Each button becomes a circle with a tiny logo in it. For example, the Twitter button turns into a blue circle with the familiar flying bird. Using icons saves space, and makes your buttons look a bit more stylish and professional (at the expense of being more difficult for less experienced people to recognize).

Text Only. You get plain text, with nothing fancy. This might suit your site if it uses a minimalist design.

Official Buttons. You use the official style conventions of the appropriate service (in other words, the Facebook button has the visual styling set by Facebook, the Twitter button has the look designed by Twitter, and so on). The drawback is that you end up with a mishmash of subtly different fonts, colors, and spacing in your sharing buttons. The advantage is that you sometimes get extra frills—for example, the Facebook button shows the number of times this post has been shared, right on the button.

10. **Optionally, edit the bit of text that appears just before the sharing buttons.**

The standard caption is "Share this." It's styled to appear in all capitals (you can't change that). But if you wanted to write, say, "Tell your peeps about me!" just click "Edit label text" and fill in your new words.

11. **Click Save Changes.**

Now you can head back to your site and check out a post. Refresh the page to see the new sharing buttons.

Changing Where Sharing Buttons Appear

Ordinarily, Jetpack puts your sharing buttons on all your posts and all your pages. But you can take control and tell Jetpack exactly where you want to see sharing buttons. You do that in the same WordPress.com page you've been using to choose and customize your sharing buttons. Just scroll down to the Options section, where you'll see the heading "Show like and sharing buttons on." There are four checkboxes you can pick:

- **Posts.** This adds the sharing buttons to single-post pages—the ones with your post content and comments section. You definitely want sharing buttons here.

- **Pages.** This adds sharing buttons to each static page (for example, the About Me page you may have on your site). Static pages usually show content that doesn't change often and isn't as newsworthy as your posts, so you might not want sharing buttons on these pages.

- **Front Page, Archive Pages, and Search Results.** This adds sharing buttons after each post, when it appears in a list of posts—for example, on your site's front page or in a page of search results. You might choose to put sharing buttons here if your home page shows complete posts rather than just excerpts. In that case, it's reasonable to assume that some visitors will do all their reading on your home page, without clicking through to the single-post page. But if your home page displays excerpts, you definitely don't want sharing buttons, because it'll seem wildly presumptuous to ask your readers to share posts they haven't even read.

NOTE It's unfortunate that Jetpack combines the Front Page, Archive Page, and Search Results options into a single setting. When you perform a search, you never see more than an excerpt of a post, so it doesn't make sense to have sharing buttons in search results, even if you do want sharing buttons on your front page. Sadly, WordPress won't let you make this distinction.

- **Media.** This adds sharing buttons to attachment pages, which display media files. For example, readers can get to this page by clicking a picture in a gallery. It's not terribly important to add sharing buttons here, because most of your readers won't go to these pages or spend much time on them.

Once you pick your sharing buttons and choose where they'll appear, click Save Changes. You can now browse to your site and give them a whirl.

FREQUENTLY ASKED QUESTION

Selectively Hiding Sharing Buttons

I don't want all my posts to be the same. Can I show sharing buttons on some but not all posts?

Yes. To do that, you need to switch on sharing for all posts, and then opt out of sharing for specific posts. (Hopefully, you want to share most of your posts, or this technique gets a bit tedious.)

The trick is an almost hidden setting that appears in the post editor. To find it, start editing a post, and then click the green Jetpack icon in the top-right corner of the post editor. This pops open a sidebar with Jetpack settings, including a "Show sharing buttons" checkbox. Turn that off, and your post will *never* show any sharing buttons. The same technique works with pages.

Remember, everyone needs to agree to show sharing buttons. If the post has the "Show sharing buttons" setting turned on, but you've told Jetpack not to show sharing buttons on posts, you won't get any buttons.

Consider an example. Imagine you have a site with 36 posts and want to allow sharing on all but three. First make sure you have Jetpack configured to show sharing buttons on all posts. Then, turn off the "Show sharing buttons" checkbox for the three posts that *shouldn't* have sharing buttons.

Letting People Like Your Site

As you've seen, visitors can use the Facebook button to share your posts. But you might prefer to let Facebookers show their appreciation for your site as a whole, using the sort of Like box shown in *Figure 12-6*.

Liking advertises your *whole* site on Facebook, not just one post. As a result, it's more likely to get people interested in your content. But that benefit comes with a cost. You need to maintain a Facebook presence for your site—a *Facebook Page* that works in tandem with your WordPress site.

NOTE You may have noticed that Jetpack has its own settings for adding a Like button. However, the Jetpack Like button isn't what you want, because it records votes only from WordPress.com users. Most of your visitors won't have WordPress.com accounts, so your site can't take full advantage of the WordPress.com voting and ranking system.

A *Facebook Page* is a public meeting spot you create on Facebook. You use it to promote something—say, a company, a cause, a product, a television show, or a band. You might already have a Facebook Page to promote your business or yourself (for example, musicians, comedians, and journalists often do). Any Facebook member can create one.

A Facebook Page is similar to a personal Facebook profile, but it's better suited to promotion. That's because anyone can visit a Facebook Page and read its content, even if they aren't Facebook friends with the page owner, or don't even have a Facebook account. Those who do have accounts can do the usual Facebook things—click Like to follow the page, post on the page's timeline, and join in any of its discussions. Generally, a personal Facebook profile is better suited to keeping up with friends or networking with business contacts, while a Facebook Page offers a better way to promote yourself, your business, or your cause to masses of people you don't know. (If you don't have a Facebook Page yet, you can quickly create one at *www.facebook.com/pages/create*.)

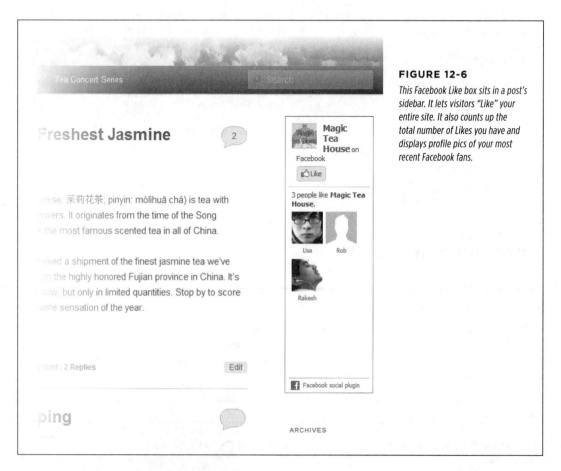

FIGURE 12-6

This Facebook Like box sits in a post's sidebar. It lets visitors "Like" your entire site. It also counts up the total number of Likes you have and displays profile pics of your most recent Facebook fans.

Once you have a Facebook Page, it's easy to add a Facebook Like box to your site. Several plugins can do the job, including the Jetpack plugin. Here's how to get set up with Jetpack:

1. **Choose Jetpack→Settings.**

   This takes you to the familiar Jetpack settings page.

2. **Click the Writing tab and scroll down to the Widgets section.**

   The Writing tab has quite a few settings, although some of them are pretty specialized (like writing posts in Markdown syntax or writing a math equation in LaTeX).

3. **Turn on the "Make extra widgets available" setting, if it isn't already on.**

   Now you have access to all of Jetpack's special widgets, including the Facebook Like box.

4. **Choose Appearance→Customize.**

   This opens the familiar theme customizer that you use to stylize your site.

5. **Click the Widgets section, and choose one of the widget areas in your theme.**

   Ideally, you should add the widget to your home page (with its list of posts), *and* to your single-post page. Some themes (like Twenty Twenty) don't include a sidebar, so you'll need to put the Like box in the footer area.

6. **Click the Add a Widget button.**

   You'll see a list of all the widgets you can choose from.

7. **Type "facebook" into the widget search box and click the Facebook Page Plugin when it appears (*Figure 12-7*).**

   This adds the Facebook Page Plugin to your sidebar. But you still need to configure a few essential details before it can go live.

FIGURE 12-7

The Facebook Page Plugin is one of the special widgets that Jetpack offers your site.

8. **Type in the Facebook Page web address.**

   The easiest way to do that is to visit your page through Facebook and then copy the address from your browser's address bar.

9. **Optionally, configure the other settings for the Facebook Like box.**

   As with any widget, you can give the Facebook Like box a title, but that's really not necessary. You can also set its width, change its color scheme, and choose whether you want to display your fans' faces (as in *Figure 12-6*), the latest posts from your page's news stream, or the latest posts from its timeline. If you opt out of all these options, you get a very compact box that includes a Like button, the number of Likes you've received, and a tiny thumbnail of the profile picture from your Facebook Page.

10. **Click Publish.**

    This adds the Like box widget to your page.

■ Keeping Readers in the Loop

The best sites are *sticky*—they don't just attract new visitors; they encourage repeat visits.

To make a site sticky, you need to build a relationship between your site and your readers. You want to make sure that even when your readers leave, they can't forget about your site, because they're still linked to its ongoing conversation. One way to do that is to notify readers about posts that might interest them. Another strategy is to tell readers when someone replies to one of their comments. Both techniques use email messages to lure visitors back to your site.

If you installed Jetpack, you automatically get a convenient opt-in system for email notifications. You just need to switch it on:

1. **Choose Jetpack→Settings from the admin menu.**

2. **Click the Discussion tab.**

3. **In the Subscriptions section, switch on "Let visitors subscribe to new posts and comments via email."**

4. **Turn on the "subscribe to site" and "subscribe to comments" options underneath.**

   Now, when someone visits your site, they'll see two new checkboxes in the Leave a Reply section (*Figure 12-8*).

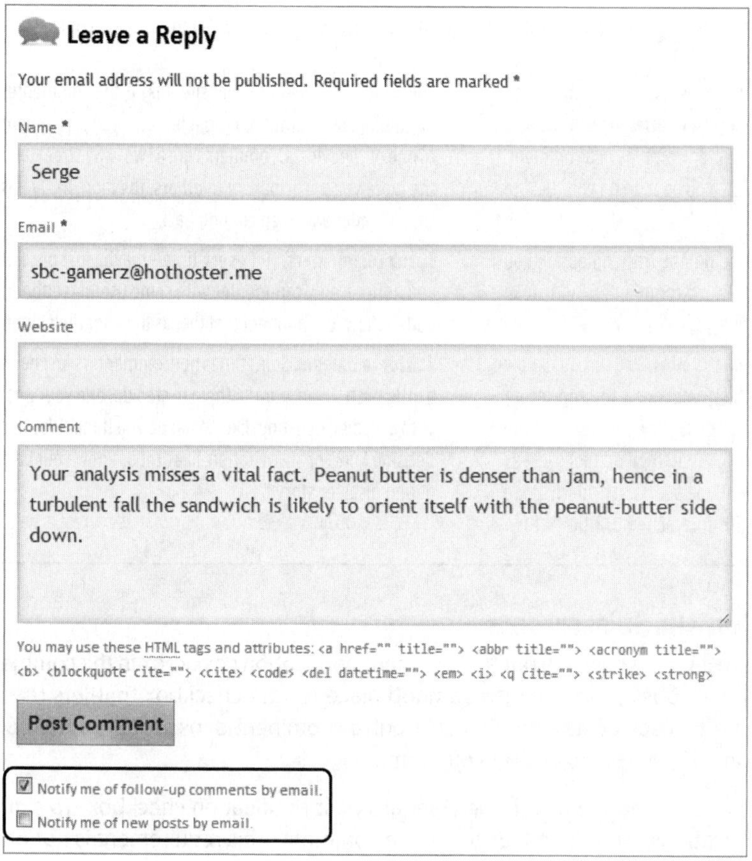

Leave a Reply

Your email address will not be published. Required fields are marked *

Name *

> Serge

Email *

> sbc-gamerz@hothoster.me

Website

Comment

> Your analysis misses a vital fact. Peanut butter is denser than jam, hence in a turbulent fall the sandwich is likely to orient itself with the peanut-butter side down.

You may use these HTML tags and attributes: `<a href="" title=""> <abbr title=""> <acronym title=""> <b> <blockquote cite=""> <cite> <code> <del datetime=""> <em> <i> <q cite=""> <strike> <strong>`

Post Comment

☑ Notify me of follow-up comments by email.

☐ Notify me of new posts by email.

FIGURE 12-8

After Serge enters his email address and writes his comment, he can sign up for WordPress notifications. WordPress lets him know when someone replies to his comments (so he can remain part of the conversation), or when you publish a new post (if he really loves your content).

UP TO SPEED

Taking Care of Your Peeps

Even on sites with thousands of comments, most readers keep quiet. Whether that's due to laziness, indifference, or the fear of being ignored, the average reader won't leave a comment. So when someone does speak up, you need to do your best to keep them in the discussion.

One way to do that is with the comment-tracking option you just read about. You can also reward commenters and stoke the conversation in several more ways:

- **Comment on your commenter's sites.** You already know that, every once in a while, you need to step into a discussion with your own comment. Visitors like to see you involved because it shows you read their opinions just as they read yours. However, if you see a particularly good comment, you can take this interaction a little further.

Follow the commenter's website link. If the commenter has a blog, stick around, read a bit, and add a comment to one of *their* posts. Comments are a two-way street, and the more you participate with others, the more likely it is that a reader will keep coming back.

- **Thank commenters.** Not every time—maybe just once. If you notice a new commenter with some useful feedback, add a follow-up comment that thanks them for their input.

- **Ask for comments.** Sometimes noncommenters just need a little push. To encourage them to step up, end your post with a leading question, like "What do you think? Was this decision fair?" or an invitation, like "Let us know your best dating disaster story."

Signing Up Subscribers

Although it makes sense to put the comment notification checkbox in the comments area of your posts, that spot isn't a good place for the checkbox that lets readers subscribe to your posts. Ideally, you'll put a prominent subscription option after every one of your posts *and* on your home page.

There's another problem with the standard post notification checkbox. To sign up for notifications, a reader needs to leave a comment. This requirement is not only a bit confusing (readers might not realize they need to write a comment, tick the site-subscription checkbox, and *then* click Post Comment), but also unnecessarily limiting.

Fortunately, Jetpack offers a better option, with a subscription widget that can sign up new followers anytime, called Blog Subscriptions. Here's how to use it:

1. **Make sure you've enabled Jetpack widgets.**

   If you're not sure, choose Jetpack→Settings, click the Writing tab, and check that the "Make extra widgets available" setting is on.

2. **Choose Appearance→Customize.**

   This launches the theme customizer.

3. **Click the Widgets section, and choose one of the widget areas in your theme.**

It's a good idea to include the subscription widget in two places: your home page and the footer area of each post (with a message like "Liked this article? Subscribe to get lots more.")

4. **Click the Add a Widget button.**

This brings up a list of all the widgets your site supports.

5. **Add the Blog Subscriptions widget.**

The easiest way to find it is to type "subscriptions" into the widget search box.

6. **Configure the Blog Subscriptions widget (*Figure 12-9*).**

You can customize several bits of information, including the widget title, the text that invites readers to sign up, and the text on the Subscribe button. For best results, keep the text in the widget brief.

You can also choose to show the total number of subscribers, in which case the subscription box adds a line like "Join 4 other followers."

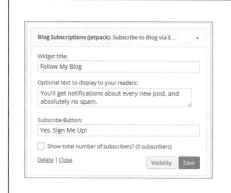

FIGURE 12-9

If you configure the subscription widget like the one on the left, your readers see a subscription box like the one on the right.

7. **Click Publish.**

Now that the subscription box is everywhere on your site, you can turn off the "subscribe to site" comment setting you used earlier (page 364). But that's up to your personal taste.

Emailing Subscribers

Occasionally, you might want to reach out to your followers and send them an email that doesn't correspond to a post. For example, you might offer a special promotion or solicit feedback on a website change. If you decide to take this step, tread carefully—if you harass readers with frequent or unwanted emails, they'll feel like they're being spammed.

If you decide to go ahead and email your followers, you first need to get their email addresses. Here's how you can get them:

1. **Choose Jetpack→Settings.**

2. **Click the Discussion tab.**

3. **Scroll down to the Subscriptions box and click "View your email Followers."**

This takes you to the WordPress.com management interface, where you can see the total number of people subscribed to your blog, and get the full list of email addresses of everybody.

If you want even more control—for example, the ability to send out your own newsletters and promotional emails—you can sign up with a professional mailing list provider like Mailchimp (*http://mailchimp.com*) or MailerLite (*http://mailerlite .com*). Every mailing list has its own WordPress plugin that you can easily find in the plugin directory. The catch is that mailing lists aren't free. At best, you'll get to sign up a couple of thousand contacts before you need to cough up a monthly payment.

PLUGIN POWER

Even Better Email Subscription Services

Jetpack gives WordPress sites a solid, straightforward subscription package. The WordPress.com servers handle all the user tracking and emailing, making your life easy. But Jetpack doesn't include any settings that let you customize the way it handles subscriptions. More advanced email and newsletter plugins (some of which will cost you a bit of cash) offer more features.

One example is the popular Subscribe2 plugin (*http://tinyurl .com/wp-sub2*). It adds the following useful features:

• **Digests.** Instead of sending readers an email after you publish every new post, Subscribe2 lets you send a single email, periodically, that announces several new posts at

once. Subscribe2 calls this message a *digest*. For example, you might choose to send subscribers a weekly digest summarizing the past seven days' posts.

• **Excluded categories and post types.** Perhaps you don't want to send notifications for every new post. Subscribe2 lets you exclude certain categories or post types (like asides and quotes) from notification emails.

• **User-managed subscriptions.** If you're willing to let readers sign up as users on your site (page 337), they can manage their own subscriptions. For example, they can subscribe to just the post categories that interest them and pick the most convenient digest option.

Publicizing Your Posts on Social Media

As you've seen, one good way to get the word out about your site is to get your readers talking and sharing on social media sites. But you don't need to wait for them to do the work for you—if you have a Facebook or Twitter account, you can use it yourself to announce new content.

This technique is often called *publicizing*, and it's not quite the same as the social sharing you learned about earlier. Sharing is when a visitor introduces new people to your content. Publicizing is when *you* tell readers about new content. The difference is that the people you tell probably already know about your site. Your goal is to get them interested enough to come back.

Publicizing is an increasingly important way to reach your readers. Twitter fanatics may pay more attention to tweets than they do to email messages. Facebook fans who won't bother to sign up for an email subscription might not mind liking your Facebook Page and getting notifications from you in their News Feeds. For these reasons, many WordPressers choose to publicize their posts.

The Jetpack plugin provides a handy Publicize feature. But first, you need to connect your site to the social media account (or accounts) you want to use. Here's how:

1. **Choose Jetpack→Settings from the admin menu.**

2. **Click the Sharing tab.**

   This is the same tab you used to add social media icons earlier.

3. **In the "Publicize connections" section, switch on "Automatically share your posts to social networks."**

   This tells Jetpack that you want to publicize your new posts on social media. But before anything can happen, you need to connect Jetpack to your social media accounts.

4. **Click "Connect your social media accounts" underneath.**

   This takes you to the WordPress.com management pages. You'll see a short list of social media services you can connect, with Facebook and Twitter at the top (*Figure 12-10*).

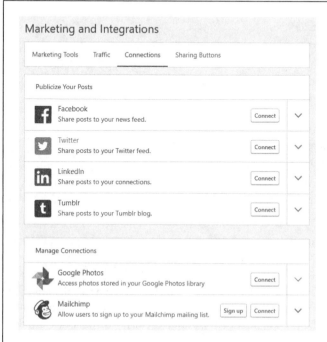

FIGURE 12-10

The Publicize feature lets you post to services like Facebook, Twitter, LinkedIn, and Tumblr. Once you pick a service, you need to log in to your social media account to complete the connection.

NOTE There's one big restriction with the Facebook publicizing feature. Currently, you can use it only if you've created a Facebook Page for your site (an ordinary personal profile isn't good enough). A Facebook Page is free to set up, and easy enough to use (but a bit of work to maintain). More details are on page 361.

5. **To use one of these services, click the Connect button next to the corresponding icon.**

   WordPress will get you to log in and confirm that your site should be allowed to post or tweet on your behalf. Then you're ready to start publicizing new posts.

The Publicize feature springs into action every time you publish a new post. However, WordPress lets you control the process before your work goes live.

When you're writing a post, you can review the publicize settings by clicking the Jetpack icon in the top-right corner of the post editor. You'll see a "Share this post" section with a list of your connected social media services (*Figure 12-11*).

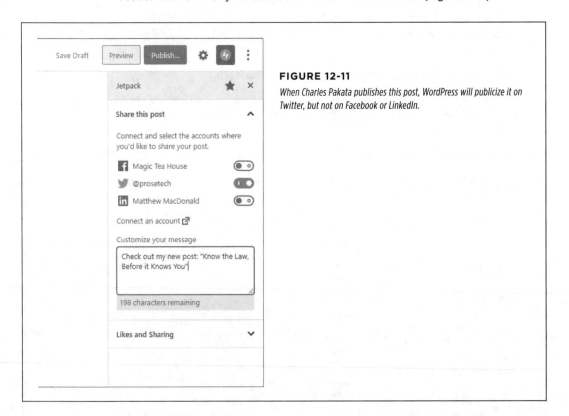

FIGURE 12-11

When Charles Pakata publishes this post, WordPress will publicize it on Twitter, but not on Facebook or LinkedIn.

There are two things you can change in the "Share this post" section:

- **Stop sharing.** If you don't want to publicize a post on a given social media platform, switch it off. You can even turn off all your linked services. But remember, this setting is for the current post only. As long as you have publicize turned on in Jetpack, every new post will opt in to social sharing.

- **Change your message.** You can edit the message that Jetpack sends out to announce your new post. Ordinarily, Jetpack uses the post title for the message, but you can substitute more descriptive text. For example, if Charles Pakata publicizes the post in *Figure 12-11* to Twitter, his followers will see the tweet shown in *Figure 12-12*.

TIP You'll also see the same section appear in the sidebar when you click Publish. However, it's easy to get into the habit of speeding through WordPress's last-minute confirmation and clicking Publish twice in a row, without reviewing the social media settings. For that reason, it's a good idea to set these settings ahead of time, like when you first start writing your post.

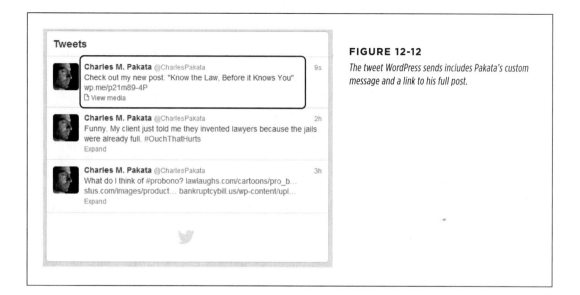

FIGURE 12-12

The tweet WordPress sends includes Pakata's custom message and a link to his full post.

Sharing Your Tweets on Your Site

You already learned how to encourage people to tweet about your site, but the integration between WordPress and Twitter runs deeper than that. If you're a Twitter-holic, WordPress has a nifty way to integrate *your* Twitter feed into your own site.

Before you dive into this feature, it's worth taking a moment to ask why you'd use it and how it fits into your site's promotional plans. There are several good reasons to use it:

- **To offer extra content.** If you're an avid tweeter, you can stuff your feed with news, tiny tips, and micro observations related to your site. Those details might be interesting to your readers even if they aren't worth a full post.

- **To make your site feel alive.** Having a Twitter feed can make your site seem more current and dynamic—provided that you tweet regularly.

- **To attract new followers.** If you display your Twitter feed on your site, some of your readers might decide to follow your feed. If they do, you'll have another way to reach them. This is particularly useful if you use Twitter to announce your blog posts using WordPress's Publicize feature (page 368).

To display a Twitter feed on your site, you could use a small, focused plugin like the popular Custom Twitter Feeds (*https://tinyurl.com/cust-twitter*). Or you could dig around in the trusty Jetpack plugin to find the Twitter Timeline widgets. Here's how:

1. **Make sure you've enabled Jetpack widgets.**

   If you're not sure, choose Jetpack→Settings, click the Writing tab, and check that the "Make extra widgets available" setting is on.

2. **Choose Appearance→Customize.**

   This loads the theme customizer.

3. **Click the Widgets section, and choose one of the widget areas in your theme.**

   Ideally, your theme will provide a sidebar widget area. Unfortunately, if you're using Twenty Twenty, you have only footer widget areas, which put your tweets at the bottom of the page.

4. **Click the Add a Widget button and add the Twitter Timeline widget.**

   The easiest way to find it is to type "twitter" into the widget search box.

5. **Fill in the Twitter Username box.**

   This is the only essential detail, because it tells the widget what feed to show. You need to use a Twitter username, which always starts with an @, as in *@CharlesPakata*.

6. **Optionally, configure the other settings of the Twitter Timeline widget (*Figure 12-13*).**

   You don't need to change anything if you don't want to. However, here are some details you might want to change:

   # of Tweets Shown lets you create a short tweet list, like the one in *Figure 12-13*. Without that, you'll get a long list of tweets and a Load More button at the end.

   No Header and **No Footer** can help make your timeline more compact, by removing the "Tweets by" message at the top or the View on Twitter link at the bottom.

   Transparent Background removes the white background so the background of your page shows through. It can help you blend the feed into your site more seamlessly.

   Timeline Theme lets you switch between a Light and a Dark option. You choose Light if your theme has a light background color, or Dark if your theme has a dark background color (in which case the feed will use light-colored text so it stands out better).

7. **Click Publish.**

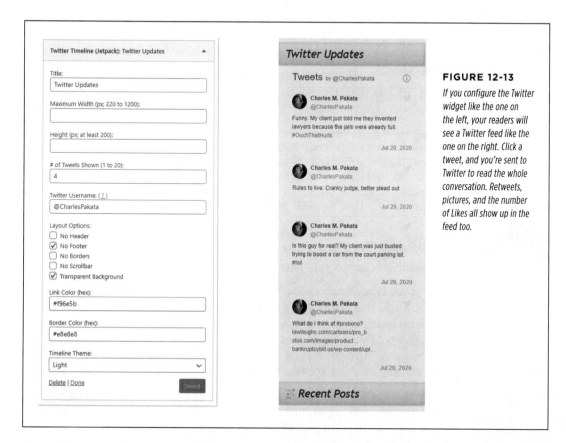

FIGURE 12-13

If you configure the Twitter widget like the one on the left, your readers will see a Twitter feed like the one on the right. Click a tweet, and you're sent to Twitter to read the whole conversation. Retweets, pictures, and the number of Likes all show up in the feed too.

Search Engine Optimization

As you've seen, you have an exhaustive range of options for getting the word out about your site. You can share your posts, publicize them, use email notifications, and tweet the heck out of everything. All these techniques share something in common—each one is a type of *social networking*, where you use connections to people you already know to reach out just a bit further.

You have another way to get people to your site, but it's more difficult and less fun. You can try to attract complete strangers when they run a web search for content related to your site. To perform this trick, you need to understand *search engine optimization* (SEO), which is the sometimes cryptic art of getting web search engines like Google to notice you.

The goal is to make your site appear in a highly ranked position for certain searches. For example, if your WordPress site covers dog breeding, you'd like web searchers to find your site when they type "dog breeding" into Google. The challenge is that for any given search, your site competes with millions of others that share the same search keywords. If Google prefers those sites, your site will be pushed farther down in the search results, until even the most enthusiastic searcher won't spot you. And if searchers can't find you on Google, you lose a valuable way to attract fresh faces to your site.

Next, you'll learn a bit more about how search engines like Google work, and you'll consider how you can help your site rise up in the rankings of a web search.

NOTE In the following sections, you'll spend a fair bit of time learning about Google. Although Google isn't the only search engine on the block, it's far and away the most popular, with a staggering 80 percent (or more) of worldwide web-search traffic. For that reason, it makes sense to consider Google first, even though most of the search engine optimization techniques you'll see in this chapter apply to all the major search engines, including Bing, Yahoo!, and even Baidu, the kingpin of web search in China.

PageRank: Scoring Your Site

To help your site get noticed, you need to understand how Google runs a web search.

Imagine you type "dog breeding" into the Google search page. First, Google peers into its gargantuan catalog of sites, looking for pages that use those keywords. Google prefers pages that include the keywords "dog breeding" more than once, and pages that put them in important places (like headings and page titles). Of course, Google is also on the lookout for sites that try to game the system, so a page that's filled with keyword lists and repetitive text is likely to get ignored.

Even with these requirements, a typical Google search turns up hundreds of thousands (or even millions) of matching pages. To decide how it should order these pages in its search results, Google uses a top-secret formula called *PageRank*. PageRank determines the value of your site by the community of websites that link to it. Although the full PageRank recipe is incredibly convoluted (and entirely secret), its basic workings are well-known:

- The more sites that link to you, the better.

- A link from a better, more popular site (a site with a high PageRank) is more valuable than a link from a less popular site.

- A link from a more selective site is better than a link from a less selective site. That's because the more outgoing links a site has, the less each link is worth. So if someone links to your site and just a handful of others, that link is valuable. If someone links to your site and *hundreds* of other sites, the link's value is diluted.

Finding Your PageRank

Because of the power of PageRank scores, it's no surprise that web authors want to know how their pages are doing. But Google won't give out the real PageRank of a web page, even to its owner.

That said, Google does allow website owners to see a *simplified* version of their PageRank score, which gives you a general idea of your site's performance. The simplified PageRank is based on the real thing, but Google updates it just twice a year, and it provides only a value from 1 to 10. (All things being equal, a website rated 10 will turn up much higher in search results than a page ranked 1.)

There are two ways to find your website's simplified PageRank. If you use the Google Chrome browser, you can add a handy browser plugin to do the job (get it at *http://tinyurl.com/wr-extension*). Another approach is to use an unofficial PageRank-checking website, like *www.prchecker.info* (*Figure 12-14*).

The simplified PageRank score isn't always accurate. If you submit a site that's very new, or hasn't yet established itself on the web (in other words, few people are visiting it and no one's linking to it), you may not get a PageRank value at all.

> **TIP** Don't worry too much about your exact PageRank. Instead, use it as a tool to gauge how your website improves or declines over time. For example, if your home page scored a PageRank of 4 last year but a 6 this year, your promotion is clearly on the right path.

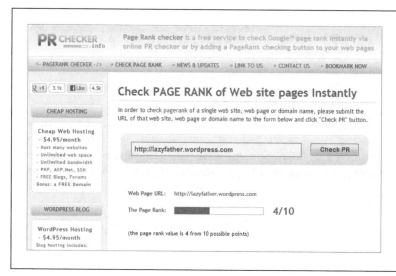

FIGURE 12-14

To see the PageRank for your home page, type in your site's address and then click Check PR. Here, http://lazyfather.wordpress.com scores a middle-of-the-road 4 out of 10.

Getting More Links

The cornerstone of search-engine ranking is links—the more people connect to you, the greater your web prestige and the more trustworthy your site seems to Google. Here are some tips any WordPresser can use to build up their links:

- **Look for sites that are receptive to your content.** To get more links, you need to reach out and interact with other websites. Offer to guest-blog on a like-minded site, join a community group, or sign up with free website directories that include your type of business.

- **Keep sharing.** The social sharing techniques you learned about in the first part of this chapter are doubly important for PageRank. Although tweets and Likes aren't as powerful as website links, Google still counts them in your favor when respected people talk about your content on social media sites.

- **Add off-site links (that point to you).** You don't need to wait for other people to notice your content. It's perfectly acceptable to post a good comment on someone else's blog, with a link that references something you wrote. Or post in a forum, making sure your signature includes

your name *and* a link to your site. The trick is to find sites and forums that share the same interests as your site. For example, if you're an artisanal cheese maker in Chicago, it makes sense to chat with people running organic food cooperatives. But be careful. There's a thin line between spreading the word about your fantastic content and spamming other people. So don't post on a forum or someone else's site unless you can say something truly insightful or genuinely helpful. If you're not sure whether to post, ask yourself this question: "If this were my site, would I appreciate this comment?"

- **Research your competitors' links.** If you find out where other people are getting their links from, you may be able to get links from the same sites. You can do this sort of research using a tool called a *backlink checker.* Multiple backlink checkers are on the web, including the free site *www.openlinkprofiler.org.* Type in a website address, and the backlink checker will find other pages that lead there. For example, searching for *www.magicteahouse .net* shows you all the sites that link to it.

Making Your Site Google-Friendly

You can't trick Google into loving your site, and there's no secret technique to vault your site to the top of the search page rankings. However, you can give your site the best possible odds by following some good habits. These practices help search engines find their way around your posts, understand your content, and recognize that you're a real site with good content, not a sneaky spammer trying to cheat the system.

Here are some guidelines to SEO that don't require special plugins or custom coding:

- **Choose the right permalink style.** Every WordPress post and page gets its own permalink. For best search results, your permalinks should include the post title, because the search engine pays special attention to the words in your web address. (Page 91 explains how to change your permalink.)

- **Edit your permalinks.** When you first create a post, you have the chance to edit its permalink. At this point, you can improve it by removing unimportant words (like "a", "and," and "the"). Or, if you use a cute, jokey title for your post, you can replace it in the permalink with something more topical that includes the keywords you expect web searchers to use. For example, if you write a post about your favorite cookware titled "Out of the Frying Pan and Into the Fire," you ordinarily get a permalink like this: *http://triplegoldcookwarereview.com /out-of-the-frying-pan-and-into-the-fire*. If you remove some words, you can shorten it to *http://triplegoldcookwarereview.com/out-of-frying-pan-into-fire*. And if you substitute a more descriptive title, you might choose *http://triple-goldcookwarereview.com/calphalon-fry-pan-review*.

- **Use tags.** Google pays close attention to the tags you assign to a post. If they match a web searcher's keywords, your post has a better chance of showing up in search results. When choosing tags, pick just a few, and make sure they clearly describe your topic and correspond to terms someone might search for (say, "artisanal cheese," "organic," and "local food"). Some search-obsessed bloggers scour Google statistics to find the best keywords to use in attracting web searchers, and use those as their tags in new posts. However, that's too much work for all but the most fanatical SEO addicts.

- **Optimize your images.** Google and other search engines let people search for pictures. When someone searches for an image, Google attempts to match the search keywords with the words that appear near the picture on a web page, and with the alternate text that describes the picture. That means people are more likely to find your pictures if you supply a title, alternate text, and a caption. Remember to use descriptive, searchable keywords when you do.

- **Make sure your site works on mobile devices.** If your theme was released in the last five years, it almost certainly has mobile support. But if you're saddled with something older, or you've built a theme yourself, make sure it adjusts itself for easy reading on mobile browsers, or use a plugin to patch the gap, like WPtouch (page 463). Otherwise, Google will penalize your site and be less likely to recommend its pages.

GEM IN THE ROUGH

Hiding from Search Engines

You don't *have* to let search engines find you. If you want to keep a low profile, choose Settings→Reading, and then turn on the setting "Discourage search engines from indexing this site." That way, your website won't appear in most search engine listings.

People will still be able to find you if they click a link that leads to your site, or if they know your site address. For that reason,

you shouldn't rely on this trick to conceal yourself if you're doing something dubious or risky—say, planning a bank robbery or cursing your employer. When you need utmost privacy, you can use WordPress's private site feature (page 336), or just keep yourself off the web.

Boosting SEO with a Plugin

You can make your site more attractive to Google and other search engines by using an SEO plugin. But be warned: most SEO plugins are a possible case of overkill for the casual WordPress site-builder. Prepare to be swamped by pages of options and search settings.

If you search WordPress's plugin repository for "SEO," you discover quite a few popular plugins. One of the best is Yoast SEO (*http://tinyurl.com/seo-yoast*). Its creator is WordPress über-guru Joost de Valk, who also blogs some useful (but somewhat technical) SEO articles at *http://yoast.com/seo-blog*.

Once you install and activate WordPress SEO, you see a new SEO menu in the admin area, packed with a dizzying array of options. You can ignore most of them, unless you want to change the way the plugin works. The following sections explain two useful features you can tap into right away.

■ CREATING AN XML SITEMAP

After installing the SEO plugin, your site gets one immediate benefit: an *XML sitemap*. This document tells Google where your content resides on your site. It ensures that all your posts get indexed, even if your home page doesn't link to them. Although you don't need to give your XML sitemap another thought, you can take a look at it—just choose SEO→General, click the Features tab, scroll down to the XML Sitemap setting, and click "See the XML sitemap."

Needless to say, WordPress SEO updates your sitemap every time you publish a new post or page.

TIP If you aren't using Yoast SEO but are using Jetpack, you can still get XML sitemaps. That's because Jetpack has its own half-hidden XML sitemap feature. To use this ability, choose Jetpack→Settings, click the Traffic tab, and turn on the "Generate XML sitemaps" setting.

■ TWEAKING SEARCH DESCRIPTIONS

The WordPress SEO plugin also gives you control over two important details: the title and description (known to web nerds as the *meta description*) of each post or page. These details are useful—even to SEO newbies—because Google displays them when it lists a page from your site in its search results. *Figure 12-15* shows an example.

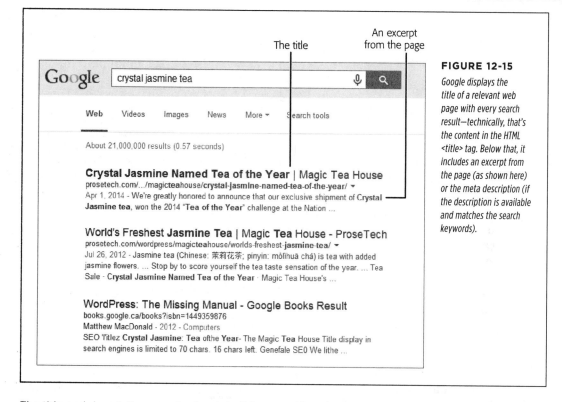

FIGURE 12-15

Google displays the title of a relevant web page with every search result—technically, that's the content in the HTML <title> tag. Below that, it includes an excerpt from the page (as shown here) or the meta description (if the description is available and matches the search keywords).

The title and description are also important because Google gives more weight to keywords in those places than keywords in your content. In other words, if someone searches for "dog breeding" and you have those words in your title, you can beat an equally ranked page that doesn't.

Ordinarily, the WordPress SEO plugin creates a good title for a post, based on a title-generating formula in the SEO→Search Appearance→Content Types page. This formula puts your post title first, followed by your site name, like this for the "crystal jasmine" post:

```
Crystal Jasmine Named Tea of the Year - Magic Tea House
```

However, you can customize the title before you publish the post by using the Yoast SEO box, which appears on the Add New Post page (*Figure 12-16*). For example, it's a good idea to shorten overly long post titles, and to replace cutesy titles with ones that clearly describe your content. You can also use the Yoast SEO box to type in a meta description.

Pick the title here. You have exactly
70 characters to work with.

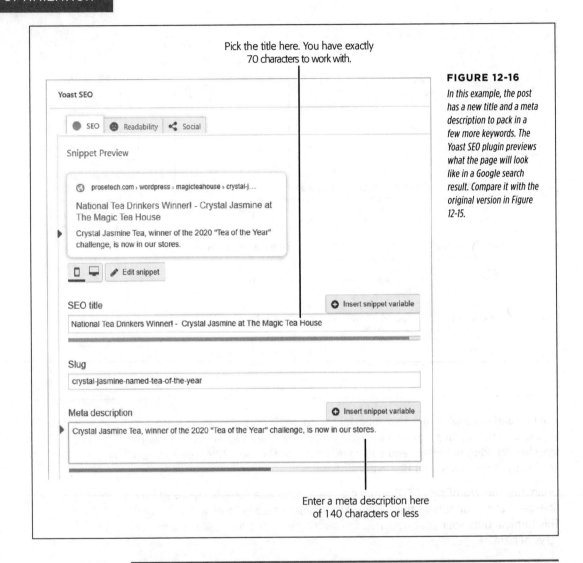

FIGURE 12-16

In this example, the post has a new title and a meta description to pack in a few more keywords. The Yoast SEO plugin previews what the page will look like in a Google search result. Compare it with the original version in Figure 12-15.

Enter a meta description here
of 140 characters or less

TIP For more ways to optimize your site for search engines using the Yoast SEO plugin, check out the detailed tutorial at *http://tinyurl.com/seo-yoast2*.

■ WordPress Site Statistics

Once you have some solid promotion tactics in place, you need to evaluate how well they perform. There's no point in pursuing a failed strategy for months, when you should be investing more effort in a technique that actually *works*. The best way to assess your site's performance, and see how it changes over time, is to collect *website statistics*.

Several popular statistics packages work with WordPress, and a range of plugins automatically add tracking code to your site. For example, one of the most popular is Google Analytics. To use it, you need to sign up with Google (*http://analytics .google.com/analytics*) and grab the free Google Analytics plugin from MonsterInsights (*https://tinyurl.com/ga-monster*). But in the rest of this chapter, you'll use a slightly simpler statistics service that's already at your fingertips. It's WordPress's own statistics-collection service, which it offers through the Jetpack plugin.

You can view your stats from the admin area by choosing Jetpack→Site Stats. You can also see an enhanced view on WordPress.com, which includes frills like a map of where your visitors live (*Figure 12-17*). To get there, click the Show Me button next to "View enhanced stats on WordPress.com."

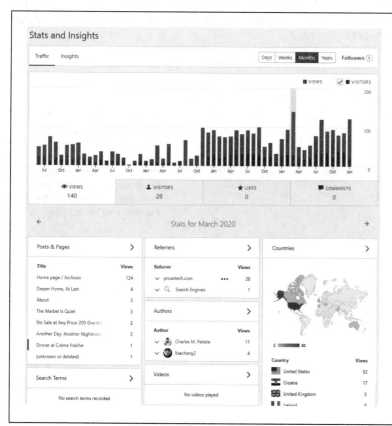

FIGURE 12-17

The Stats and Insights tab has a lot of information jockeying for your attention. Here are the details for The Real Estate Diaries site.

The obvious question is, now that you have all this raw data, what can you do with it? Ideally, you'll use site statistics to focus on your strengths, improve your site, and keep your visitors happy. You should resist the temptation to use it as a source of endlessly fascinating trivia. If you spend the afternoon counting how many visitors hit your site from Bulgaria, you're wasting time that could be better spent writing brilliant content.

The following sections present four basic strategies that can help you find useful information in your statistics, and use that insight to improve your site without wasting hours of your time.

Strategy 1. Find Out What Your Readers Like

If you know what you're doing right, you can do a lot more of it. For example, if you write a blog with scathing political commentary, and your readers flock to any article that mentions gun control, you might want to continue exploring the issue in future posts. (Or, to put it less charitably, you might decide to milk the topic for all the page views you can get before your readers get bored.)

To make decisions like that, you need to know what content gets the most attention. A Facebook Like button may help you spot popular posts, but a more thorough way to measure success is to look at your traffic. On the Stats page, focus on the Top Posts & Pages box, which shows you the most read posts and pages over the past couple of days (*Figure 12-18*).

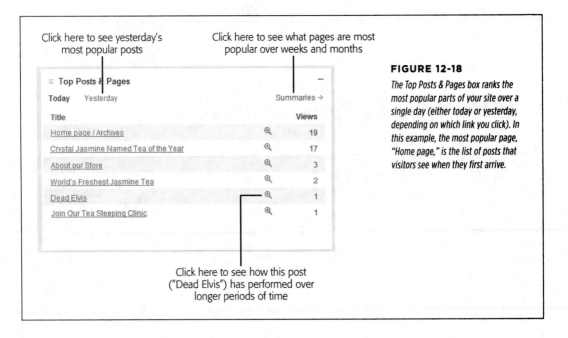

Click here to see yesterday's most popular posts

Click here to see what pages are most popular over weeks and months

Click here to see how this post ("Dead Elvis") has performed over longer periods of time

FIGURE 12-18

The Top Posts & Pages box ranks the most popular parts of your site over a single day (either today or yesterday, depending on which link you click). In this example, the most popular page, "Home page," is the list of posts that visitors see when they first arrive.

The Top Posts & Pages box gives you a snapshot of the current activity on your site, but to make real conclusions about what content stirs your readers' hearts, you need to take a long-term perspective. To do that, click the Summaries link. Now WordPress lets you compare your top pages over the past week, month, quarter, year, or all time. Just keep in mind that bigger timeframes are often biased toward older articles, because they've been around the longest.

If you analyze a site on WordPress.com, you can also check out the Tags & Categories section. It shows you the categories and tags that draw the most interest. You can form two conclusions from this box: popular categories may reflect content your readers want to keep reading, and popular tags may indicate keywords that align with popular search terms (see Strategy 3 on page 384).

Strategy 2. Who's Giving You the Love?

A visitor can arrive at your site in three ways:

- By typing your address into their browser (or by using a bookmark, which is the same thing)

- By following a link from another site that points to you

- By performing a search and following a link in the search results page

The first type of visitor already knows about you. There's not much you can do to improve on that.

The second and third types of visitor are more difficult to predict. You need to track them so you can optimize your web promotion strategies. In this section, you'll focus on the second type of guest. These people arrive at your site from another website, otherwise known as a *referrer*.

If you followed the link-building strategies laid out on page 376, the social sharing tips from page 352, and the publicizing techniques described on page 368, you've created many routes that a reader can take to get to your site. But which are heavily traveled, and which are overgrown and abandoned? To find out, you need to check the Referrers box, which ranks the sites where people come from, in order of most to least popular (*Figure 12-19*).

Once you know your top referrers, you can adjust your promotional strategies. For example, you may want to stop spending time and effort promoting your site in places that don't generate traffic. Similarly, you might want to spend more effort cultivating your top referrers to ensure you keep a steady stream of visitors coming to your site.

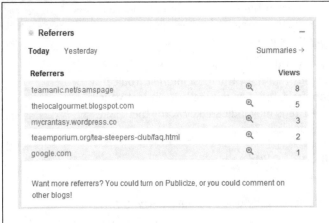

FIGURE 12-19

Use the Referrers box to see where your visitors come from. It shows you the referring sites from a single day. You can click a specific referrer to get more information, or you can click Summaries to examine your top referrers over longer periods of time.

Strategy 3. Play Well with Search Engines

In any given minute, Google handles well over a million search queries. If you're lucky, a tiny slice of those searchers will end up at your site.

Webmasters pay special attention to visitors who arrive through search engines. Usually, these are new people who haven't read your content before, which makes them exactly the sort of people you need to attract. But it's not enough to know that visitors arrive through a search engine. You need to understand what *brought* them to your site, and to understand that, you need to know what they were searching for.

The Search Engine Terms box can help you find out (*Figure 12-20*). It lists the top queries that led visitors to your site for a single day (or, if you click Summaries, over a longer period of time).

If you use SEO to find what you think are the best keywords for tags, titles, and descriptions (see, for example, page 380), the Search Engine Terms box helps you determine if your efforts are paying off. And even if you don't, it gives you insight into hot topics that attract new readers—and which you might want to focus on in the future.

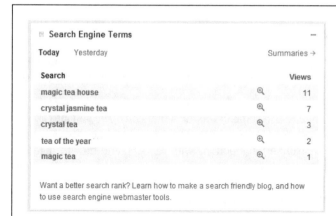

FIGURE 12-20

Here are the keywords that led searchers to the Magic Tea House. Notice that you may not see the common, short keywords that you expect (like "tea," by itself). That's because the more general a keyword is, the more sites there are competing for that keyword, and the less likely it is that a searcher will spot your site.

Strategy 4. Meet Your Top Commenters

On the WordPress.com version of the stats page, you can tap into one more set of useful statistics. Take a look in the Comments box to see which of your visitors left the greatest number of comments and which posts stirred the most conversation (*Figure 12-21*).

FIGURE 12-21

Comments are the lifeblood of a WordPress site. A site with a thriving Comments section is more likely to attract new visitors and to keep existing ones. By examining the Comments box, you can see who deserves the most credit for keeping your conversations alive.

The most interesting information is the top commenters. These people are particularly valuable, because their input can start discussions and keep the conversation going.

Once you identify your top commenters in the past week or month, you can try to strengthen your (and therefore your site's) relationship with them. Make an extra effort to reply to their comments and questions, and consider making a visit to their blog, and commenting on their posts. If they stick around, you might even offer them the chance to write a guest post for your site or to become a contributor.

The Last Word

The greatest site isn't worth much if no one visits it. But promotion is a hard job in any industry, and advertising a website in a world with several billion of them isn't any easier.

In this chapter, you learned to use some of the tools that can make you into a more effective promoter. You've learned how to help get your content out on social media platforms, how to build stronger connections with readers, and how to measure the traffic on your site. None of these techniques can help you if you don't make a quality site. But if you build a site that's stocked with solid content or provides a service people want, these promotional tools will ensure that you make the most of your work.

Editing Themes to Customize Your Site

A s you've traveled through this book, you've taken a look at every significant feature that WordPress offers and used those capabilities to build a variety of sites. However, you've always played by the rules, picking themes from the theme gallery, installing plugins from the plugin directory, and sticking to the safety of the WordPress admin area. But a whole other world of possibilities awaits those who can color outside the lines.

The key to unlocking more flexibility and building a truly unique WordPress site is to create your own theme. As you know, a theme is a mash-up of HTML markup, formatting rules, and PHP code. Ordinarily, WordPress hides these details from you. Instead, the people who create the themes and plugins your site uses handle all this, while you focus on writing fab content and adjusting settings in the admin area. But if you decide to cut loose and customize your own themes, you step into a different world. Be forewarned: this world can seem dizzyingly complex. But you don't need to understand every detail. Instead, you simply need to find the parts of a theme you want to change and work on those.

In this chapter, you'll start your journey by taking a close look at how themes work, and you'll learn how to make small alterations that can have big effects. First, you'll try modifying styles. Then, you'll crack open a theme's template files to change the code inside.

■ The Goal: More Flexible Blogs and Sites

Before you begin fiddling with themes, you need to have a clear idea of *why* you'd want to do so. What do you hope to gain by changing a theme that you can't get by using a good, preexisting theme with just the right combination of WordPress settings and plugins?

There are several good reasons:

- To get something *exactly* the way you want it. Maybe you need your WordPress site to use the official corporate colors of your business, or match other non-WordPress pages.

- To make your site unique (so it doesn't look the same as every other site that uses the same theme). After all, only so many themes exist, and if you pick a good one, it's a safe bet that you're following in the footsteps of thousands of other webmasters.

- To create a site that doesn't look "bloggy." For example, maybe you don't want date information on all your posts.

It's important to understand that you don't necessarily need to customize your themes to get these results. Some themes are more flexible than others. For example, some themes might allow you to keep the layout you love while tweaking the colors and typefaces you don't. But other themes force you to buy a professional version to get customization options, and many themes just don't provide much customizability. For them, your only option is to make more drastic changes.

Most WordPress sites have some characteristics that make them feel more bloggish and less like the sort of complex, traditional websites that rule the world of business and ecommerce. For example, WordPress sites usually display dated posts in reverse-chronological order. But what if you want to build a product showcase with posts that are actually product profiles, and you don't want those profiles to include any date information at all, like the site shown in *Figure 13-1* (top)?

Another limitation is that WordPress treats all posts the same way. What if you want to create an e-magazine with a custom-made home page for each news category you cover (Sports, Current Affairs, Lifestyle, and so on)? WordPress has that sort of extensibility built into its template system. But to use it, you need to be able to create your own theme templates.

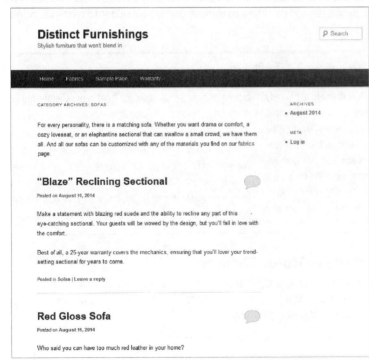

FIGURE 13-1

Top: Not many clues tell you that WordPress's blogging engine powers this ecommerce store.

Bottom: Here's the same site without a custom theme. It still functions (more or less), but it feels like a blog.

Taking Control of Your Theme

Before you can create a brilliantly customized WordPress theme, you need to know a bit more about how themes work behind the scenes.

When you first met themes in Chapter 5, you discovered that every theme consists of a combination of files. These files, which work together to create all the pages on your WordPress site, fall into three basic categories:

- **Stylesheets.** These files contain style rules that format different parts of your site, such as headlines, sidebar headings, and links. They use the much-loved CSS standard, which will be familiar to anyone who's dabbled in web design.

- **Templates.** These files contain a mix of ordinary HTML and PHP code. Each template is responsible for a different part of your site—for example, there's a template for the list of recent posts, the page header, the footer, the single-post view, and so on.

- **Resources.** These are other files that your theme's templates might use, often to add distinctive touches. Examples include image files (for fancy graphical bullets, header details, backgrounds, and so on) and JavaScript code (to create animated effects or run a featured image slideshow).

At a bare minimum, every theme needs two files: a single stylesheet named *style.css*, which sets the colors, layout, and fonts for your entire site; and a template named *index.php*, which creates the list of posts on your home page. Most themes have a few more stylesheets and many more templates, but you'll get to that in a moment.

Getting Comfortable with Themes

Editing a theme can be a bit intimidating. You need to be ready to crack open the template files that run the WordPress show. To understand what's inside, you should know at least a little about HTML markup—comfortable enough to find your way through the tangle of angle brackets and elements that live in every web page. It also helps to know some CSS (that's the Cascading Style Sheets standard that formats every modern web page). If you don't, you'll still be able to feel your way around with the introduction you'll get in this chapter, but be prepared for a steep learning curve.

If your web skills aren't quite up to snuff, plenty of good online resources can help perk them up. Read the basic overview at *www.htmldog.com*, or run through the videos and exercises on Khan Academy (*https://tinyurl.com/khan-html*). You don't need to understand everything, but a little background will make the whole process more manageable.

How WordPress Stores Themes

On your WordPress site, the *wp-content/themes* folder holds all your themes. For example, if your website address is *http://magicteahouse.net*, the Twenty Twenty theme will be in the following folder:

```
http://magicteahouse.net/wp-content/themes/twentytwenty
```

Each theme you install gets its own subfolder. So if you install seven themes on your website, you'll see seven subfolders in the *wp-content/themes* folder, even though you use only one theme at a time.

All of a theme's stylesheets and template files reside inside the theme's folder (*Figure 13-2*). Most themes also have subfolders. For example, they might tuck JavaScript files into a subfolder named *js* and image files into a subfolder named *images*. You don't need to worry about these details as long as you remember that themes are a package of files and subfolders you need to keep together in order for your site to function properly.

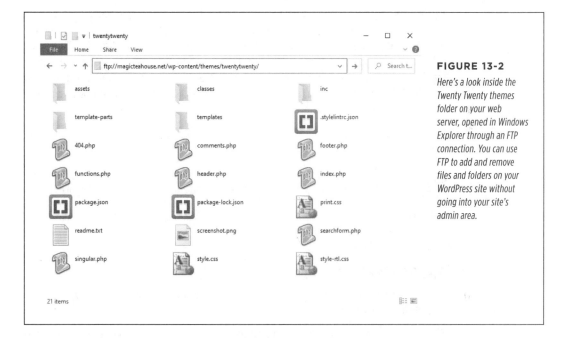

FIGURE 13-2

Here's a look inside the Twenty Twenty themes folder on your web server, opened in Windows Explorer through an FTP connection. You can use FTP to add and remove files and folders on your WordPress site without going into your site's admin area.

Style.css: How a Theme Identifies Itself

The *style.css* file is the starting point for every theme. In most themes, it's a huge file packed with formatting instructions. For example, the Twenty Twenty theme's *style.css* file weighs in with nearly *two thousand* lines of formatting magic.

The *style.css* file defines a few essential pieces of information about the style itself, using a *theme header* at the beginning of the file. Here's a slightly shortened version of the header that starts the Twenty Twenty *style.css* file. Each distinct bit of information is highlighted in bold:

```
/*
Theme Name: Twenty Twenty
Text Domain: twentytwenty
Version: 1.1
Requires at least: 4.7
```

```
Requires PHP: 5.2.4
Description: Our default theme for 2020 is designed to take full advantage of
the flexibility of the block editor. Organizations and businesses have the...
Tags: blog, one-column, custom-background, custom-colors, custom-logo...
Author: the WordPress team
Author URI: https://wordpress.org/
Theme URI: https://wordpress.org/themes/twentytwenty/
License: GNU General Public License v2 or later
License URI: http://www.gnu.org/licenses/gpl-2.0.html
*/
```

WordPress brackets the header with two special character sequences: It starts with */* and ends with *%* the CSS comment markers. As a result, browsers don't pay any attention to the header. But WordPress checks it and extracts the key details. It uses this information for the theme description you see in the Add Themes page. It also tracks the version number and checks the theme URL for new versions. (Although it says "URI" in the theme file, a URI is the same as a URL when you deal with content on the web, as in the case of themes.)

If your theme lacks these details, WordPress won't recognize that it's a theme. It won't show it in the Appearance→Themes page, and it won't let you activate it on your site.

> **NOTE** Many themes include stylesheets beyond *style.css*. However, these stylesheets add extra features or handle special cases—for example, they provide alternate color schemes, deal with old browsers, handle languages that write text from right to left, and format the content in the editing box on the Add New Post page so that it provides the most realistic preview possible.

The Theme Editor

You can edit your current theme in the WordPress admin area. Start by choosing Appearance→Theme Editor.

The first time you go to the theme editor, WordPress shows a warning explaining that changes here can damage your theme (*Figure 13-3*). The safe alternative is to create a child theme, which you'll get to in the next section. But for now, it's safe to click "I understand" to carry on and take a peek under the hood of your theme.

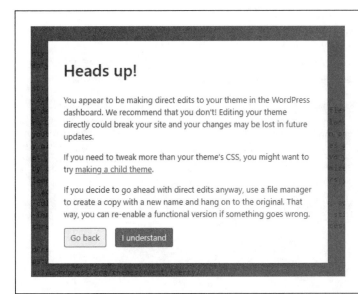

FIGURE 13-3

In a real site, you don't edit your theme without making a copy first. However, it's still safe to use the theme editor to look inside your theme files and plan your changes.

WordPress splits the theme editor into two parts. On the left is a giant text box that shows the contents of the currently selected file. On the right is a sidebar that lists all the files in your theme. Although you can use the theme editor to look at most of the files in your theme, WordPress starts you out by showing the most commonly modified file. That's the theme's main stylesheet, *style.css* (*Figure 13-4*).

The *style.css* file shows the formatting instructions that create your theme's unique visual feel. By understanding what's happening in *style.css*, you can tweak colors and fonts and other details that a theme might not ordinarily let you change. It's also the easiest part of a theme to change. But before you go down that road, you need to make sure that you aren't in danger of totally scrambling your theme.

To be safe, you need to create a child theme.

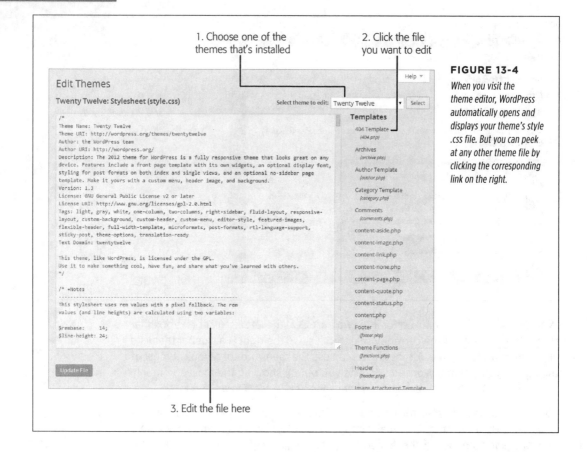

1. Choose one of the themes that's installed

2. Click the file you want to edit

3. Edit the file here

FIGURE 13-4

When you visit the theme editor, WordPress automatically opens and displays your theme's style .css file. But you can peek at any other theme file by clicking the corresponding link on the right.

Protecting Yourself with a Child Theme

Before you start mucking around with one of your themes, you should think about the long-term effects. Someday, probably not long from today, the person who created the theme you're editing will release a new and improved version that you'll want to use on your site. Here's the problem: if you install a theme update, you'll wipe out all the edits you made to your theme files. Editing themes is enough work without having to do it over and over again.

Fortunately, there's a solution. You can create a *child theme*, which takes the current theme as a starting point and lets you slap your customizations on top of it. You don't change the original theme (known as the *parent theme*) at all. Instead, you selectively edit the templates and stylesheets and save those altered files in the child theme folder. These new files override the same-named templates and stylesheets in the parent theme, so you get the features you want in your site. And when you update the *parent* theme at some future date, your customizations stay in place, because WordPress stores the child theme as a separate group of files. (If this sounds confusing, here's a hamburger analogy: the parent theme is the hamburger, and your child theme adds the toppings. When you update the parent theme, you get a brand new hamburger, but you keep all the toppings you added. They're just slapped on top of the new sandwich.)

When you customize a theme, you should *always* start by creating a child theme. It's the safest approach, and the only way to guarantee that your work will survive future changes.

The Anatomy of a Child Theme

The easiest way to create a child theme is with a plugin. But before we get to that, it helps to take a closer look at how child themes work.

Every child theme is built out of two ingredients:

- A new theme folder in the *wp-content/themes* section of your website. For example, you might call this folder *twentytwenty-child* if it's based on the Twenty Twenty theme. (By convention, theme folders have no spaces or capital letters.)

- A properly configured *style.css* file. This file will link itself to the parent theme through the theme header.

Eventually, you'll put other customized files in your child theme. But to get started, all you need is *style.css*.

Unlike the *style.css* file in the parent theme, *style.css* in the child theme can start out super simple. Here's an example:

```
/*
Theme Name: Twenty Twenty Reboot
Description: This is a customized version of the Twenty Twenty Theme for the
Magic Tea House.
Author: Katya Greenview
Template: twentytwenty
*/
```

You've already seen the first three details (theme name, description, and author). You can use any name you want, put in any descriptive text you want, and name any author you want (although usually the author is you).

But the *template* setting is a new and important detail. It points to the parent theme's folder. When WordPress sees the template line shown here, it knows to look in a folder named *twentytwenty* (in the themes section of your site). When it does, it finds the files for the familiar Twenty Twenty theme.

The *style.css* file also needs to include the parent styles that format the layout and appearance of your site. But WordPress won't automatically grab them the parent theme, unless you tell it to. One quick and easy way to do that is with the *@import* command:

```
@import url("../twentytwenty/style.css");
```

This line grabs all the styles from the Twenty Twenty theme and applies them to the child theme. You can put this line immediately after the header in the *style.css*, and you're done. There's no need to add any style rules, except those that you decide to create to customize the parent theme.

Now that you understand how a child theme is made, you could make one on your own. As long as you have a theme folder with a proper *style.css* file, WordPress will spot it and add it to theme gallery so you can activate it on your site.

However, you do need to be comfortable creating folders on your website by hand and uploading files. If you don't have these webmastering skills yet, there's a simpler approach—you can use a plugin that does the heavy lifting for you.

Creating a Child Theme with a Plugin

If you search the plugin library for "child theme," you'll find several that offer to help you create child themes. One of the most reliable is Child Theme Configurator (*https://tinyurl.com/childconfig*). Here's how to use it:

1. **Choose Plugins→Add New from the admin menu.**

   This brings you to the plugin section of your site.

2. **Install the Child Theme Configurator plugin.**

   The easiest way to find it is to search for the exact name, "child theme configurator."

3. **Activate the Child Theme Configurator.**

   Once it's activated, the Child Theme Configurator adds itself to the Tools menu.

4. **Choose Tools→Child Themes.**

   This is a new menu command, courtesy of Child Theme Configurator. It opens a multitabbed page with a pile of options.

5. **Select the parent theme you want to use from the theme list (*Figure 13-5*).**

   To use a theme, you need to have it already installed on your site. To follow along with the examples in this chapter, use the Twenty Twenty theme. It's a good starting point for theme-customization practice because it's clean, straightforward, and adaptable. If you decide to start with a different theme, all the concepts described in this chapter still apply, but the exact class names and formatting details will vary.

6. **Click Analyze to verify the theme.**

   Sometimes themes are built in nonstandard ways that make it more difficult to use them to create a child theme. Child Theme Configurator performs a quick analysis to see if any likely problem points exist in the theme you've chosen (*Figure 13-6*).

FIGURE 13-5

This site has four themes installed. You pick the one you want to use as a parent theme.

FIGURE 13-6

Child Theme Configurator reports that you can make a child theme using Twenty Twenty as its parent. It also helpfully points out that some styles are in another file (the ones used to format blog posts if the reader chooses to print them), and so they can't be overridden in the usual way in style.css.

7. **Choose a name for the child theme folder.**

Child Theme Configurator adds *-child* to the end of the parent theme name, so *twentytwenty* becomes *twentytwenty-child*. This is the standard WordPress convention, so there's no need to change the name.

8. **Scroll down to the Click to Edit Child Theme Attributes box and click it.**

This reveals the theme information that's used to make the theme header. If you want, you can adjust some of these details, like the theme name (say, change "Twenty Twenty Child" to "Twenty Twenty Reboot").

9. **Optionally, choose to copy the menus and widgets in your child theme.**

As you may remember, sometimes settings don't stick when you switch themes. Menus and widget arrangements are one of the things you lose, but if you click the helpful Copy Menus checkbox, you can transfer these details to the child theme (with the caveat that sometimes you'll *still* lose some of your theme settings, depending on the theme).

10. **Click Create New Child Theme.**

The Child Theme Configurator plugin creates the child theme folder on your web server and creates a new *style.css* file inside it. However, you still need to activate the child theme before it comes into effect.

11. **Choose Appearance→Themes to visit the theme gallery.**

Scroll down until you find the new child theme you created (*Figure 13-7*). It's a good idea to preview your child theme (click Live Preview) before you activate, just to make sure there aren't any lurking problems. If everything looks right, sail on!

FIGURE 13-7

When it starts its life, a child theme (like the Twenty Twenty Reboot child theme shown here) is identical to its parent theme.

12. **Activate your child theme.**

Everything should look the same as it did before. Take a moment to check your widgets and menus. If your theme has other types of customizations (like the fancy dynamic home pages you saw on page 238), you may need to return to the theme customizer and reapply any customizations that didn't make the transition to your child theme.

Changing Your Child Theme

Once you've created a new child theme, you're ready to make it your own. The best place to start is the theme editor. Choose Appearance→Theme Editor to take a look (*Figure 13-8*).

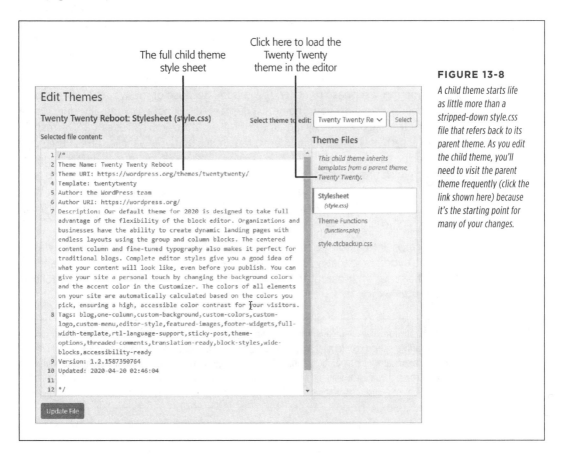

FIGURE 13-8

A child theme starts life as little more than a stripped-down style.css file that refers back to its parent theme. As you edit the child theme, you'll need to visit the parent theme frequently (click the link shown here) because it's the starting point for many of your changes.

You'll start out with just two files in your child theme (and one backup). The short *style.css* file has the header you learned about and *no* style rules. This is where you'll put your style customizations. Some of the details you alter include colors, fonts, spacing, and borders. You can also hide design elements you don't want to see.

You'll also see *functions.php*, which is a template file where themes put custom bits of code.

NOTE When you create a child theme with Child Theme Configurator, it uses a small bit of code in the *functions.php* file to grab the styles from the parent. That means it doesn't need to use the *@import* command in the *style.css* file.

Ready to try out a simple change? Scroll right to the bottom of the simple *style .css* file (past the */* marker that indicates the end of the header) and add this style:

```
body {background:hotpink;}
```

Click the Update File to save the modified file. Then, refresh your site. On most themes (including Twenty Twenty), this rule is all you need to turn the background behind every page to an unappealing shade of hot pink. But don't worry—you can easily reverse the change by removing this line and saving *style.css* again.

If you don't see the hot pink background appear right away, the problem is probably that your browser hasn't noticed the change because of the way it *caches* (stores a temporary copy of) your theme's stylesheet. It will notice eventually, but to save you the trouble of endlessly hitting the refresh button, you can use a *hard refresh*—a more assertive way of telling your web browser that you want to refresh the page and all its resources *right now*. The way you perform a hard refresh differs depending on which browser and operating system you're using. Usually, you need to hold down either Shift or Ctrl (Command on a Mac) while you click the refresh button. For a more specific overview of hard refreshes in every browser, check out *https:// tinyurl.com/hard-refresh*.

The real trick to making changes to your child theme is knowing the right styles to use. To understand that, you need to know a bit more about the stylesheet standard and the styles your parent theme uses. You'll tackle both challenges in the following section.

TIP Child themes are a great way to customize your site permanently, but they're also a handy tool for *temporarily* changing your formatting. For example, if you're running the Magic Tea House site, you might decide to hold a special winter promotion. During this promotion, you plan to make sweeping changes to your site's color scheme. If you use a child theme to redecorate, you can quickly go back to your original theme when the promotion ends—just choose Appearance→Themes and activate the parent theme.

■ Decoding the Style Rules in Your Theme

Every theme is slightly different, but most include a gargantuan *style.css* file stuffed full of hundreds of formatting instructions. Reading the average *style.css* file is not for the faint of heart. As explained earlier, the typical theme includes hundreds or thousands of formatting instructions. Fortunately, although it may be hard to digest everything that's happening in a stylesheet, it's not too difficult to understand how each rule works. In this section, you'll take a closer look at CSS, the standard that sets the rules for every stylesheet on the internet, including the ones in WordPress themes.

Decoding a Basic Style Rule

The first step to understanding CSS styles is to take a look at a few rules and get familiar with their syntax.

Every stylesheet is a long list of rules, and every rule follows the same format, which looks like this:

```
selector {
  property: value;
  property: value;
}
```

Here's what each part means:

- **The selector** identifies the type of content you want to format. A browser hunts down all the elements on a web page that match that selector. There are many ways to write a selector, but one of the simplest (shown next) is to identify the elements you want to format by their element name. For example, you could write a selector that picks out all the level-1 headings on a page.

- **The property** identifies the type of formatting you want to apply. Here's where you choose whether you want to change colors, fonts, alignment, or something else. You can have as many property settings as you want in a rule—the preceding example has two.

- **The value** sets a value for the property. For example, if your property is *color*, the value could be light blue or queasy green.

Now here's a sample rule, of the sort you'll find in a theme like Twenty Twenty:

```
body {
  background: #f5efe0;
  line-height: 1;
}
```

This tells a browser to find the web page's *<body>* element, which wraps all the content on the page. The browser applies two formatting instructions to the element. First, it changes the background color. (You'll be excused for not knowing that *#f5efe0* is an HTML color code that indicates a very light shade of peach.) Second, it sets the line spacing to a normal value of 1. (A higher value would add more space between each line of text.)

You need several skills to decode a style rule like this. First, you need to know the basics of HTML, so you can understand what the *<body>* element is and what it does in a web page. Second, you need to know what style properties are available for you to tweak. Style rules let you change color, typeface, borders, size, positioning, and alignment. To understand the sample rule shown previously, you need to know that CSS defines a background property that lets you change the color behind an element. Third, you need to know what values are appropriate for a property—for example, you set a page's background color by using an HTML color code. (In the case of colors, you can pick the color you want and get its HTML color code from a color-picking site like *www.colorpicker.com*.) WordPress will do its best to prompt you with a basic autocomplete feature (*Figure 13-9*).

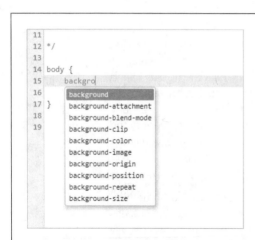

FIGURE 13-9

As you type, WordPress suggests matching style properties and values. If you want to insert one of them, click it in the list or press Enter when it's highlighted.

> **TIP** You can get stylesheet help from a book like *CSS: The Missing Manual* by David Sawyer McFarland (O'Reilly). Or, if all you need is an overview of the style properties you can change and their acceptable values, check out the stylesheet reference at *https://tinyurl.com/moz-css*.

Multiple Rules and Media Queries

Sometimes a stylesheet contains multiple rules for the same element. In such a situation, the browser combines all the property settings into one super-rule. For example, you can break the stylesheet rule you just saw into two pieces, like this:

```
body {
  background: #f5efe0;
}
```

```
body {
  line-height: 1;
}
```

In this case, the effect is the same. Big stylesheets may use this approach to break down complex rules and better organize them. The styles can be arranged in any order.

Many stylesheets include *media queries*—one or more blocks of conditional rules that spring into action when certain browser conditions are met. You can recognize a media query by the fact that it starts with the code *@media*. Here's an example:

```
@media screen and (min-width: 960px) {

  /* Conditional styles go in here. */

}
```

This translates into the following instruction: "If this page is currently being displayed on a screen (not sent to a printer or another device), and the width of the page is at least 960 pixels, then run the following styles." This is how a theme like Twenty Twenty shifts the layout and adjusts the menu on smaller screens and smartphones (page 136).

Class and ID Selectors

The preceding stylesheet rules target the *<body>* element. This type of rule is called an *element rule*, because it applies to a specific element on a page. For example, if you write a rule that formats *<h1>* headings, every first-level heading on the page gets the same formatting.

CSS supports other types of selectors, and WordPress themes use them heavily. One of the most popular is the *class selector*, which starts with a period, like this:

```
.entry-header {
  margin-bottom: 24px;
}
```

This rule formats any element that has a class named *entry-header* applied to it. If you look at the HTML markup for one of your posts, you'll find an element tagged with the *entry-header* class, like this:

```
<header class="entry-header">
  <h1 class="entry-title">Magic Tea House's Grand New Opening</h1>
  <div class="comments-link">
    <a href="http://magicteahouse.net/grand-new-opening/#respond"
    title="Comment on Magic Tea House's Grand New Opening">
```

```
      <span class="leave-reply">Leave a reply</span>
   </a>
 </div>
</header>
```

Here, the theme uses the *entry-header* class to wrap the header section of each post, which includes the post title and the "Leave a reply" comments link.

It's important to realize that the word "entry-header" doesn't mean anything special to WordPress or to a browser. It's simply a naming convention the Twenty Twenty theme uses (as do many other themes).

NOTE Sometimes you'll see a CSS rule that combines element selectors and class selectors. So the selector *h1.entry-title* refers to any level-1 heading that uses the *entry-title* class.

There's one more selector that's similar to class selectors, called an *ID selector*. It starts with a # character. Here's an example that uses an ID named *site-footer*:

```
#site-footer {
  background-color: #fff;
  border-color: #dedfdf;
  border-style: solid;
  border-width: 0;
}
```

In Twenty Twenty, this rule is for the footer that appears at the very bottom of each page of your site. Ordinarily, it contains copyright information and the text "Powered by WordPress."

Here's the snippet of HTML that gets its formatting from the *site-footer* rule:

```
<footer id="site-footer" role="contentinfo" class="header-footer-group">

  <div class="section-inner">
    <div class="footer-credits">
      <p class="footer-copyright">&copy;2020 Magic Tea House</p>
      <p class="powered-by-wordpress">
        <a href="https://wordpress.org/">Powered by WordPress</a>
      </p>
    </div>

    ...
  </div>
</footer>
```

The difference between class selectors and ID selectors is that a class selector can format a number of elements (provided they all have the same class applied to them), while an ID selector targets just a single HTML element (because two elements can't have the same ID). A WordPress page will have only one footer, so it makes sense to use an ID selector for it, even though a class selector would work just as well. Twenty Twenty includes a *#site-header* rule for the same reason.

NOTE Some stylesheets use ID selectors heavily, while others do all their work with class selectors. It's really a matter of preference. You simply need to follow the convention your theme uses.

This is where things can get a bit head-spinny. Understanding the syntax of CSS is one thing, but editing the styles in a theme means knowing which class and ID names that theme uses, and what elements are associated with those names. You'll get some pointers on page 411.

Combining Selectors

You can *combine* selectors to create even more powerful formatting rules, so long as you separate them with a comma. WordPress then applies the style rule to any element that matches any *one* of the selectors.

Here's an example that changes the alignment of three HTML elements—the *<caption>* element used with figures, and the *<th>* and *<td>* elements used with tables:

```
caption, th, td {
  text-align: left;
}
```

You can also create more complex rules that match elements *inside* other elements. This is called a *contextual selector*, and you build one by combining two ordinary selectors, separated by a space. Here's an example:

```
.comment-content h1 {
  ...
}
```

This selector matches every *<h1>* element inside the element with the class name *comment-content*. It formats the heading of every comment, while ignoring the *<h1>* elements that appear elsewhere on the page.

Here's another example that formats every link (represented by the *<a>* element) inside the main menu, without affecting ordinary links elsewhere on the page:

```
.primary-menu a {

  ...
}
```

WordPress loves contextual selectors—in fact, most of the Twenty Twenty theme's style rules use contextual selectors. If you haven't seen them before, they may take some time to decipher. Just remember that a browser works on the selectors one at a time. It starts by finding an element that matches the first selector, and then it looks inside that element to match the second selector (and if the rule includes another selector, the browser searches inside *that* element too).

Here's another example to practice your CSS-decoding skills:

```
.singular .entry-header {
  background-color: #fff;
  padding: 4rem 0;
}
```

Got it? This selector looks for an element with the class name *.entry-header* inside an element with the class name *.singular*. Why did the creators of Twenty Twenty decide to make this two-level rule? Sometimes it's just a good way to be more specific and to better indicate the structure of the elements on the site. But often it's a technique that's used to apply more targeted formatting.

For example, the creators of Twenty Twenty want to show the post header in different places, without using the same formatting. The *.singular* class is used on the single-post page (when you're reading a full post). However, post headers also appear in the list of posts on the home page, in category pages, in search results, and so on. In these other cases, the *.entry-header* class will be used, but the *.singular* class won't be used.

But for the most part, the specifics of how a stylesheet is structured won't matter to you. When you want to modify your theme's CSS, you'll find the style rule that controls the detail you want to change and you'll change its formatting, without asking too many questions.

■ Changing the Twenty Twenty Styles

The Twenty Twenty theme is filled with styles, and you can override any of them. For example, say you decide that you don't want the standard menu bar to appear on your site. The menu isn't a widget; it's a built-in part of the Twenty Twenty theme. If you don't create your own menu, Twenty Twenty uses an automatic menu (page 213).

You could assign a menu with no items, which works well enough while the site is expanded, but not as well when you shrink the menu. (You still get the Menu button with the three dots, only now it will pop open an empty menu.)

But here's where your child theme and CSS knowledge pay off, because you can hide the menu by using the right style rules (*Figure 13-10*). Here's what you need to add to the *style.css* file:

```
.primary-menu-wrapper {
  display: none;
}


.header-inner .toggle {
  display: none;
}
```

This works because the *.primary-menu-wrapper* class is attached to a *<nav>* element that holds the menu for your site. The *display:none* instruction tells browsers to collapse this element, and its contents, into nothingness. But that's not quite enough. You still get a mobile menu (the button with three dots) appearing when the browser window is small enough. To hide that, you need a second rule that targets the *.header-inner* class.

NOTE When you add a style rule to a child theme, it *overrides* any conflicting rule in the parent theme, but it allows style settings that don't conflict. For example, if you change the *color* of your post title in the child theme, that won't affect the *font* the parent theme applies to the title.

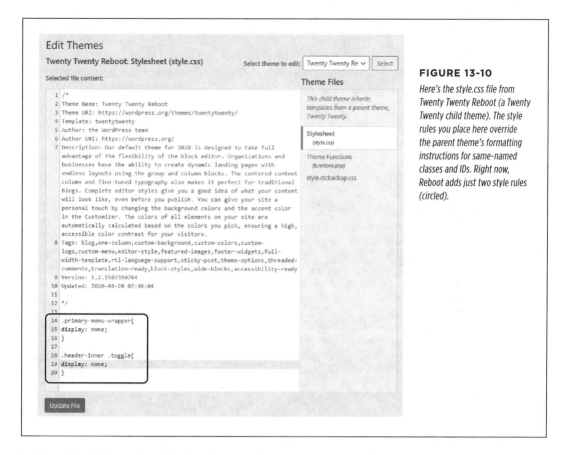

FIGURE 13-10

Here's the style.css file from Twenty Twenty Reboot (a Twenty Twenty child theme). The style rules you place here override the parent theme's formatting instructions for same-named classes and IDs. Right now, Reboot adds just two style rules (circled).

Incidentally, this isn't the only way to write these rules. Because there are several layers of nested elements, you have different options about which ones to hide. As always, your best starting point is often a Google search of the WordPress support forums. If you use the former (search for "hide the menu twenty twenty"), you'll find a topic on the latter discussing menus in more detail (*https://tinyurl.com/2020hide*).

If you're comfortable with CSS (or you're using one of the CSS resources mentioned on page 390), you'll have no trouble understanding the display property. However, you might have more trouble finding the ID and class names of the elements you want to change. Table 13-1 provides a cheat sheet to some of the key elements in the Twenty Twenty theme.

GEM IN THE ROUGH

Adding CSS in the Theme Customizer: Another Approach

A child theme is the tried-and-true method for safely customizing a theme. If you want to make just one simple change (like hiding a menu) and not take your customizations any further, there's a simpler alternative. You can use the *additional CSS* feature in the theme customizer.

The advantage of using the theme customizer is that you don't need to make a child theme. The disadvantage is that certain changes aren't supported or are more difficult to implement. You can't use fancy Google fonts or edit templates, as you'll learn to do later in this chapter. You also get just a tiny box

to hold all your CSS. Overall, the additional CSS feature is best for minor tweaks rather than complete theme customizations.

To use the additional CSS feature, fire up the theme customizer with Appearance→Customize. Next, click the Additional CSS tab at the bottom of the panel. An empty box appears, where you can type in your CSS, just as you enter it in the theme editor. You won't see the theme header or any other styles, because these are all safe and sound in the *style.css* file. The additional CSS box just holds your *extra* CSS. When you're finished, click Publish to update your site and see the effect.

TABLE 13-1 *Class names in the Twenty Twenty theme*

STYLE SELECTOR	CORRESPONDS TO...
#site-header	The header section at the top of every page. It includes the site title, site description, and navigation menu. (For many themes, it also holds a header image.)
.site-title	The site title in the header (for example, "Magic Tea House").
.site-description	The byline that appears under the site title in the header (for example, "Tea Emporium and Small Concert Venue").
.primary-menu-wrapper	The standard horizontal menu, with all its items in an HTML list.
.header-inner .toggle	The mobile menu, which is shown when the window is sized narrow enough.
.search-toggle-wrapper	The search button in the normal desktop view.
#site-content	All the page content between the header and the footer.
.footer-widgets-wrapper	The section of the footer that holds all the widgets.
.widget	Any widget in a widget area.
.widget-title	The optional title that appears above a widget.
#site-footer	The final footer on every page, which starts with a solid horizontal line and appears after the footer widgets.
.entry-title	The title of any post. Post titles appear in several places (including the home page, the single-post page, and the search page). You probably don't want to format titles all the same way, so you need to combine a class name with another selector that's used on the <body> element, as shown in the next three examples.

STYLE SELECTOR	CORRESPONDS TO...
.blog .entry-title	The title of any post in the main list of posts.
.single .entry-title	The title of any post on a single-post page.
.archive .entry-level	The title of any post when listing them on a category, tag, or date page. (You can further differentiate between these three by using *.category* or *.tag* or *.date* in place of *.archive* in your style rule.)
.entry-content	The content in the post. As with the *entry-title* class, this applies to the post content no matter where it appears, unless you combine this selector with another one.
.post-meta	A few pieces of information about the post, which appears just under the title. This includes who published the post, the publication date, and how many comments it has.
.comments	The comments area after a post.
.comment-author	The name of the person who left the comment.
.comment-metadata	Information about when the comment was left.
.comment-content	The comment text.
.bypostauthor	A comment left by you (the author of the post).
.logged-in	A class added to the *<body>* if the current visitor is logged in as a website user (page 319). For example, you can use the selector *body.logged-in* to change the formatting of the page for registered users.

Although Table 13-1 is tailored to the Twenty Twenty theme, many other WordPress themes follow a similar structure. The standard year themes are particularly consistent. And WordPress itself is responsible for adding class names like *.blog*, *.single*, and *.logged-in* to the *<body>* element to flag key details about the current state of a page. However, other themes are free to change many of these details in ways both subtle and maddening. For most class names, there are no guarantees, so every theme customization task must begin with a process of exploration.

Once you get the right class or ID name, you can target the exact visual ingredient you want to alter. For example, to change the font, color, and size of the text in a blog post in a single-post page, you can add a style rule like this:

```
.entry-content {
    font-family: "Times New Roman";
    color: red;
    font-size: 21px;
}
```

Type it into the editor (Appearance→Theme Editor), hit Update File, and refresh the website to see the change (*Figure 13-11*).

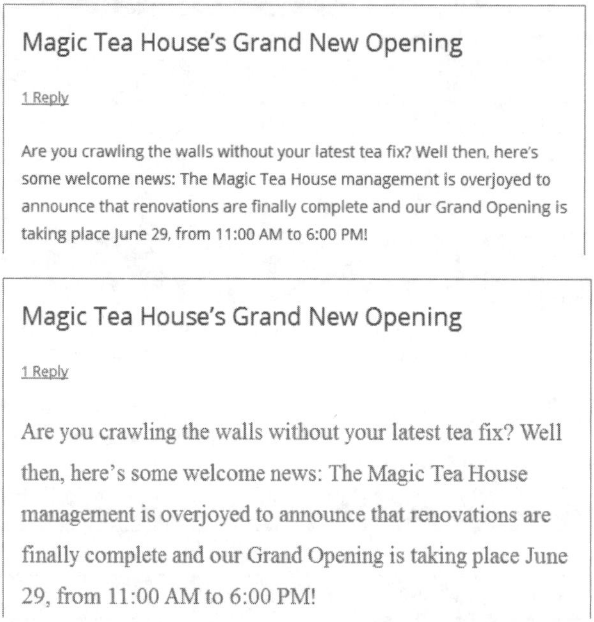

Magic Tea House's Grand New Opening

1 Reply

Are you crawling the walls without your latest tea fix? Well then, here's some welcome news: The Magic Tea House management is overjoyed to announce that renovations are finally complete and our Grand Opening is taking place June 29, from 11:00 AM to 6:00 PM!

Magic Tea House's Grand New Opening

1 Reply

Are you crawling the walls without your latest tea fix? Well then, here's some welcome news: The Magic Tea House management is overjoyed to announce that renovations are finally complete and our Grand Opening is taking place June 29, from 11:00 AM to 6:00 PM!

FIGURE 13-11

Top: An ordinary font in a simple theme.

Bottom: This revamped post, with its large red-lettered text, doesn't exactly look better, but it does look different.

It's a simple recipe: find the class name or ID name for the element you want to change, add some style properties, and your page gets an instant makeover.

Here's another example that makes your comments ridiculously obvious, with a yellow background, bold text, and a gray border. Comments left by other people aren't given the same enhancements.

```
.bypostauthor .comment-body {
  background-color: LightYellow;
  font-weight: bold;
  padding: 5px;
}
```

This rule could use just the *.bypostauthor* selector on its own, but that grabs the whole comment area, including the space immediately underneath the previous comment. The *.comment-body* class is a smaller section inside that holds all the comment information and text. You could use a similar technique to target just the comment text (*.bypostauthor .comment-content*) or commenter's name (*.bypostauthor .comment-author*).

Puzzling Out the Styles in a Theme

Half the battle in editing *style.css* is figuring out how to write your selector. Often you won't know the class or ID name of the page element you want to change, so you need to do a bit of detective work.

The best starting point is to scour the HTML that WordPress creates for your page. Many browsers make this process easier with an Inspect Element feature that lets you reveal the HTML in a specific part of a page (*Figure 13-12*). If that doesn't work, you can wade through the complete HTML for the page. To do that, right-click the page and choose a command with a name like View Source. You'll need to search the HTML for the piece of content you want to change. To get started, hit Ctrl-F (Command-F on a Mac), and type in a bit of the text that's near the part of the page you want to change. For example, to change the comments section, you might search for "Leave a Reply" to jump to the heading that starts off the section.

Once you find the right place—roughly—the real hunt begins. Look at the elements just before your content, and check the *class* and *id* attributes for names you recognize or that seem obvious. Pay special attention to the *<div>* element, which HTML pages use to group blocks of content, like sidebars, posts, menus, headers, and footers. You'll often find one *<div>* nested inside another, which lets the theme apply a layered tapestry of style settings (which is great for flexibility, but not so good when you're trying to understand exactly what rule is responsible for a specific formatting detail).

TIP   WordPress.com has a few helpful videos that show how to pinpoint individual elements in a web page, using different web browsers. You can watch these micro-overviews at *http://tinyurl.com/css-inspection*.

Once you have a potential class or ID name, it's time to experiment. Pop open the theme editor and add a new style rule that targets the section you identified. Do something obvious first, like changing the background color with the *background-color* property. That way, you can check your site and immediately determine if you found the right element.

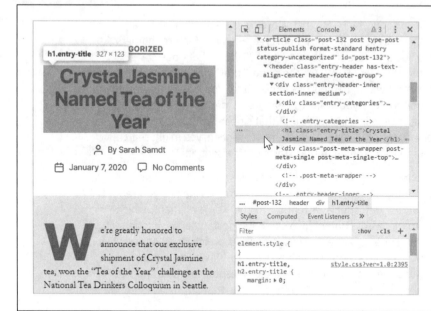

FIGURE 13-12

In most browsers, there's an easy way to find the class name of an element without searching through all the HTML on a page. Just right-click the part of the page you want to examine and choose "Inspect element." A panel opens with the corresponding bit of HTML markup selected. That quickly tells you that the post title is an <h1> element with the class name entry-title (but you already knew that).

Using Fancy Fonts

One of the most common reasons to edit the styles in a template is to change an element's font. In fact, it's often the case that all you need to do to turn a popular theme into something uniquely yours is to change some of the typefaces.

Originally, HTML pages were limited to a set of *web-safe fonts*. These are the typefaces every web surfer has seen, including stalwarts like Times New Roman, Arial, and Verdana. But web design has taken great strides forward in recent years with a CSS feature called *embedded fonts*. Essentially, embedded fonts let you use almost any typeface you want on your web pages. But first you need to upload the font to your web server in the right format and write some convoluted CSS.

Happily, you can sidestep these problems by using a web font service like Google Fonts. There, you can pick from a huge gallery of attractive typefaces. When you find a font you want, Google spits out the CSS style rule for you. Best of all, Google *hosts* the font files on its high-powered web servers, in all the required formats, so you don't need to upload anything to your site.

To use a Google font on a WordPress site, follow these steps:

1. **Go** *http://fonts.google.com*.

   Google displays a long list of fonts (*Figure 13-13*). At the time of this writing, there were roughly 1,000 typefaces to choose from.

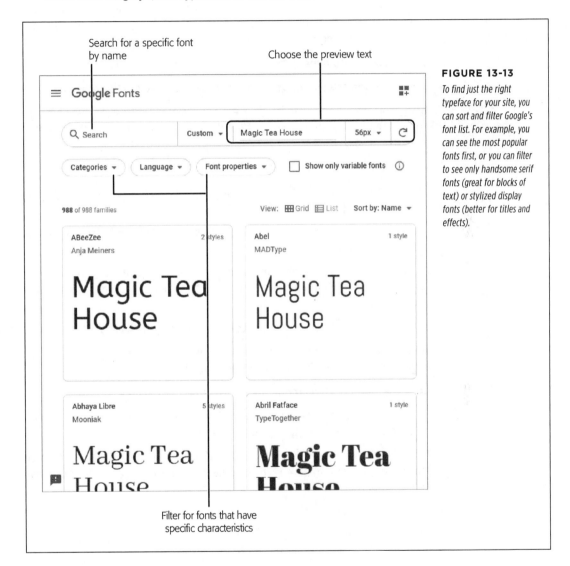

Search for a specific font
by name

Choose the preview text

Filter for fonts that have
specific characteristics

FIGURE 13-13

To find just the right typeface for your site, you can sort and filter Google's font list. For example, you can see the most popular fonts first, or you can filter to see only handsome serif fonts (great for blocks of text) or stylized display fonts (better for titles and effects).

2. **Optionally, change your font preview settings.**

   You can type in your own preview text and choose the size of the preview text to help you make a better decision about which font is best for your needs.

3. **Optionally, filter the list for specific types of fonts.**

Click the Categories box to choose a specific type of font. *Serif* and *sans-serif* fonts are best for text, while *display* fonts have plenty of eye-catching (and wacky) choices for titles and other visual elements. You can also filter for *hand-written* (cursive) fonts or *monospace* fonts (where every letter has the same width, as in code listings).

Click Properties to filter for fonts with a specific thickness or slant.

4. **Scroll through the list of fonts. When you see a font you like, click it.**

Google shows a more detailed preview and a list of every style this font has. For example, many fonts have different weights (thicker and bolder styles) and italic styles.

5. **For each style you like, click "+ Select this style."**

Google opens a panel on the right side with a list of the typefaces you've picked so far.

6. **If you want to pick more fonts, click your browser's back button to return to the font listing.**

As you add more styles from more fonts, Google keeps adding them to the sidebar. When you have everything you want, continue to the next step to get your styles.

7. **In the sidebar, click Embed, and then click *@import* (Figure 13-14).**

To use a font, you need to copy Google's font styles into *your* stylesheet. The easiest way to do that is with the *@import* command.

8. **Copy the *@import* line and paste it into your *style.css* file.**

Leave out the *<style>* tags that Google shows, because those are needed only if you're defining a stylesheet inside an HTML page.

You can add as many *@import* commands as you want. However, the most efficient approach is to pick all the Google fonts you want at once and use just one *@import* command that has them all.

9. **Create a style that uses the font.**

Once you import the Google stylesheet, you can use your new font, by name, wherever you want. Just set the *font-family* property.

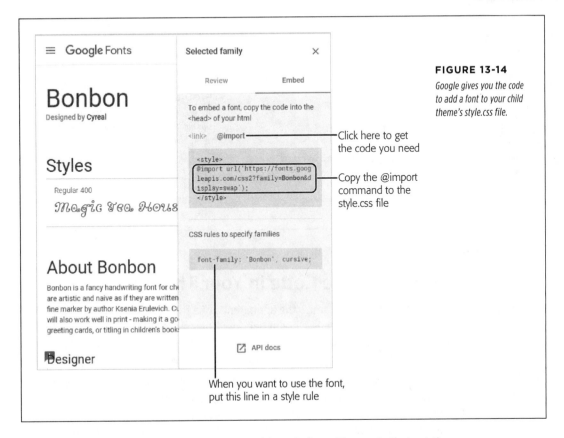

FIGURE 13-14

Google gives you the code to add a font to your child theme's style.css file.

Here's a complete *style.css* file that uses this technique. The parts that set the custom font are highlighted in bold:

```
/*
Theme Name: Twenty Twenty Reboot
Description: This is a customized version of the Twenty Twenty Theme for the
Magic Tea House.
Author: Katya Greenview
Template: twentytwenty

*/
@import url(http://fonts.googleapis.com/css?family=Bonbon);

.entry-header .entry-title {
  font-family: 'Bonbon', 'Times New Roman';
  font-size: 30px;
}
```

Figure 13-15 shows this font in action in a post.

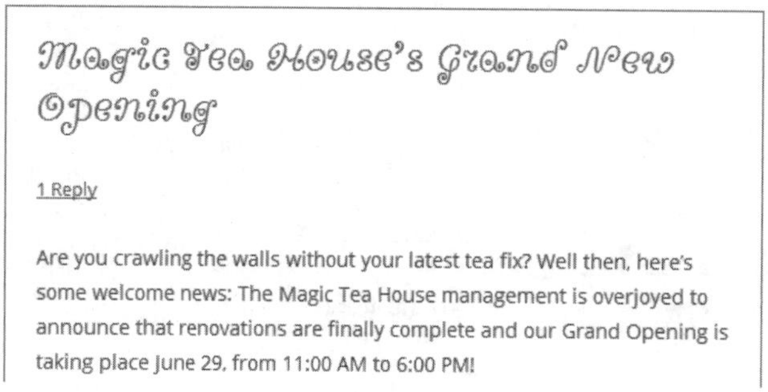

Magic Tea House's Grand New Opening

1 Reply

Are you crawling the walls without your latest tea fix? Well then, here's some welcome news: The Magic Tea House management is overjoyed to announce that renovations are finally complete and our Grand Opening is taking place June 29, from 11:00 AM to 6:00 PM!

FIGURE 13-15

The Bonbon font is perhaps not the best for this site, but it's impossible to deny that it makes for eye-catching post titles.

Editing the Code in Your Theme

When you want to customize the appearance of a theme, the first place you should look is the *style.css* file. But if you need to make more dramatic changes—for example, revamp the layout, change the information in the post list, or add new widget areas—you have to go further. Your next step is to consider the theme's *template files*.

A typical theme uses anywhere from a dozen to 50 templates. If you crack one open, you see a combination of HTML markup and PHP code. The PHP code is the magic ingredient—it triggers the specific WordPress actions that pull your content out of the database. Before WordPress sends a page to a visitor, it runs all the PHP code inside it.

Writing this code is a task that's well beyond the average WordPress website owner. But that's not a problem, because you don't need to write the code yourself, even if you're building a completely new theme. Instead, you take a ready-made page template that contains all the basic code and *edit* that file to your liking. Here are two ways you can do that:

- **Change the HTML markup.** Maybe you don't need to change the code in the template file at all. You might just need to modify the HTML that wraps around it. After all, it's the HTML (in conjunction with the stylesheet) that determines how your content looks and where it appears.

- **Modify the PHP code.** You start with a template full of working code. Often you can carefully modify this code, using the WordPress documentation, to change the way it works. For example, imagine you want the list of posts on the home page to show fewer posts, include just post titles or images instead of content, or show posts from a specific category. You can do all this by adjusting the code that's already in the home-page template.

Of course, the more thoroughly you want to edit the PHP, the more you need to learn. Eventually, you might pick up enough skills to be an accomplished PHP tweaker.

UP TO SPEED

Learning PHP

The actual syntax of the PHP language is beyond the scope of this book. If you want to develop ninja programming skills, plenty of great resources for learning PHP are available, whether you have a programming background or are just starting out. Don't rush off just yet, however, because although learning PHP will definitely help you customize a WordPress theme, it may not help you as much as you expect.

Learning to customize a WordPress template is partially about learning PHP (because it helps to understand basic language details like loops, conditional logic, and functions). But it's

mostly about learning to use WordPress's functions in PHP code. For that reason, you'll probably get more practical value out of studying WordPress functions than learning the entire language, unless you plan to someday write dynamic web pages of your own.

To get started with WordPress functions, check out the function reference at *http://tinyurl.com/func-ref*. To learn more about PHP, start with the basics with the tutorial at *www.w3schools.com/php*.

Introducing the Template Files

Every theme uses a slightly different assortment of templates. The WordPress staple Twenty Twenty uses a fairly typical set of about 30 templates.

You can recognize a template by the *.php* at the end of its filename. Although template files hold a mix of HTML and PHP, the *.php* extension tells WordPress that there's code inside that the WordPress engine needs to run before it sends the final page to a browser.

Even though a template is just a mix of HTML and PHP, understanding where it fits into your site can be a bit of a challenge. That's because every page WordPress stitches together uses several template files.

For example, imagine you click through to a single post. WordPress has a template, called *singular.php*, that creates the page on the fly. However, *singular.php* needs help from a host of other templates. First, it inserts the contents of the *header.php* template, which sits at the top of every page in a WordPress site. The *header.php* file takes care of basics, like linking to the stylesheet, registering some scripts, and showing the header section, complete with the top-level menu.

After the header, the *singular.php* template calls for help from another template, *content.php*, which pulls in several more templates. For example, *entry-header.php* builds the title and header section of the post, and *featured-image.php* handles the display of the featured image. And after the post is finished, *navigation.php* shows the links to the previous and next post, and *comments.php* shows the familiar comments area.

Finally, when *content.php* has finished doing all its work, the *singular.php* template rounds out the page by calling two final templates into action. The *footer-menus-widgets.php* template handles the widgets area at the bottom of the page and any social menus. Lastly, the *footer.php* template shows the closing line and ends the page.

If you're going cross-eyed trying to follow this template assortment, *Figure 13-16* shows how it all breaks down with a bird's-eye view of a single post.

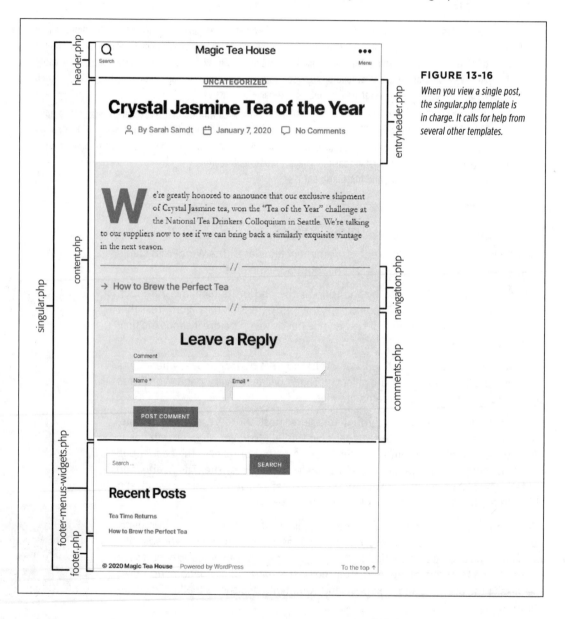

FIGURE 13-16

When you view a single post, the singular.php template is in charge. It calls for help from several other templates.

At first glance, this system seems just a bit bonkers. How better to complicate life than to split the code for a single page into a pile of template files that need to work together? But in typical WordPress fashion, the design is actually pretty brilliant. It means that a theme designer can create a single template file that governs a repeating site element—like a header—and use it everywhere, without needing to worry about duplicating effort or being inconsistent.

When you edit theme styles, your first challenge is finding the right style rule to change. When you edit *templates*, the first challenge is finding the right template file to modify. Table 13-2 can help you get started. It describes the fundamental templates that most themes, including Twenty Twenty, use. Don't worry about digesting absolutely all this information now—just keep it in your back pocket for those times when you need to figure out what a template file does.

TABLE 13-2 *Important WordPress templates*

TEMPLATE FILE	DESCRIPTION
index.php	This is a theme's main template, and the only one that's absolutely required. WordPress uses it if there's no other, more specific template to use. Most themes use *index.php* to display a list of posts on the home page.
category.php, tag.php, date.php, author.php	Lists posts when you browse a category, a tag, a date range, or a specific author. If you don't use these templates, WordPress falls back to *index.php*. Twenty Twenty doesn't use these templates, but you can add them. You can also create templates that target specific categories, tags, and authors, by adding the name to the end of the template, like *category-tea.php* (the category with "tea" as its simplified name).
singular.php, single.php, and page.php	Displays a single post or page. Many themes have separate templates for posts and pages, in which case *single.php* is the template for posts, and *page.php* is the template for pages.
header.php	Shows the banner that appears across the top of every page. Often *header.php* includes a navigation menu.
footer-menus-widgets.php	Holds the widgets that appear under the post but above the footer.
footer.php	Shows the footer that stretches across the bottom of every page.
sidebar.php	Twenty Twenty doesn't include a sidebar with widgets, but many themes do (including some year themes). Usually, sidebar templates have names like *sidebar.php* or *sidebar-right.php*.
content.php	Shows the content of a post or page. Some themes create different content templates for different types of posts and pages (like *content-aside.php* or *content-page.php*).
entry-header.php	Shows the full content of a post or page, with the help of several other templates.
featured-image.php	Shows the featured image at the beginning of a post or page.
navigation.php	Shows the navigation links that let you skip from one post to another.
comments.php	Shows the comments section after a post or page.
404.php	Shows an error message when the requested post or page can't be found.

■ Changing a Twenty Twenty Template

By this point, you've digested a fair bit of WordPress theory. It's time to capitalize on that by editing a template file.

In this example, you'll try something that seems simple enough. You want to remove the "Powered by WordPress" message that appears at the bottom of every page on your site, just under the footer widgets (*Figure 13-17*). (It's not that you're embarrassed by WordPress. You just can't help but notice that none of the other big-gun WordPress sites have this message slapped on their pages. Sometimes, being professional means being discreet.)

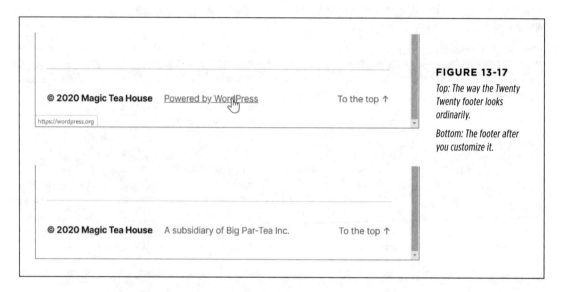

FIGURE 13-17

Top: The way the Twenty Twenty footer looks ordinarily.

Bottom: The footer after you customize it.

> **NOTE** Even if you don't want to hide the footer text in your own theme, this pattern—opening a template and searching for something to remove—is a common one. For example, you might use it to stop showing date or author information in you posts, if you've decided those details aren't relevant.

Step 1. Find the Template File

Start by examining the list of templates in Table 13-2. In this case, the *footer.php* file is the obvious candidate. It creates the entire footer section, widgets and all, for every page.

Step 2. Create a Copy of the Template File

Once again, you need to start with a child theme (page 394). If you don't, you can still customize the footer, but your hard-won changes will vaporize the moment you install a theme update.

Here's where things get a bit more awkward. As you know, WordPress templates are really a collection of many template files. To change a template, you need to figure out the changes you want to make to your pages, and then find the template file responsible for that part of the page (*singular.php, comments.php*, and so on). Then you add a new version of that template file to the child theme. That new template will override the one in the parent theme.

You do this by copying the template file you want to edit from the parent theme folder to your child theme folder, and then making your changes. In this example, that means you need to copy the *footer.php* file in the *twentytwenty* folder and paste it into the *twentytwenty-child* folder.

If you're a seasoned webmaster, you can do this using an FTP program. Otherwise, you'll probably want to do the work with the Child Theme Configurator you used earlier. Here's how:

1. **Choose Tools→Child Themes.**

    This brings you to the same Child Theme Configurator page you used to create your theme.

2. **Click the Files tab.**

    This shows you the template files in your parent theme and the template files you've copied over. Right now, you'll have just one template file, *functions.php*, which the Child Theme Configurator adds when you generate the child theme.

3. **In the Parent Templates section, put a checkmark next to each template you want to customize (*Figure 13-18*).**

    In this example, you want only *footer.php*. Don't pick files you don't want to customize, because that will prevent you from getting theme updates for those files.

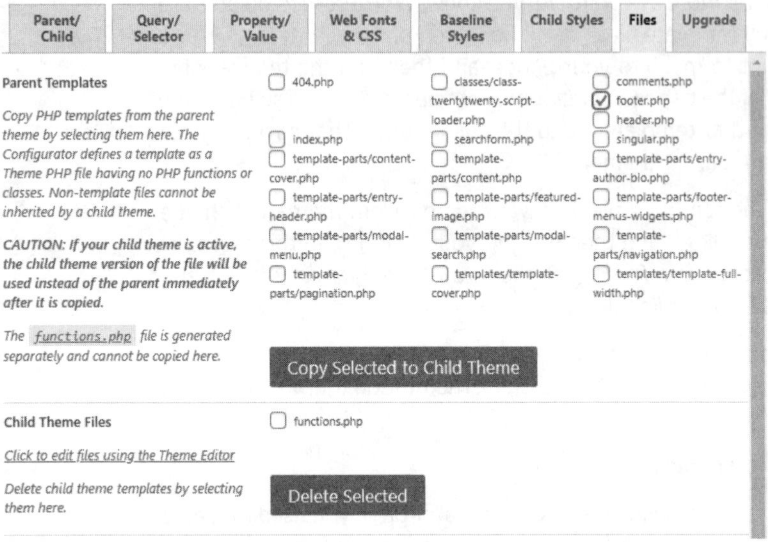

FIGURE 13-18

The Child Theme Configurator can copy the footer.php from the parent theme to your child theme, where you'll customize it.

4. **Click Copy Selected to Child Theme.**

    The Child Theme Configurator makes a copy of the template and adds it to your child theme.

5. **Choose Appearance→Theme Editor.**

    Now your theme has a *style.css* file, a *functions.php* file, *and* the *footer.php* template.

    Click Theme Footer to view the full markup for the footer template (*Figure 13-19*). You can make changes and click Update File to save your altered template. The only problem is that it will take a bit of digging before you understand how the template works.

NOTE The child theme's version of a template completely overrides the parent theme's copy of the template. In this example, that means WordPress ignores the original version of the *footer.php* file, which is in the original theme folder (twentytwenty). You can refer to it anytime to check the code or copy important bits, but WordPress won't run the code anymore unless you delete the child theme's version.

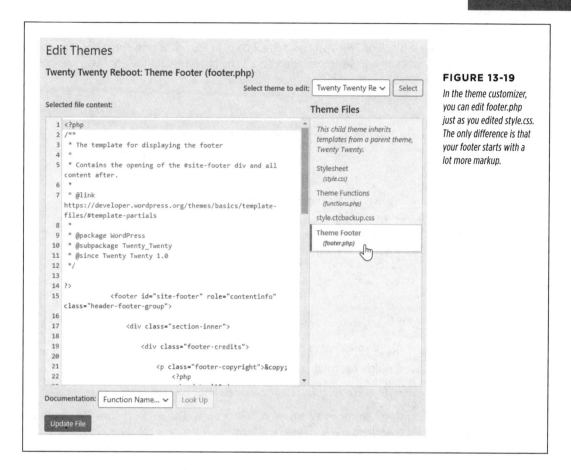

FIGURE 13-19

In the theme customizer, you can edit footer.php just as you edited style.css. The only difference is that your footer starts with a lot more markup.

Step 3. Examine the Template File

The *footer.php* file is one of WordPress's simpler template files. But even simple templates have a fair bit of boilerplate to wade through.

In this section, you'll look a closer take at the contents of *footer.php*. You don't always need to take this step, but it's good practice when you're just starting out and trying to make sense of WordPress's template system.

If you've written web pages before, you probably know that programming code, like JavaScript, needs to be carefully separated from the HTML on the page. The same is true for PHP, although it has its own special syntax. Every time a block of PHP code appears, it starts with this:

```
<?php
```

Similarly, the following character sequence marks the end of a block of PHP code:

```
?>
```

You can see this system at work at the very beginning of every template file, where there's a block of PHP that has no real code at all, just a long comment. The comment lists some basic information about the template. Here's what you see at the beginning of *footer.php*:

```php
<?php
/**
 * The template for displaying the footer.
 *
 * Contains the opening of the #site-footer div and all content after.
 *
 * @link https://developer.wordpress.org/themes/basics/template-files/
#template-partials
 *
 * @package WordPress
 * @subpackage Twenty_Twenty
 * @since Twenty Twenty 1.0
 */


?>
```

The next line identifies the beginning of the footer content. This section has the ID *site-footer*, which you might remember from Table 13-1:

```
<footer id="site-footer" role="contentinfo" class="header-footer-group">
```

The following lines open two new *<div>* sections. The ID *footer-credits* hints that you're approaching the area with the footer text you want to alter:

```
<div class="section-inner">


    <div class="footer-credits">
```

The template files use indenting to help you keep track of what element is inside another. And there's one other trick. At the end of many sections, a comment reminds you of what it is. For example, scroll down to the end of the footer credits section and you'll see this:

```
</div><!-- .footer-credits -->
```

The *</div>* ends the *<div>* element, as per usual. But the HTML comment is just a helpful extra. It explains that this just-finished *<div>* uses the *.footer-credits* class, so you can figure out where you are in the template.

What you're looking for right now is the pair of paragraphs that hold the copyright information for your site and the "Powered by WordPress" message. They look like this:

```
<p class="footer-copyright">&copy;
  <?php
echo date_i18n(_x( 'Y', 'copyright date format', 'twentytwenty' ));
  ?>

  <a href="<?php echo esc_url( home_url( '/' ) ); ?>"><?php bloginfo( 'name'
); ?></a>
</p><!-- .footer-copyright -->

<p class="powered-by-wordpress">
  <a href="<?php echo esc_url( __( 'https://wordpress.org/', 'twentytwenty' )
); ?>">
    <?php _e( 'Powered by WordPress', 'twentytwenty' ); ?>
  </a>
</p><!-- .powered-by-wordpress -->
```

The markup looks a bit complicated, because several PHP code blocks are embedded inside the markup. The first block grabs the current year for the copyright message. The next one adds a link to your home page. Another PHP block inserts the name of your website to make the link text (like "Magic Tea House").

But the part you want is this link, which adds the "Powered by WordPress" message and links it to the WordPress.org home page:

```
  <a href="<?php echo esc_url( __( 'https://wordpress.org/', 'twentytwenty' )
); ?>">
    <?php _e( 'Powered by WordPress', 'twentytwenty' ); ?>
  </a>
```

You could delete the line altogether, or you can replace it with some text of your own, like this:

```
  A subsidiary of Big Par-Tea Inc.
```

Now click Update File to save your new footer.

This completes the example and drives home a clear point: even the most straight-forward theme-editing tasks require a bit of slogging.

WARNING The theme editor can be an unforgiving tool. WordPress does not store past versions of your template files. You can always delete your template (in which case the parent's version takes over again), but you can't roll back to an older version of your child template. To make sure you don't run into trouble, get into the habit of making basic, bare-minimum backups. Before you make a big change to a theme file, copy all the content from the editing box in the theme editor, and paste it into a blank text file on your computer.

FREQUENTLY ASKED QUESTION

Updating a Child Theme

What happens when I update a theme that uses customized templates?

Child templates don't work in exactly the same way as child styles. Child styles extend style rules already put in place by the parent theme. Even if the parent gets a new, updated *style. css* file, the child styles remain, and WordPress applies them on top of the parent styles.

But page templates don't extend parent templates; they replace them. As soon as you add *footer.php* to your child theme,

WordPress starts ignoring the *footer.php* in the original theme. That means that if you update the parent theme and change the original *footer.php* file, no one really notices.

This is probably the safest way to handle theme updates, because there really isn't a way that WordPress could combine two versions of a template. However, if you plan to customize all the templates in a theme, you may as well build a completely separate theme of your own.

■ The Last Word

In this chapter, you covered some serious ground. First, you peeked under the hood and learned a bit more about how themes are structured. Then you learned how to create your own child themes, to give you a safe space to make changes and customizations.

Once you have a child theme, there's no limit to what you can change, but the learning curve is steep. In this chapter, you dipped your toe into relatively simple examples. You learned to change colors and alter fonts. You also saw how to hunt down tiny details in templates. These examples are great when you have an almost perfect theme and want to fix just one issue. But if you're willing to go deeper into the WordPress universe, you can deconstruct and rebuild everything about your theme. One good starting point for tinkerers is the WordPress Theme Developer Handbook at *https://developer.wordpress.org/themes.*

Often WordPressers first delve into a theme to make a tiny alteration. But they rarely stop there. The ability to transform a site, often by changing just a single style rule or modifying a single line of code, is addictive. If you catch the bug, you'll want to customize every theme you touch, and you won't be able to rest until you get exactly what you want for your site.

More Tools for Professional Sites

As you already know, WordPress began its life as a nifty tool for building a blog. In the two decades since, it's exploded into an all-purpose way to make almost any type of site.

This transformation is exciting, but it also means that life with WordPress can quickly get complicated. Once you've wrapped your mind around the core WordPress features, you need to add the right plugins, set up the right backup and security practices, fine-tune your site's performance, and consider buying a professional theme. This adds up to a lot of extra details that you need to master.

In this book, you've already started to branch out beyond the standard WordPress tools. As you've travelled through the past 13 chapters, you've dipped into plenty of popular third-party plugins. In this chapter, you'll look at some more. You'll set up regular backups so your site won't disappear if your web host crashes. You'll use caching to perk up sluggish sites. And you'll learn how to give your site ecommerce shopping powers.

■ Disaster Proofing with Backups

You probably don't spend much time worrying about the safety of the files on your website. After all, even small web hosting companies take reliability seriously. They use systems that have a high level of *redundancy*—web servers with multiple hard drives, for example, and groups of computers that work together so that a hardware failure in one won't sideline an entire website. They often use backups to put the files they host in a separate storage location, so they can recover them after catastrophes, like floods and fires.

These measures are important, but they aren't enough to fully protect your Word-Press site. Unless you do your own backups, your site is exposed to serious risks that your web hosting company can't prevent. For example, an attacker could break into your administrator account and sneak some advertising or malware into your site. In some cases, these attacks are stealthy enough that you won't notice any effect for weeks. Even if your web host has a backup, it's likely to be a copy of your infected site, making it useless. Other problems can occur, too—and the web is full of backup horror stories. Maybe your web host's backups go mysteriously missing, just before a serious outage. Or sudden bankruptcy puts your web host out of business, taking your website with it.

To make sure your site can weather any crisis, you need to make your own regular backups. You could do backups by hand, periodically downloading your content to your own personal computer for safekeeping. But inconvenient backup strategies rarely work, because in real life people get busy or forget to do them—usually just before an unexpected disaster hits.

A better idea is to use an automated backup service. This service has a simple job. Every day (ideally, when your site isn't too busy), the backup service makes a copy of your site's content and transfers it to another server elsewhere in the world. Best of all, this process happens automatically, without requiring any thought from you.

Understanding a WordPress Backup

A complete WordPress backup consists of two parts:

- **A backup of your database.** As you learned in Chapter 1, WordPress stores every post, page, comment, and stitch of content in a database on your web server. This is the most important part of your site, because without content, all you have is an empty shell.

- **A backup of your files.** These files include the contents of your media library, including every picture and resource you uploaded to your WordPress site, the theme files that tell WordPress how to lay out your content, and any plugins you've installed.

Sometimes you might decide to back up just the database. There are a few reasons for this—maybe you have limited space and the files take too much room, or maybe you've backed up the files recently and you don't think they've changed. But the best bet is to grab everything and ensure you have a complete snapshot of your site. That way, you're also guaranteed that you can restore your backup from scratch on a fresh, blank web server, with a minimum of fuss.

NOTE You might wonder why you back up theme files when they're freely available. Won't that just waste space? But there are two reasons. First, as you become a more advanced WordPress designer, you may begin customizing the themes (as you saw in Chapter 13). Second, it's always possible that the particular version of a theme you're using will disappear from the web or be replaced by a version that doesn't work quite the same.

Copying Your Files by Hand

Can't I just log in to my web host account and copy my website files?

Not exactly. Yes, if you're handy with an FTP program, you can back up your files anytime. All you do is browse to your web hosting account and copy the contents of your WordPress website folder to your home computer. If you use Windows, this is as simple as firing up Windows Explorer, pointing it to your web host's FTP site, logging in, and dragging the folder with your WordPress site onto your desktop.

However, don't make the mistake of thinking that this is a complete backup. Even if you copy every file you see, you still won't get the contents of your WordPress database, which is the beating heart of your WordPress site. To get that, you need a tool that can access the database, extract its contents, and put it in a file. (This tool also needs to be able to do the reverse, copying the data from a backup file into a new database, in case you need to re-create your site.) Unless you're a MySQL database guru, your best bet is to use a WordPress backup plugin to help you out.

Using the UpdraftPlus Plugin

Plenty of decent backup plugins exist for WordPress. (Common choices include VaultPress, BackWPup, BackupBuddy, WP-DB-Backup, and BackupWordPress.) Most of these plugins aren't free. Some of them bundle their own storage services. And often you'll spend more for yearly backup and storage costs that you do for web hosting.

But in the past few years, one backup plugin has rocketed ahead in popularity. Today, UpdraftPlus (*http://tinyurl.com/updraftplus*) is the most popular WordPress backup plugin, with millions of active installs. It's full-featured, fairly priced, and a good starter choice, because it offers a reasonably capable free version.

To get started with UpdraftPlus, you do the usual: visit Plugins→Add New page, search for UpdraftPlus by name, install it, and activate it. Once UpdraftPlus is running on your site, you have two options to back up your site:

- **Manually.** You perform an instant backup in the admin area, and then download the backed-up data in one big ZIP file for safekeeping.

- **Scheduled.** You schedule a backup to run at certain times (for example, once a day), and have the backed-up ZIP file placed in a storage location you specify.

You'll try out both options in the following sections.

Creating an Instant Backup

The first thing you should do after you install UpdraftPlus is try a manual backup. That way, you'll know if the plugin works properly, and you'll have at least one complete backup of your website to start you off. Here's how:

1. **Choose Settings→UpdraftPlus Settings.**

   This brings you to the control center for UpdraftPlus. This single tabbed page controls every type of backup.

2. **Click the Backup/Restore tab.**

   This is where you can launch manual backups and review the backups you've made so far (*Figure 14-1*).

UpdraftPlus Backup/Restore

UpdraftPlus.Com | Premium | News | Twitter | Support | Newsletter sign-up | Lead developer's homepage | FAQs | More plugins · Version: 1.16.23

| Backup / Restore | Migrate / Clone | Settings | Advanced Tools | Premium / Extensions |

Next scheduled backups:

Files: Database:
Nothing currently scheduled   Nothing currently scheduled   Backup Now

Time now: Thu, April 30, 2020 19:42   Add changed files (incremental backup) ...

Last log message: Download most recently modified log file

The backup apparently succeeded and is now complete (Apr 30 19:43:16)

FIGURE 14-1

Although UpdraftPlus is crammed with settings, you can't miss the big blue Backup Now button.

3. **Click Backup Now.**

   UpdraftPlus pops open a window with a few more settings (*Figure 14-2*).

4. **Choose what you want to back up.**

   Usually, you'll want to choose both the "Include your database" and the "Include your files" options. That way, your backup is complete, and has everything you need to restore your site on a brand-new server if disaster strikes.

5. **Optionally, choose to make this backup permanent.**

   If you choose "Only allow this backup to be deleted manually," UpdraftPlus will keep this backup copy until you decide to delete it. Otherwise, UpdraftPlus follows a revolving backup cycle. You'll see how to configure the backup cycle on page 432. But out of the box, UpdraftPlus keeps just two backups at a time and tosses the oldest ones out first.

6. **Click Backup Now.**

   You'll need to wait a moment while UpdraftPlus creates a backup job for your web server. (You'll see a "Requesting start of backup" message.)

Once the backup starts, UpdraftPlus will gather your site's information, stuff everything inside a few ZIP files, and then copy those files to a safe place on your web server. While it works, you'll see a progress bar ticking away.

TIP You don't need to stay on the UpdraftPlus settings page while it backs up—or even in the admin area. Once your backup job is scheduled, it will take place safely in the background, whether you're there watching it in your browser or not. And when it's complete, UpdraftPlus sends your admin email a quick report.

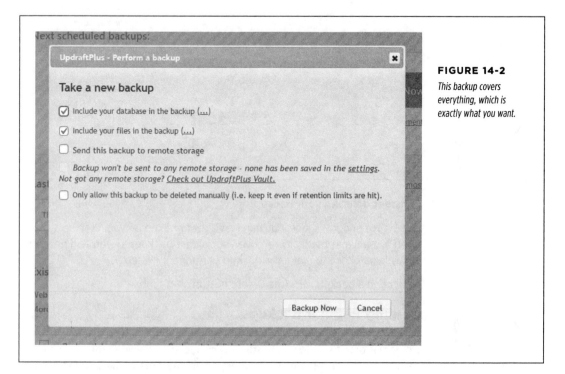

FIGURE 14-2

This backup covers everything, which is exactly what you want.

7. **Scroll down to review your backups.**

   At the bottom of the Backup/Restore tab is a list of recent backups (*Figure 14-3*). If you're curious, UpdraftPlus puts all these files in the *wp-content* folder of your website, in a dedicated subfolder named *updraft*.

8. **Download the backed-up files to your computer.**

   To download the database, you click the Database button in the "Backup data" column and then click "Download to your computer." You'll be rewarded with a compressed file that has all your data. Keep it safe.

   UpdraftPlus makes the process of downloading manual backups somewhat time-consuming. If you want to download everything (and you should), you need to click each button in the "Backup data" column. In other words, you click Database to download a copy of your website's database, Plugins to download all the files for the currently installed plugins, Themes to download your currently installed themes, Uploads to capture your media files, and Others to grab other types of files. By the end, you'll have five compressed files in your computer's Downloads folder.

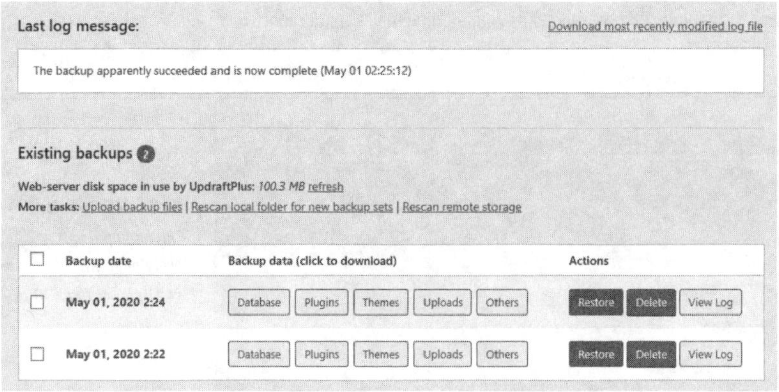

Last log message: Download most recently modified log file

The backup apparently succeeded and is now complete (May 01 02:25:12)

Existing backups ⓘ

Web-server disk space in use by UpdraftPlus: *100.3 MB* refresh

More tasks: Upload backup files | Rescan local folder for new backup sets | Rescan remote storage

☐	Backup date	Backup data (click to download)					Actions		
☐	May 01, 2020 2:24	Database	Plugins	Themes	Uploads	Others	Restore	Delete	View Log
☐	May 01, 2020 2:22	Database	Plugins	Themes	Uploads	Others	Restore	Delete	View Log

FIGURE 14-3

This website has two manually created backups that were made minutes apart. The "Backup data" column has buttons for the database and different types of files, which shows that these were complete backups.

9. **Consider deleting your backup.**

If you have enough space, you can leave your backup on your web server. UpdraftPlus will automatically remove your oldest backups as you add new ones (unless you chose to make the backup permanent in step 5).

To remove a backup, click the Delete button next to it.

Scheduling Regular Backups

You can do a perfectly good job of disaster-proofing your site with manual backups. The problem is that it's up to you to start every backup, download the final product, and keep your backed-up files somewhere smart. As time passes, you might find yourself forgetting to make backups or download the copies, which leaves your website at risk. And even if you do remember, clicking all those buttons takes time.

The solution is to tell your plugin to do the backup work for you, at regularly scheduled intervals. Here's how it works with UpdraftPlus:

1. **Choose Settings→UpdraftPlus Settings; then click the Settings tab.**

The Settings tab holds everything you'll want to configure in UpdraftPlus.

2. **Choose the backup frequency.**

UpdraftPlus lets you set a different schedule for your database and your file backups, which is handy. Database backups are more important, and the content in your database changes more frequently. For most sites, it makes sense to back up the database more frequently (say, once a day) than you back up the files (say, once a month).

To set the frequency, change the current schedule setting from Manual (which means backups are performed only when you do them yourself) to anything else. Your options range from every hour to once a month (*Figure 14-4*).

TIP Keep in mind that making backups takes computer effort. If you are trying to copy the full content of your database every hour, you may slow your site's performance. When choosing a backup frequency, you need to balance the risk of losing data (say, a day's worth of posts) with the risk of slowing down your site. You can get a feeling for how big your backups are by doing a manual backup and looking at the size of the downloaded files. And if you're still in doubt, talk with your web host.

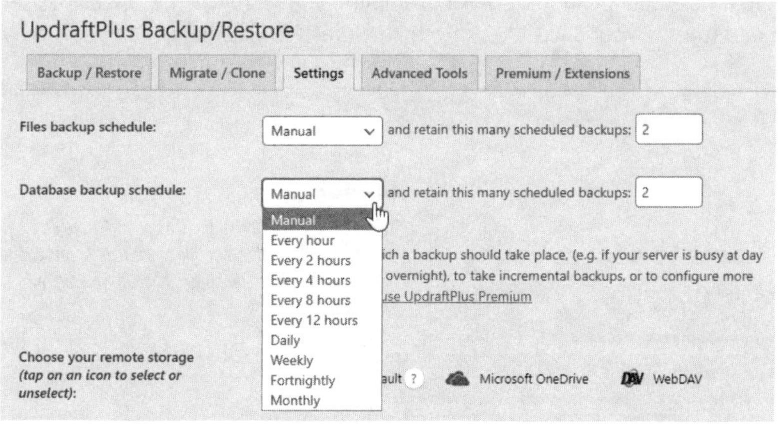

FIGURE 14-4

At first, both database and file backups will be manual-only, as shown here.

3. **Choose the number of backup copies you want to keep.**

   The standard option keeps two copies of each type of database, but you can expand that to hold more. For example, if you decide to perform daily database backups, choosing 14 means you'll always have two weeks to fall back on. Your choice depends on what storage service you choose (in the next step) and how much space is available.

TIP A blended backup strategy often makes the most sense. Every once in a while (say, once a month), make a manual backup and keep that on your computer forever. The rest of the time, let UpdraftPlus manage a short-term backup cycle. That way, you're protected against short-term disaster (you install a bad plugin and your site crashes) and long-term issues (you realize some important posts were accidentally deleted six months ago).

4. **Choose a remote storage location.**

   UpdraftPlus gives you a variety of options to choose from, ranging from its own paid storage service (UpdraftVault), to more common services like Dropbox (*www.dropbox.com*), which is free for the first 2 GB of storage; Google Drive (*http://drive.google.com*), which is free for the first 15 GB; and Amazon S3 (*http://aws.amazon.com/s3*), which costs pennies per gigabyte.

If you don't already have a paid account with one of these services, our recommendation is Google Drive. Its initial allotment of 15 GB will hold plenty of backups, although you need to share it with other Google services you might be using, like Gmail, Google Photos, and Google Docs.

5. **Authenticate your storage location.**

For example, if you choose Google Drive, you'll see a link that you need to click to give UpdraftPlus access. Click that, log in to Google, and agree to grant it permission to use your storage (*Figure 14-5*). A few services don't require a separate authentication step. For example, if you're using Amazon S3, you simply need to type your secret key into the UpdraftPlus settings.

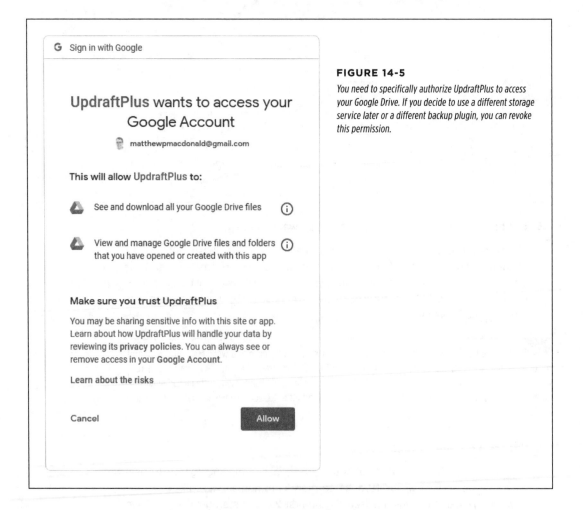

FIGURE 14-5

You need to specifically authorize UpdraftPlus to access your Google Drive. If you decide to use a different storage service later or a different backup plugin, you can revoke this permission.

6. **Optionally, fine-tune which files are included in your backups.**

You can choose to include or ignore plugins, themes, and uploads. You can also add custom rules that exclude folders you don't want to back up. UpdraftPlus adds a few preset rules to make sure it isn't wasting time and space backing up backups. It excludes the *updraft* folder, as well as folders that are commonly used by other backup plugins.

7. **Click Save Changes.**

Although you can't set exact backup times in the free version of UpdraftPlus, you can monitor when your next backup is going to take place. Click the Backup/Restore tab and you'll see the next scheduled file backup and the next scheduled database backup (*Figure 14-6*).

Once UpdraftPlus has made its backup, you'll see it appear in the recent backup list, alongside any manual backups you've made. (UpdraftPlus adds an icon next to the backup to show which storage service it uses.) You can even download or delete your remote backups from the UpdraftPlus page, just as you do with manual backups.

POWER USERS' CLINIC

Why You Might Need the Premium Edition

The free edition of UpdraftPlus has everything you need to create reliable backups. But a bunch of features *aren't* available unless you buy the premium edition:

- **More complex schedules.** The free edition doesn't allow you to set exact times, back up to more than one storage service, or overlap different types of backup schedules.

- **Encryption.** If you're worried about the safety of your backed-up site, encryption will scramble it securely with a key. Then, even if someone breaks into your storage, they won't be able to read your data.

- **Incremental backups.** Use these to tame huge backups.

With an incremental backup, UpdraftPlus copies only the changed files, reducing the amount of space the backup needs, and the workload of your web server.

- **Updraft cloning.** Cloning is an interesting feature that lets you restore your backup to a temporary spot on an UpdraftPlus web server. You can use this to test your backups or experiment with changes (like adding plugins) that might disturb your site.

UpdraftPlus sells its plugin in different packages. For a one-site Personal license, you need to pay $70 (at the time of this writing) and another $42 per year. Minimally more expensive plans let you use UpdraftPlus on several sites at once.

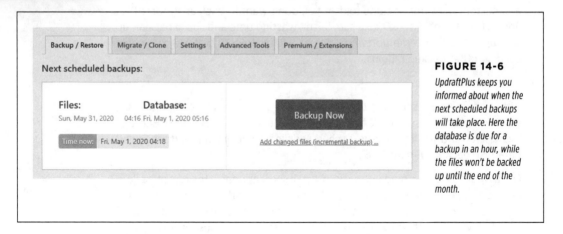

FIGURE 14-6

UpdraftPlus keeps you informed about when the next scheduled backups will take place. Here the database is due for a backup in an hour, while the files won't be backed up until the end of the month.

Restoring a Site

You need to configure your backup schedule only once. Once you have it sorted out, you can relax in the newfound glow of WordPress security.

But what happens if you actually need to *use* your backup—say, to bring a dead website back to life?

Let's start with the easiest scenario: rolling back to an older version of your site. Accomplishing this is easy. Just head back to the familiar Settings→UpdraftPlus Settings page, click the Backup/Restore tab, and scroll down to the list of recent backups. When you see the backup you want to use, click the Restore button next to it. You can choose to restore just a part of your backup (for example, just the database or theme files).

Life is slightly more complicated if you want to restore to an older backup that you've downloaded to your computer but isn't online any longer. In this situation, you need to upload your backup to your site first. Just above the list of backups in UpdraftPlus, click the "Upload backup files" link. You can then drag your backup files into an upload box. Remember to get all the ZIP files that belong to the backup set (and don't even bother trying files that were archived by another backup program, because UpdraftPlus won't understand them).

The biggest restore job is re-creating your site on a new web server. You'll need to do this if your site is vaporized by a truly serious incident. Begin by installing a fresh copy of WordPress and the UpdraftPlus plugin. Next, connect it to your remote storage service, or upload your backup files. Once your backup appears in the list, you're ready to bring it back to life. Click Restore to copy everything over and resurrect your site.

TIP   If you're running a mission-critical site, it's worth trying this out at least once. Create a new, temporary WordPress site on your web host in a different folder. Or use the Local tool from Chapter 3 to run a copy of WordPress on your computer. Then, try to create a duplicate version of your real site working from your backup. If you succeed, you know you can trust your backup strategy to keep your site safe.

GEM IN THE ROUGH

Using Your Backup Plugin to Experiment with New Changes

Once your WordPress site is established, it gets harder to change things. You might want to add a new plugin, experiment with a different menu organization, or even try out a customized version of your theme. But all these changes are risky, because they affect the live copy of your site that everyone can see. If a change goes wrong, you'll be stuck trying to patch your site back to the way it used to be, while the entire world sees an embarrassing mess.

One solution to the problem is to use a staging service, which some higher-end WordPress web hosts provide (page 42). But if you don't have that, you can make your own no-frills staging system with a backup plugin like UpdraftPlus.

To begin, create a full backup of your test site. Then use that to create a new, duplicate version of your site, either on your web host or on your computer with Local. Then you can let loose with all your experiments. When you're finished, you have several options for applying the changes to your live site. If they're simple (like installing a new plugin), you can repeat them on the live site. If they're more complex, you may be able to use your backup plugin to do a partial restore. For example, if you're working on a new theme, you can back up your theme files on your test site and then restore them to your real site.

Keep in mind that while you work, your real site could be changing—for example, someone might be posting new content or writing new comments. For that reason, you need to be very careful about restoring the test copy of the database to your live site. There's no way to combine databases, and you don't want to accidentally erase new content. To be extra safe, create *another* backup of your live site just before you do a restore, just in case you find out you're missing something.

■ Better Performance with Caching

As you learned at the very beginning of this book, WordPress websites are powered by code. When a visitor arrives at one of your pages, the WordPress code grabs the necessary information out of your database, assembles it into a page, and sends the final HTML back to your guest's browser. This process is so fast that ordinary people will be blissfully unaware of all the work that takes place behind the scenes.

But even the fastest web server can't do all that work (run code, call the database, and build a web page) instantaneously. When someone requests a WordPress page, it takes a few fractions of a second longer to create it than it would to send an ordinary HTML file. Normally, this difference isn't noticeable. But if a huge crowd hits your site at the same time, the WordPress engine will slow down slightly, making your entire website feel just a bit laggy.

This is where *caching* comes into the picture. The basic idea is this: the first time someone requests a page on your site, WordPress goes to work, running its code and generating the page dynamically. But once it delivers the page, a caching plugin stores the result as an ordinary HTML file on your web server. Now here's the ingenious part: the next time a visitor asks for the same page, the caching plugin sends back the previously generated HTML, sidestepping the usual page-generating process and saving valuable microseconds (*Figure 14-7*). This shortcut works no matter how many people visit your site—as long as the plugin has an ordinary HTML copy of the finished page, it uses that instead of creating a new copy all over again. Caching plugins can use other tactics, too, such as caching just part of the content, compressing the cached data, and discarding cached copies after a certain amount of time.

> **NOTE** Caching takes extra space on your website, because it stores extra HTML files. However, these files are rarely big enough to worry about. Caching plugins don't attempt to cache your site's pictures (or the other resources you uploaded to the media library).

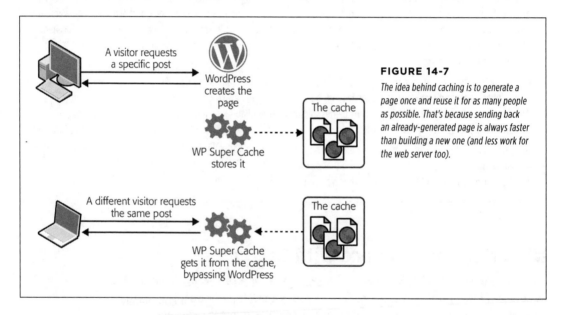

FIGURE 14-7

The idea behind caching is to generate a page once and reuse it for as many people as possible. That's because sending back an already-generated page is always faster than building a new one (and less work for the web server too).

The WP Super Cache Plugin

The world is awash in WordPress caching plugins. Some web hosts even preinstall them on new WordPress sites so they can handle more traffic for more people at once. My web host preinstalls LiteSpeed Cache, but you may come across W3 Total Cache (the current market leader), WP Rocket, WP Fastest Cache, or countless others. In this section, you'll use the WP Super Cache plugin (*http://tinyurl.com /wpsupercache*). It's developed by Automattic, the same company that does most of the development on the core WordPress software, which increases the odds that it will remain reliable for years to come.

To get started with WP Super Cache, search for the plugin by name, install it, and activate it in the usual way. When you first activate WP Super Cache, it won't do anything. You need to explicitly turn on caching first:

1. **Choose Settings→WP Super Cache.**

   Like many advanced plugins, WP Super Cache has a multitabbed page that's stuffed with settings (*Figure 14-8*).

2. **Click the Caching On option.**

   The only reason you would keep caching off is if you're trying to resolve an odd conflict (see the box on page 441).

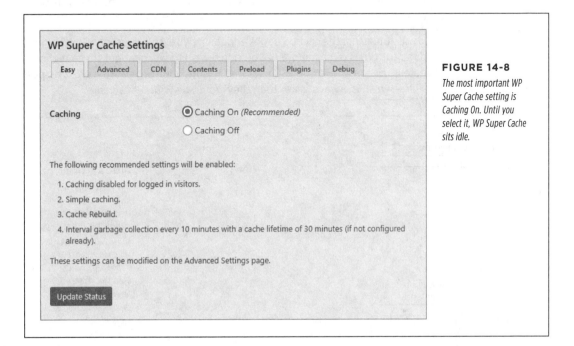

FIGURE 14-8

The most important WP Super Cache setting is Caching On. Until you select it, WP Super Cache sits idle.

3. **Click the Update Status button.**

   Now WP Super Cache is ready to work. Before you forget about it completely, it's good to run a quick test.

4. **Open a new, private browser window, and visit a few pages on your site.**

   WP Super Cache won't cache anything until someone visits your site. It's not enough for you to do it while you're logged in, because WP Super Cache won't cache those pages either. Instead, you need to pretend to be an ordinary visitor. You can do that in several ways—load a new browser (where you aren't signed in), use another computer or device (like a phone), or just launch a private browser window where you aren't logged in (for example, open the menu in Chrome and pick "New incognito window").

   No matter which approach you use, take the time to load up a few different posts or pages.

5. **In the WP Super Cache settings page, click the Contents tab.**

   Here you'll see a summary of what's currently stored in the cache.

6. **Click the "Regenerate cache stats" link.**

   WP Super Cache will report the number of cached files it created (*Figure 14-9*). You can't tell that these pages are being cached when you see them in a browser, because they don't look any different. But WordPress will send them to your visitors more quickly, bypassing most of the processing the WordPress engine ordinarily does.

   You'll notice that the cached pages appear in two lists, representing two slightly different types of caching. But you don't need to worry about these technical details, because WP Super Cache makes sure your pages are as fast as possible—which is always faster than they would be without caching.

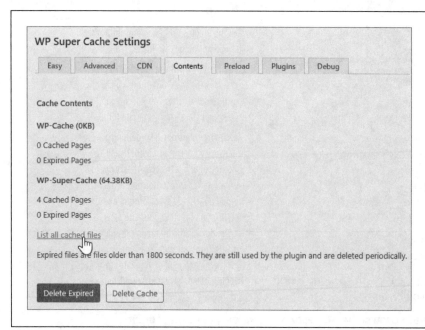

FIGURE 14-9

This site currently has four cached pages. To see them, click the "List all cached files" link. You can also use the Delete Cache button to discard every cached page, which is important in rare cases—for example, if you've changed something about the design of your site and WP Super Cache has somehow failed to notice.

7. **Optionally, you can adjust more settings in the Advanced tab.**

   Once you verify that your site still works properly, you can tweak a few WP Super Cache settings. Click the Advanced tab and look for settings that have "(Recommended)" next to them, in italics. This indicates a setting that improves the way WP Super Cache works for most people, but may cause problems in rare situations (and, for that reason, may be initially switched off). One example is the "Compress pages so they're served more quickly to visitors" setting, which improves performance for most people but causes trouble with some web hosts that don't support compression properly.

There are advanced options you can control, like the amount of time a page is kept in the cache and how often WP Super Cache checks for expired pages. However, you won't usually fiddle with any of these details unless you're a blackbelt caching guru.

> **NOTE** WP Super Cache is almost always smart enough to know when it needs to regenerate the cached copy of a page. For example, if you update a post or a visitor leaves a new comment or you change your menu, WP Super Cache takes notice and dumps the affected pages from the cache. However, in rare cases, WP Super Cache may fail to notice a change and keep serving old content. (This is most common when you're testing changes to your site's design.) If this happens, you can quickly fix the problem by manually clearing the cache. Just click the Delete Cache button, which is always visible in the toolbar at the top of the page when you're logged in to your site.

WORD TO THE WISE

WP Super Cache's Effect on Other Plugins

Caching changes the way your website works, and it can have unexpected side effects on other plugins. In fact, when you look at *plugin conflicts*—when two plugins won't work nicely together on the same site—one of the offenders is often a caching plugin.

Here are some tips to help you steer clear of plugin conflicts:

- **Check the documentation for your plugins.** WP Super Cache is popular enough that other plugin creators often test their plugins with it. To see if the plugins you use need special settings to get along with WP Super Cache, look up the plugin in WordPress's plugin directory (*http://wordpress.org/plugins*) and check the FAQ tab.

- **Visit the Plugins tab in the WP Super Cache Settings.** A few plugins get special attention from WP Super Cache. If you have one of them, you can tell WP Super Cache to change the way the cache works to avoid disrupting the other plugin. To do that, go to Settings→WP Super Cache,

click the Plugins tab, find your plugin, and then click the Enabled option next to it.

- **Delay caching until you're ready to go live.** Switching on caching is the very last thing you should do with your WordPress site, after you polish your theme, tweak your layout, and start using your site for real.

- **Learn to troubleshoot.** If something goes wrong, you need to be ready to track it down. Usually, the most recently activated plugin is the culprit—try disabling it and seeing if your site returns to normal. If that doesn't work, you need to deactivate every plugin, and then activate them one at a time, testing your site after each step. Also, be on the lookout for theme versus plugin conflicts, which are less common but occasionally occur. If you change your theme and part of your site stops working, you can troubleshoot the problem by switching off all your plugins and then activating them one at a time.

Faster Images with a Content Delivery Network

A caching plugin like WP Super Cache doesn't try to cache pictures and other types of media files. This is because picture files don't require much web server work. They aren't created on the fly, like a page or post. There's no code to run, so caching them doesn't save the web server's time or effort. And picture files are big, so they'll fill up a cache in no time, pushing everything else out.

But another type of caching does work for pictures. It's not as essential as the standard WordPress caching you've just seen, but it can still be useful, particularly for image-heavy sites like photoblogs and portfolios. This is called *browser caching*—caching that happens on the visitor's computer.

The idea behind browser caching is to make sure a visitor doesn't have to keep downloading the same picture over and over again. Why would a browser download the same picture more than once? There are several reasons it happens:

- The same picture appears on more than one page.
- A picture appears on a page and in some sort of gallery display on the home page.
- A visitor visits the same page more than once, perhaps on different days.

With browser caching, the browser stores a copy of the picture locally, on the computer or device that's viewing your site. Then, when the visitor reloads the same page later, or goes to another page that uses the same picture, the browser uses the copy it already has on hand (*Figure 14-10*).

NOTE The biggest difference between browser caching and server caching (the type that WP Super Cache does) is that browser caching asks every visitor to hold on to their own separate cache. Server caching keeps one big pool of cached files on the web server, which it shares with everyone. Non-geek translation: server caching gives you more bang for the buck, but browser caching is often worthwhile too.

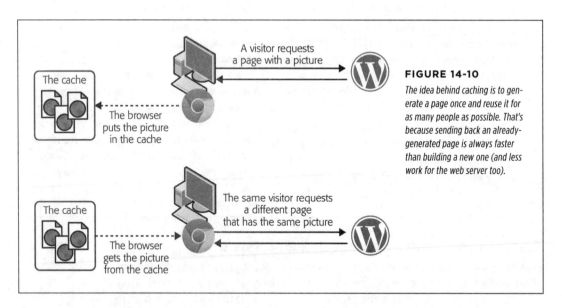

FIGURE 14-10

The idea behind caching is to generate a page once and reuse it for as many people as possible. That's because sending back an already-generated page is always faster than building a new one (and less work for the web server too).

Browsers are already in the habit of caching files—they do it automatically and without being asked. But to reuse a cached copy of a picture on different pages, the browser needs to recognize that the new page is using the *same* picture. One option is to use a *content delivery network*, or CDN—one or more web servers that work together to deliver files more quickly.

WP Super Cache has a CDN feature, but it requires you to sort out all the hosting details. A much easier approach is to use the Jetpack plugin, which has an automatic, free CDN feature built in. Best of all, Jetpack's CDN feature uses Automattic's web servers to distribute your pictures. Because Automattic is a sprawling, multibillion-dollar web company, it has an army of servers distributed across the globe. That means that when visitors hit your site, they'll probably get pictures faster when they're sent from Automattic's servers than if they were coming from your web host's servers.

Here's how to enable Jetpack's CDN feature:

1. **Make sure Jetpack is activated.**

   If you've somehow avoided setting up Jetpack until now, you can get all the details about it in Chapter 9.

2. **Choose Jetpack→Settings, and click the Performance tab.**

   This page has options for buying Jetpack's search and video add-ons, but it doesn't have much else. You'll find what you need in the "Performance & speed" section.

3. **Switch on "Enable site accelerator."**

   This turns on both the settings underneath:

   Speed up image load times. This enables caching for your pictures.

   Speed up static file load times. This enables caching for some style and script files that WordPress uses. This means that if a browser downloads these files after visiting one WordPress website, they won't need to redownload them when they visit another WordPress website that uses the same plugin. Right now, Jetpack caches only files that are used by the core WordPress software, Jetpack, and the WooCommerce plugin you'll learn about next.

■ Adding Ecommerce to Your Site

So far in this chapter, you've been exploring ways that plugins can improve a site. First you used them to protect yourself with backups. Then you used them to enhance performance for large crowds of visitors. But now you're about to take a look at a different type of plugin—one that doesn't just improve your how your site works behind the scenes, but transforms the way it presents itself. You're going to add ecommerce.

If you're running a business, it's easy to use a WordPress site to promote yourself. You can send customers to your local store, advertise your services, and even create a product catalog (with the right theme). But what if you want to go a step further, and use your site to *sell* things?

Selling products online might sound like a high hurdle, but with the right plugin, it's easy. And thanks to online money handlers like PayPal, you don't need to worry about credit card processing or opening a business account at the bank.

A Quick Overview of PayPal

PayPal strikes a simple but compelling deal. You tell PayPal what you want to sell, and it manages the payment process for you. PayPal accepts whatever type of credit card or bank transfer a customer wants to use, and forwards the money on to you (minus its fees—more on that in the box on page 445). When you're ready, you can transfer the PayPal money you make to a bank of your choosing.

NOTE PayPal isn't the only way to make bank on a WordPress website, but it was the first and remains the most popular. Other options—called *payment gateways*—include Stripe, Square, Amazon Pay, and more. Most ecommerce plugins support several payment gateways.

You might already have a PayPal account, but odds are it's a personal one—suitable for buying other people's goods but not much else. Before you can use PayPal to collect money on your site, you need a premier account or a business account.

NOTE All PayPal accounts are free to set up. PayPal makes its money on the commission it takes when you make a sale.

If you already have a personal PayPal account, you can upgrade to a premier or business account quickly by visiting *www.paypal.com/upgrade*. If you don't have an account, you can sign up for a new one by visiting *www.paypal.com* and clicking the Sign Up link.

When you create a new PayPal business account, you need to specify a few details about your business (like the type of business you run, the business name, and the business address). What's more notable is what you *don't* need. Unlike opening a business account at a bank, you won't be asked for a business license, financial statements, or any other type of documents or certification.

Some of these moneymaking schemes are easier than others. (Asking for money: easy. Selling premium content through a membership: more difficult.) Depending on the plugin you're using, you might need to pay for a premium version to get the features you need. But it's impossible to argue with the fact that selling your wares online is easier today than it has ever been before.

In the following pages, you'll take a look at a couple of WordPress plugins that make ecommerce happen. First, you'll see how to ask up front for a one-time payment. Next, you'll see how to implement a basic shopping cart using the premier WordPress ecommerce plugin, WooCommerce.

The Three Types of PayPal Account

The first decision you need to make when you sign up with PayPal is the kind of account that's right for you. PayPal gives you three options:

- **Personal account.** This type of account lets you use PayPal to buy items on sites like eBay. You can also *accept* money transfers from other PayPal members without having to pay any fees. However, there's a significant catch—personal accounts can't accept credit card payments, so they won't work on an ecommerce site.

- **Premier account.** This type of account gives you an easy way to run a small business. You can still make payments to others, and you can accept any type of payment that PayPal accepts, including both credit and debit cards. However, PayPal charges you for every payment you receive, an amount that varies by sales volume but ranges from 1.9 percent to 2.9 percent of the payment's total value (with a minimum fee of 30 cents). That means that on a $25 sale, PayPal takes about $1 off the top. If you accept payments in another currency, you surrender an extra 2.5 percent. To get the full scoop on fees and to see the most current rates, refer to *www.paypal.com/fees*.

- **Business account.** This type of account has the same features and fees as a premier account, with two key differences. First, it lets you do business under your business name (instead of your personal name). And second, it supports multiple users. For that reason, a business account is the best choice if you have a large business with employees who need to access your PayPal account to help manage your site and its finances.

One you have a PayPal account, you're ready to use it on your site. You can choose from several approaches:

- **Donations.** You can ask for a one-time money contribution, whether it's a thank-you "tip" for good content or a donation toward a charitable enterprise.

- **Memberships.** You can charge a fee before your visitors get access to specific content or downloads, whether it's an ebook or your band's music recording.

- **Subscriptions.** You can sign customers up for a weekly, monthly, or yearly charge. PayPal handles all the billing.

- **Products.** You can create a full ecommerce-style store with a shopping cart, shipping options, and a checkout process.

Asking for a Donation

Begging for coin is awkward, but it may work. If your site offers genuinely useful advice, you can ask for tips with something like this: "Like what you read? Buy me a coffee!" If you're writing in support of a charitable cause or if your writing entails danger or significant sacrifice, you can ask for support, as in "Donate today to support independent journalism."

Whatever the case, the easiest approach is to set up a PayPal account and add a plugin. Full-featured plugins like WooCommerce (which you'll look at in the next section) can certainly do the job. But you'll do just as well with something simple and streamlined, like the basic PayPal Donations plugin (*http://tinyurl.com/paypal-d*).

Once you install and activate the PayPal Donations plugin, you need to configure it with your PayPal details. To do that, choose Settings→PayPal Donations in the admin menu (*Figure 14-11*).

Every PayPal plugin needs one key piece of information: your email address, because that's how PayPal directs money to you (and not someone else). You can also set other details, such as a recommended donation amount, the button style, the currency you're using, the page where the visitor should be sent after they donate, and so on.

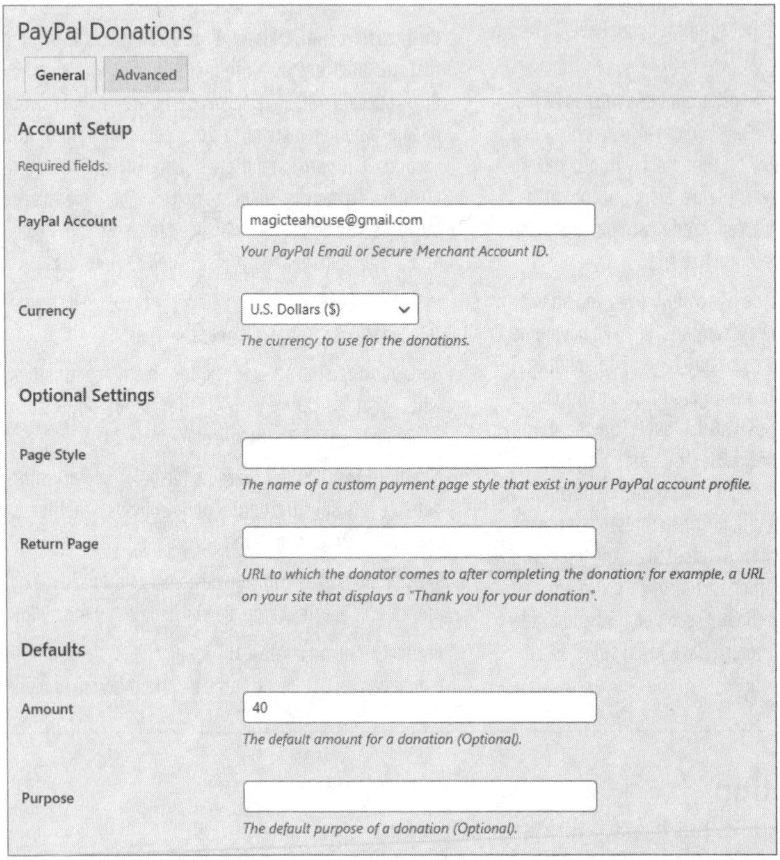

PayPal Donations

General | Advanced

Account Setup

Required fields.

PayPal Account

magicteahouse@gmail.com

Your PayPal Email or Secure Merchant Account ID.

Currency

U.S. Dollars ($)

The currency to use for the donations.

Optional Settings

Page Style

The name of a custom payment page style that exist in your PayPal account profile.

Return Page

URL to which the donator comes to after completing the donation; for example, a URL on your site that displays a "Thank you for your donation".

Defaults

Amount

40

The default amount for a donation (Optional).

Purpose

The default purpose of a donation (Optional).

FIGURE 14-11

You can set several options for a PayPal donation. The email address is the only essential detail.

> **TIP** It's possible that the small set of options the PayPal Donations plugin offers won't be exactly what you want, in which case you can try one of many similar plugins, like Quick PayPal Payments or WP Easy PayPal Payment Accept. They work in almost exactly the same way but have slightly different choices. Although none of these plugins is clearly better than the others, some let you set a required amount, present more than one payment option in a list, or tweak the appearance of the PayPal button.

Once you've configured your PayPal plugin, it's time to insert the Donate button. Here's how:

1. **Start editing a post or page.**

Go to wherever you want to put your Donate button.

2. **Click Add Block and choose Shortcode.**

   To get the PayPal Donations plugin to show your button, you need to use a slightly old-fashioned integration option called a *shortcode*. A shortcode is a magic square-bracketed code that summons a plugin into action when you don't have a block that can do the job.

3. **In the Shortcode box, type "[paypal-donation]."**

   This shortcode tells the PayPal Donation plugin to insert the Donate button. The only catch is that you won't see the button while you're writing the post. The shortcode works only in the published version of the page (or the preview).

4. **Preview or publish the page.**

   You'll see a distinctively PayPal-ian Donate button (*Figure 14-12*). If you click this button, you'll be sent to PayPal's servers to finish the transaction.

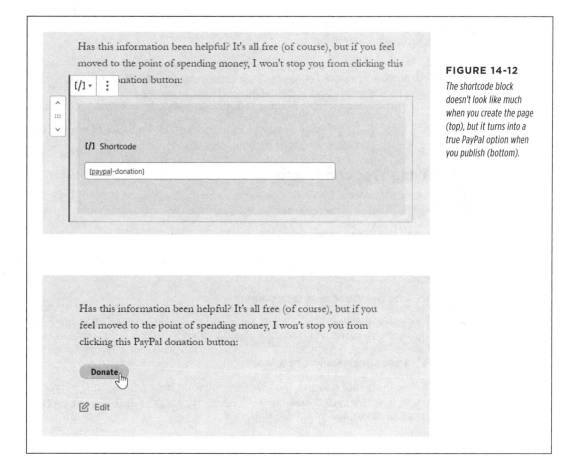

FIGURE 14-12

The shortcode block doesn't look like much when you create the page (top), but it turns into a true PayPal option when you publish (bottom).

Introducing WooCommerce

The PayPal Donations plugin gives you a bare-bones way to ask for a one-time money payment. There's much more you can do with PayPal on a WordPress site, but you'll need to step up to a more comprehensive plugin first. The clear choice is the wildly popular *WooCommerce*, which—at last measure—powered roughly 10% of the world's ecommerce websites.

WooCommerce began its life as a handy plugin by WooThemes, an independent theme designer. But in 2015, Automattic (the company that created WordPress) bought WooThemes and WooCommerce, making them nearly official ingredients in the WordPress kitchen.

The drawback to WooCommerce is that it's complex. WooCommerce itself is just one plugin, but it works in tandem with many more optional plugins (known as *extensions*) to support sophisticated features like subscriptions, shipping procedures, barcodes, reporting, preorders, PDF product catalogs, mailing lists, and just about everything else a business could think of. There are dozens and dozens of WooCommerce extensions. And if you want to go full ecommerce website, you probably want to pair WooCommerce with a WooTheme that can create a proper product-browsing experience, like a mini Amazon.

> **NOTE** WooCommerce is 100% free. However, if you start building a site with WooCommerce, you may find that you want a specialized theme and additional extensions that have a yearly cost.

In this chapter, you'll get only a taste of WooCommerce. We'll dive in and build a basic shopping cart for selling teas on the Magic Tea House site.

Installing WooCommerce

The installation process for WooCommerce is a bit more elaborate than for the other plugins you've used in this chapter. Here's how it goes:

1. **Install the WooCommerce plugin.**

   This is the same as always—search for it by name on the Plugins→Add New page, and add it to your site.

2. **Activate the WooCommerce plugin.**

   Here's where things take a turn. As soon as you activate WooCommerce, it launches a multistep setup wizard (*Figure 14-13*).

3. **Fill in your business information.**

   This includes your address, the currency you want to be paid in, and the type of products you're selling—real physical goods (like tea) or digital products (like a tea maker's recipe ebook) or both.

4. **Click "Let's go!"**

WooCommerce enthusiastically takes you to the next step, where you'll tell it the kind of payments you want to accept.

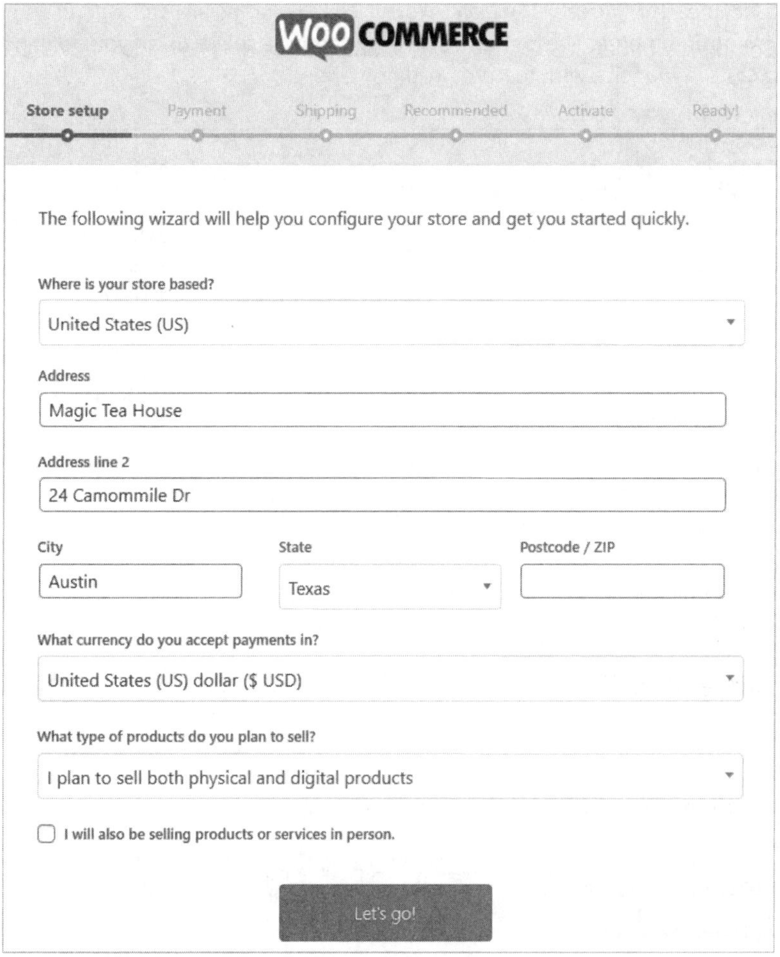

FIGURE 14-13
The first step in the WooCommerce setup asks you to tell it a bit about your business.

5. **Choose your payment options and click Continue.**

At first, WooCommerce gives you just a few choices (*Figure 14-14*).

The two most common payment choices are Stripe and PayPal accounts. (If you created a PayPal account earlier in this chapter, it's a great, hassle-free way to work with WooCommerce.) WooCommerce even offers to set up a Stripe or PayPal account *for* you, although that means it will need to gather even more information. Alternatively, you can choose to take checks, bank transfers, or cash on delivery. WooCommerce calls these choices offline payments, because they all happen outside WooCommerce.

You can turn on just one payment option or as many as you want. You can also leave them all switched off and configure your payment options afterward. Depending on your choices, WooCommerce may install a companion plugin (as it does in *Figure 14-14*). Don't worry—these plugins are still a genuine part of WooCommerce. The creators of WooCommerce split it up so you don't need to clutter your site with features you don't need.

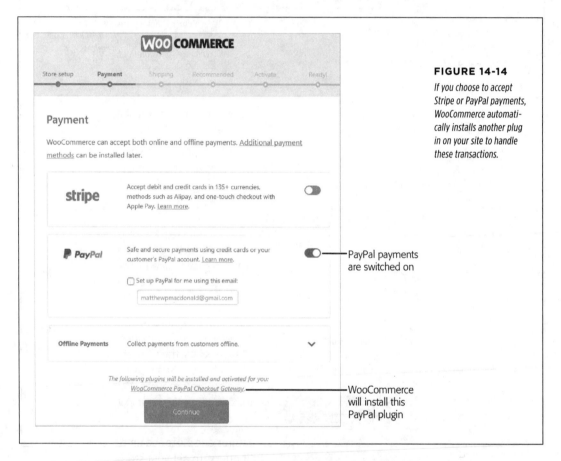

FIGURE 14-14

If you choose to accept Stripe or PayPal payments, WooCommerce automatically installs another plug in on your site to handle these transactions.

6. **Fill in your shipping information and click Continue.**

   If you've opted to sell real, physical items, now's the time to tell WooCommerce how you plan to ship them. You can choose to offer free shipping or charge a flat shipping rate. Real-time quotes from shipping providers like UPS are another option, but that requires yet another WooCommerce extension. We recommend that you keep a simple fixed rate for now, because you can alter all these details after you get WooCommerce integrated into your site.

7. **To keep things simple for now, uncheck all the optional extensions and click Continue.**

   WooCommerce recommends a pile of useful plugins. Some of them may duplicate features you already have from other plugins on your site.

   It's easier to understand WooCommerce if you get a handle on the basic package first. Be particularly careful about the Storefront Theme option—if you install that, WooCommerce activates Storefront and deactivates your current theme.

NOTE The Storefront Theme has plenty of features that make it a good fit with WooCommerce, but it starts out looking pretty plain. To get the slick results you really want, you'll probably need to buy an extension that lets you customize it or a Storefront child theme. To learn more, check out *https://woocommerce.com/storefront*.

---

8. **Choose whether you want to use Jetpack.**

   WooCommerce makes it seem like you *need* to use Jetpack, but it's really just a recommendation. (Both Jetpack and WooCommerce are owned by Automattic.)

   If you want Jetpack and don't already have it, click the big Continue with Jetpack button. If you don't, no worries—just hunt for the tiny "Skip this step" link at the bottom of the page.

9. **In the final step, click Visit Dashboard.**

   When you arrive at the end of the setup wizard, you'll see a bunch of buttons prompting you to sign up for a newsletter, start making products, or review more WooCommerce settings. But the easiest approach is just to head back to the admin area, where you can start reviewing all the new stuff WooCommerce has added.

The first thing you'll notice is the greatly expanded admin menu (*Figure 14-15*). WooCommerce adds three new menu headings, each of which is filled with links. The WooCommerce menu lets you manage settings, install extensions, track orders and customers, build reports, and run promotions. And if that doesn't seem overwhelming enough, you also get a Products menu for managing the products you want to sell, and an Analytics menu chock full of links for analyzing the financial performance of your business.

Don't panic! You can investigate WooCommerce's many features one piece at a time, and there's no need to become a master of every feature it offers. To get your store running, you need just a few simple ingredients: a product or two, a place to show them, and a shopping cart that lets customers buy them.

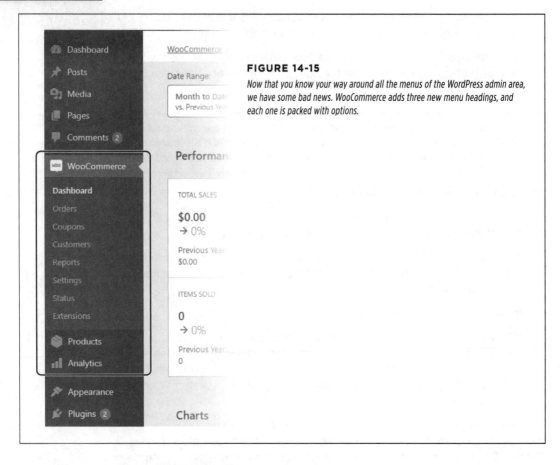

FIGURE 14-15

Now that you know your way around all the menus of the WordPress admin area, we have some bad news. WooCommerce adds three new menu headings, and each one is packed with options.

Adding Products

The heart of WooCommerce is *products*, the things you want to sell on your site. Here's the clever thing about WooCommerce—each product is actually a special type of post. Like a post, it can have a title (the product name), some content (the product description), and a featured image (the product picture). But WooCommerce also attaches extra bits of information to every product. At a bare minimum, each product gets a price, but you can also add a barcode number, the size and weight for shipping, and so on.

To create a new product, choose Products→Add New. You can then fill in the product page with the essential details (see *Figure 14-16*), and click Publish to put it on your site.

The product name

The product page

The product description

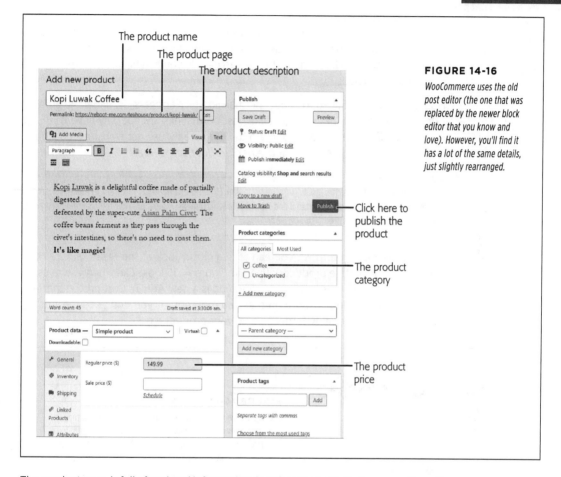

FIGURE 14-16

WooCommerce uses the old post editor (the one that was replaced by the newer block editor that you know and love). However, you'll find it has a lot of the same details, just slightly rearranged.

The product page is full of optional information, but there's plenty that you won't need to use. To create the coffee product shown in *Figure 14-16*, here's what we added:

- **A product name and full description.** There's also a box for a shortened description at the bottom of the page, but this example keeps things simple.

- **A product image.** This is really a renamed featured image. You set it in a box on the right, near the bottom of the page (not shown in *Figure 14-16*). Without a proper product image, you'll get a no-picture placeholder, which doesn't look very professional.

- **A product category.** You can assign categories and tags to products just as you do to posts. In this example, the Kopi Luwak Coffee product is in the Coffee category. WooCommerce keeps your product categories separate from your post categories (and your product tags separate from your post tags). That way, there's no confusion. If you want to create or edit a bunch of product categories, go to the Products→Categories page.

- **A regular price.** You don't need to add a price, but if you don't, WooCommerce won't show the buttons that let visitors buy your product. The price is the only detail this product uses in the "Product data" box. Poke around in this section and you'll find all sorts of extra details you can fill in, like shipping and inventory details.

Notice one other detail—the product permalink, which appears just under the product title. Product permalinks work just like post permalinks, except they have */product* in them. So if you create a product on the magicteahouse.net website, you'll get a permalink like *http://magicteahouse.net/product/angel-tea*.

TIP You can customize the last part of the permalink (the product name) if you want to make it short and snappy. Just click the Edit button next to it and type in your replacement. For example, we replaced the full product name *kopi-luwak-coffee* with the shorter *kopi-luwak*.

Once you publish your post, you can take a closer look at the result. *Figure 14-17* shows what the coffee product defined in *Figure 14-16* looks like when you're using the Twenty Twenty theme. The main addition is a new Add To Cart button that lets people buy the product (more on that in a moment). If you don't see the Add To Cart button, you probably forgot to give your product a price.

FIGURE 14-17

The Twenty Twenty theme works pretty well for products, although the author name and date information at the top isn't that useful. You can get more product display options with the Storefront theme, if you're ready to redesign your site.

You can create as many products as you want. To see your whole product list, choose Products→All Products.

Showing Products

Once you've created a product, you might wonder how to showcase it on your site. There are many approaches:

- Create a link in a post that leads readers to an associated product page.

- Create a menu with links to your product pages.

- Create a menu with links to your various product categories.

- Use one of the many blocks included with WooCommerce to drop a mini product box right inside another post or page.

- Use one of the many widgets included with WooCommerce to show some or all of your products in a widget area. (This works best on themes that have a big sidebar area.)

- Use the Storefront theme to get more choices for showing big product catalogs on your home page.

- Go to the /shop page (as in http://magicteahouse.net/shop). You'll see a basic list of all your products that WooCommerce creates automatically. You can customize this page in the theme customizer (using the sidebar, go to the section named WooCommerce→Product Catalog).

In other words, WooCommerce lets you stay as simple or get as complicated as you want when it comes to displaying your products. If you're interested in offering just a couple of products for the occasional quick sale, it's easy. If you want to build your whole site around a browsable product catalog for an ecommerce store, that works too, but you'll probably want to shift themes and pile on the WooCommerce extensions.

Using the Shopping Cart

Once you've created a product or two with WooCommerce, you're ready to make your sale. As in virtually every ecommerce site that's been created in the past 20 years, this is a two-step process. First, your customer adds the item to a shopping cart (and, optionally, many more). Then, the customer goes through the checkout process.

For example, say a customer clicks the Add to Cart button on your product page. WooCommerce helpfully shows an "added to your cart" message (*Figure 14-18*), and offers to take the customer to the cart. But what happens if the customer chooses to browse to another page instead?

What you really need is a site-wide command that can take viewers to their shopping cart. You could add a View Cart or Checkout link in your menu. You just need to point it to the */cart* page (for example, *http://magicteahouse.net/cart*). But that feels a little clumsy. You can do better.

A more sophisticated option is to show a mini shopping cart preview—for example, a tiny shopping cart icon that shows the number of items in the cart. To do that, you need to add the WooCommerce Cart widget somewhere on your site. *Figure 14-19* shows the Cart widget in the Twenty Twenty footer.

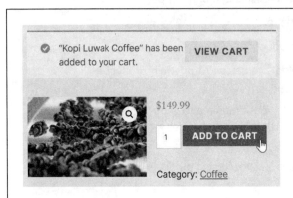

FIGURE 14-18

When you add an item to the cart, WooCommerce shows a confirmation message and the View Cart button.

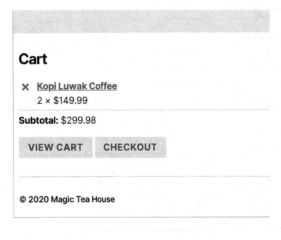

FIGURE 14-19

Unfortunately, Twenty Twenty lets you put widgets only in the footer. Other, more compact themes with sidebars and header widgets can make your website look more store-like.

When a customer buys one of your products, PayPal handles the checkout process and then notifies you by email. At this point, you need to deliver the goods (for example, by shipping them out or sending them electronically). Shortly after the transaction, the money appears in your PayPal account. You can then transfer it to a bank account or use it to buy stuff on other PayPal-equipped websites.

Even though the Magic Tea House example could use a bit more polish, it does the job perfectly well. If you click View Cart, you'll see an itemized list of what you've picked. If you click Checkout, you'll be asked for essential customer information (name, postal address, email address) and be prompted to send the payment through PayPal (or whatever payment service you picked). All this data is then funneled into the WooCommerce section of the admin area. Choose WooCommerce→Customers to see who's signed up on your site, and WooCommerce→Orders to see what they've bought. You'll also get email notifications telling you that you've received an order and it's time to prepare it for shipment. (To adjust email notification settings, visit WooCommerce→Settings and click the Emails tab.)

There's plenty more to customize in WooCommerce. You can make custom cart and checkout pages, calculate taxes, and run promotions. Most of all, you can add extensions to turn your website into a full-on extension of your business. When you're tired of tinkering in the admin area, you can study WooCommerce's voluminous documentation at *http://tinyurl.com/woo-doc* to learn more.

◼ A Final Grab Bag of Useful Plugins

WordPress offers thousands of plugins, but seasoned webmasters approach them with caution. After all, you can't afford to install any old plugin you stumble across. At best, a pile of bad plugins will slow down your site. At worst, they can open up security vulnerabilities or trigger conflicts that break your site.

In other words, it's important to choose your plugins wisely. Use just those that you need. And when you want a plugin, find the best one of its kind. In this book, you've seen plenty best-of-breed plugins, from Akismet (for controlling spam) to Yoast SEO (for optimizing your search ranking) to Jetpack (for doing a little bit of everything). And in this chapter, you tried out a few more quality plugins: Updraft-Plus, WP Super Cache, and WooCommerce. But before the book ends, it's time for a shout-out to a few more plugins you haven't seen yet. These plugins just might solve a problem or save you from a future headache. And they've all been tested over time on thousands of sites.

Search for Bad Links with Broken Link Checker

WordPress goes out of its way to avoid problems with links. When you add a link, WordPress lets you pick from a list of your site's pages and posts, so you can't mistype the address. And here's another safety feature: if you change the permalink of a published page or post (page 91), WordPress keeps the old link alive, and helpfully redirects people who use it to the proper, new location.

But these safeguards won't solve all your linking problems. There are plenty of ways that you can still end up with broken links that go nowhere. Here are some examples:

- You link to a post and then delete that post. Now the original link is broken—click it, and you'll get the infamous 404 "not found" error page.

- You link to another website, and that website deletes (or renames) the page.

- You add a picture in a post and then delete the picture from your media library.

- You add some embedded content from another website (like a YouTube video), and that content is removed.

Broken link problems aren't too hard to fix. You can change the link or just remove it altogether. But the problem is that broken links usually happen long after you've published a post or page. That means you won't know when a problem crops up. This issue—where your connections to other sites and pages slowly deteriorate—is called *link rot*.

The Broken Link Checker (*https://tinyurl.com/broken-link*) is a simple WordPress plugin that fights link rot. Once you install it, it begins scanning your site, examining each link, and checking whether it points to a real, working endpoint in the web. It even checks for links in comments. To take a look at all the broken links it can find, choose Tools→Broken Links. You'll get a report like the one shown in *Figure 14-20*.

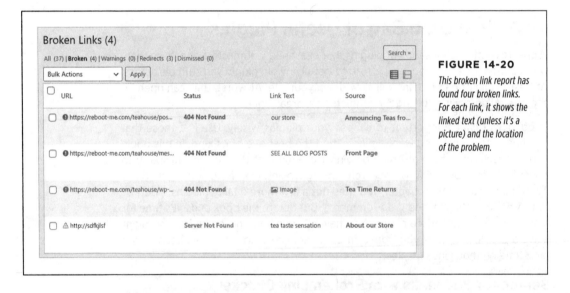

FIGURE 14-20

This broken link report has found four broken links. For each link, it shows the linked text (unless it's a picture) and the location of the problem.

TIP You can use the Broken Link Checker to get a complete list of links (broken *and* working) culled from all the posts and pages on your site, which is pretty nifty. To see the full list, click All above the link list (*Figure 14-20*).

The Broken Link Checker doesn't perform just a single check. As long as it's active on your site, it performs periodic scans. You can control its link-checking schedule, and request that it emails you whenever it detects a problem, by adjusting the settings on the Settings→Link Checker page.

Fight Database Junk with WP-Optimize

As you live with your site, it slowly accumulates useless bits of extra data. Over a few months, there's nothing to worry about, but after a couple of years, this extra junk can hog up space and slow down your site. To slim down, you can use the handy WP-Optimize plugin (*http://tinyurl.com/wp-opti*), which is built by the same people who develop the UpdraftPlus backup utility you used earlier.

WP-Optimize has three core features:

- **Database cleaning.** WP-Optimize has optimizations it can use to de-bloat your database. It can remove the records for spam comments, post revisions, old drafts, and more.

- **Image compression.** Large images take longer to travel over the web to your visitors. With aggressive compression, you can squash them to smaller, more efficient sizes.

- **Caching.** WP-Optimize has its own caching feature, which you might consider if you haven't already installed WP Super Cache.

The premium version of WP-Optimize also supports scheduling, which means it can perform regular database cleanups. But you'll do perfectly well if you use WP-Optimize manually every year or so.

To perform a database scan, choose WP-Optimize→Database. You can choose from a short menu of optimizations, but you'll do best if you stick with WP-Optimize's recommendations. It turns on the quickest, most useful optimizations, but turns off the slower options, which are marked with a tiny exclamation mark icon (*Figure 14-21*).

At the top of the list of optimizations is the setting "Take a backup with UpdraftPlus." If you have UpdraftPlus installed, you can choose this option and WP-Optimize will wait for UpdraftPlus to perform a complete backup before it changes anything. If you don't have UpdraftPlus, it's up to you to use your favorite backup plugin to make a copy of your site.

WARNING Before you optimize anything, make sure you create a full backup of your site. You never know when something could go wrong—say, a power failure in the middle of a sensitive operation on your database.

When you're ready to run your cleanup, click "Run all selected optimizations." WP-Optimize will scan through the database and make its changes. This operation is usually pretty fast, because WP-Optimize doesn't need to inspect your data—it just quickly cleans out certain types of records.

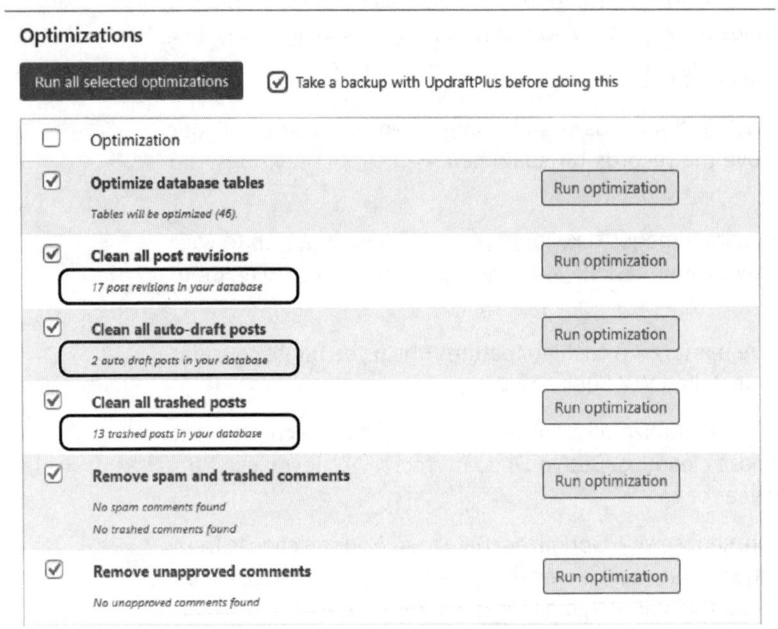

FIGURE 14-21

If you look carefully, you'll see that WP-Optimize details some facts about your database under each option. In this tiny Word-Press site, WP-Optimize reports that it can remove 17 post revisions, 2 automatically saved drafts that you probably don't want, and 13 posts that are already in the trash. There are currently no spam comments.

WP-Optimize can perform a similar magic to shrink big images, by replacing them with smaller lower-quality copies. Usually, the lower-quality copies look the same (unless you peek really closely at your pixels), so it's an optimization worth making.

To try out WP-Optimize's image compression, choose WP-Optimize→Images. Scroll to the bottom, pick some or all your images, and click the "Compress the selected images" button to start the job.

As it works, WP-Optimize takes each image, sends it to a remote image-squashing service (the site *http://reSmush.it*), and swaps in the altered version on your site. However, WP-Optimize also keeps your original, higher-quality images just in case you don't like the result. Ordinarily, it removes these backups after 50 days, but you can change this timeline—or delete them yourself—if you peek into the "Show advanced options" section.

Change How Your Site Looks on Mobile with WPtouch

In the old days (a few years ago), WordPress sites didn't always look good on tiny phone displays. Sometimes they'd stick to their normal, full-size layout, and force phone users to zoom and scroll around awkwardly. But these days, almost every theme is able to adjust itself to look decent on small screens and mobile devices.

However, sometimes you'll find a theme that you love on the desktop but aren't as enthusiastic about on a mobile device. Maybe the mobile version is too plain, wastes too much space, or just doesn't put its menus and widgets in the places where you think they belong. Maybe it doesn't give you an option to create a cut-down mobile version of your menu. Maybe you want a simpler post list on your home page that shows titles only, so you aren't stuck scrolling for days.

If you're in this situation, you don't need to trash your theme. Instead, you can use a plugin to change what the theme looks like on mobile devices. And the best plugin for the job is WPtouch (*http://tinyurl.com/wptouch*).

Most themes use mobile versions that rearrange themselves when the browser window gets really small. For example, if you visit a website that uses Twenty Twenty and you shrink your browser window, it switches to the same compressed layout you see on a smartphone. But WPtouch doesn't work this way. Instead, it recognizes smartphones and touch devices based on their *user agent*—a small bit of descriptive information their browsers send to the web server. When WPtouch realizes that a mobile device is making a request, it steps in and switches to its slick mobile theme. The WPtouch mobile theme overrides your standard theme completely. Even if your current theme has its own mobile-specific version, WPtouch replaces it.

Cleverly, WPtouch uses the same theme customizer you already know to tweak its custom mobile theme. To see how this works, load up the theme customizer (Appearance→Customize) and look for the new Switch to Mobile Theme button at the top of the sidebar. You can use this to jump back and forth between your official theme and WPtouch's mobile replacement (*Figure 14-22*).

The WPtouch mobile theme departs from the average WordPress theme in several important ways:

- **WPtouch ignores widgets when it creates a mobile version of your site.** Most other themes keep your widgets but sometimes put them somewhere else (like after the main content instead of in a sidebar).

- **WPtouch doesn't display any content in its post listings.** It simply shows the title, date, and number of comments for each post. Mobile surfers need to tap a title to read the post. Once again, this differs from themes like Twenty Twenty, which list all the content and force mobile readers to scroll the day away.

- **WPtouch doesn't load all your posts at once.** Instead, the page ends with a Load More Entries link. Click it, and the page fetches a new batch of posts and adds them to the bottom of the page.

Dig around a bit, and you can change WPtouch's color scheme and typefaces. But to get more features for tailoring menus, changing the layout, and designating certain content as mobile-only or desktop-only, you'll need to shell out for the premium version.

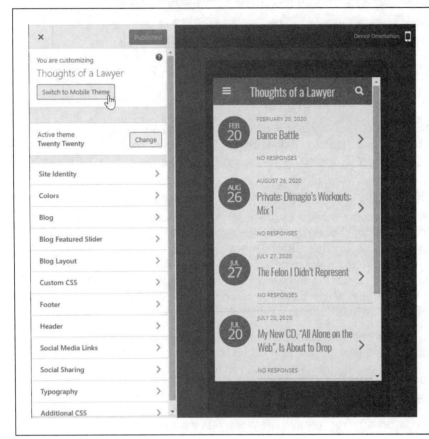

FIGURE 14-22

When you're looking at the mobile version of this site (provided by WPtouch), there's no way to tell that the desktop version uses the Twenty Twenty theme. WPtouch's mobile theme keeps the list of posts simple, highlighting the title, date, and number of comments, but leaving out the content.

Tighten Security with Limit Login Attempts

Security isn't the sort of thing you can tack on to a website after you've built it. And that's why we've been careful to follow good security practices throughout this book, like choosing secure passwords and avoiding mystery plugins. But you can take one simple yet important step to prevent *brute-force* attacks, where malicious internet programs try to hack into your site by trying randomly generated passwords over and over again. To stop these bots, you can limit the number of failed login attempts that your site is willing to tolerate. One simple but effective plugin that does this job is Limit Login Attempts Reloaded (*https://tinyurl.com/limit-logins*).

Unlike most of the plugins you've seen in this chapter, Limit Login Attempts Reloaded does just one thing—and as a result, it's refreshingly straightforward. There are just a few settings to review, and you can see them all by choosing Settings→Limit Login

Attempts. You can set how many failed logins are allowed from a given computer ("allowed retries," usually four), how many minutes the lockout stays in effect (initially, that's 20), and when it should ramp up its lockout time (to 24 hours, after four subsequent lockouts). You can also request that Limit Login Attempts Reloaded sends you a notification email to let you know after a certain number of lockouts.

The Limit Login Attempts Reloaded plugin gives you a quick and easy way to patch a potential security gap. But WordPress also has far more advanced security plugins that you might want to experiment with if you have a large site running a big business. One of the best known is Sucuri Security (*https://tinyurl.com/sucuri-p*), a massive plugin that helps the paranoid lock down their sites. Sucuri Security has an automated malware scanner, which means it can detect malicious software if it gets onto your site. It can also catch subtle attacks like template files that have been tampered with and changed from their official versions. But its most significant feature is security *auditing*—the system it uses to record potentially significant events, like users logging in and plugins being installed. Sucuri Security's security auditing is complicated and could slow down a small site that has a modest web hosting plan. But if you have the resources for it, Sucuri Security offers a higher level of protection.

■ The Last Word

Every WordPress fan has a short list of favorite plugins to use for building a site. In this chapter, you met plugins that manage backups and improve performance. You saw others that could fix common problems like bloated databases, broken links, and awkward mobile sites. You even dipped a toe into the world of professional ecommerce, all without leaving the comfort of WordPress's warm embrace.

The next step is up to you. If you want to keep in touch with the worldwide WordPress community, you have plenty of options. Follow the latest WordPress news on its official blog (*https://wordpress.org/news*). Join a meetup in your local city to talk with other WordPress lovers in real life (*https://www.meetup.com/pro/wordpress*). And check out the page of links from this book, complete with follow-up sites for future exploration, at *http://prosetech.com/wordpress*.

Index

O'REILLY®

There's much more
where this came from.

Experience books, videos, live online
training courses, and more from O'Reilly
and our 200+ partners—all in one place.

Learn more at oreilly.com/online-learning